T0298654

Effective Waste Management and Circular Economy

Effective Waste Management and Circular Economy: Legislative Framework and Strategies is an invaluable resource for researchers, policymakers, implementers and PhD, graduate and Under Graduate level students in universities and colleges analysing the legal framework, strategies in waste management, circular economy adoption, use of mathematical and statistical modelling in setting waste management strategies, sanitation and Hygiene in waste management. While huge wastes are wasted by dumping, there is a potential for resource circulation by enforcing legislative framework for effective resource utilisation and creating business opportunities. The circularity of resources in waste streams can contribute to a more secure, sustainable and economically sound future through the followings:

- Effective legal framework, strategies and policy instruments,
- Adoption of circular economy and recycling technologies,
- Support of IoT and appropriate decision making and modelling,
- Adoption of alternatives to plastics and other hazardous materials,
- Economic feasibility as business case, commercialisation, generating employment.

This book addresses most of the above issues in a lucid manner by experts in the field from different countries, which are helpful for the related stakeholders, edited by experts in the field.

Sadhan Kumar Ghosh, Professor at Jadavpur University, internationally well-known expert working in varied interdisciplinary fields including waste management having research collaboration in 40 countries.

Sasmita Samanta, Pro-Vice Chancellor, KIIT Deemed to be University, Bhubaneswar, Odisha, India having research experience in management and academic administration.

Harish Hirani, Director at CSIR-CMERI, Durgapur, having wider fields of research in IIT Delhi with a number of research collaboration.

Carlos Roberto Vieira da Silva, Director, Presidente, ABRELPE, Sao Paulo/SP – Brazil and Presidente, International Solid Waste Association, the Netherlands has experience in working on a number of international projects.

The Circular Economy in Sustainable Solid and Liquid Waste Management

Series Editor:

Dr. Sadhan Kumar Ghosh,

Professor, Mechanical Engineering, & Chief Coordinator, Centre for Sustainable Development and Resource Efficiency Management, Jadavpur University, Kolkata, India President, International Society of Waste Management, Air and Water (ISWMAW)

Biomethane through Resource Circularity: Research, Technology and Practices
Edited by Sadhan Kumar Ghosh, Michael Nelles, H.N. Chanakya, and Debendra Chandra Baruah

The Circular Economy in Construction Industry
Sadhan Kumar Ghosh, Sannidhya Kumar Ghosh, Benu Gopal Mohapatra, and Ronald L. Mersky

Effective Waste Management and Circular Economy:
Legislative Framework and Strategies
Sadhan Kumar Ghosh, Sasmita Samanta, Harish Hirani, and Carlos Roberto Vieira da Silva

For more information about this series, please visit: https://www.routledge.com/The-Circular-Economy-in-Sustainable-Solid-and-Liquid-Waste-Management/book-series/CESSLWM

Effective Waste Management and Circular Economy
Legislative Framework and Strategies

Edited by

Prof. Dr. Sadhan Kumar Ghosh

Professor, Mechanical Engineering, & Chief Coordinator, Centre for Sustainable Development and Resource Efficiency Management, Jadavpur University, Kolkata 700032, India
Founder & President, International Society of Waste Management, Water and Water (ISWMAW)

Prof. Dr. Sasmita Samanta

Pro-Vice Chancellor, KIIT Deemed to be University, Bhubaneswar, Odisha, India

Prof. Dr. Harish Hirani,

Director, CSIR-CMERI, Durgapur, India

Mr. Carlos Roberto Vieira da Silva

Director, Presidente, ABRELPE, Sao Paulo/SP - Brazil & Presidente, ISWA - International Solid Waste Association, Netherlands

CRC Press
Taylor & Francis Group
Boca Raton London New York

CRC Press is an imprint of the
Taylor & Francis Group, an **informa** business

First edition published 2023
by CRC Press
6000 Broken Sound Parkway NW, Suite 300, Boca Raton, FL 33487-2742

and by CRC Press
4 Park Square, Milton Park, Abingdon, Oxon, OX14 4RN

CRC Press is an imprint of Taylor & Francis Group, LLC

Library of Congress Cataloging-in-Publication Data
Names: Ghosh, Sadhan K. (Sadhan Kumar) editor. | Samanta, Sasmita Rani, editor. | Hirani, Harish, editor.
Title: Effective waste management and circular economy : legislative framework and strategies / edited by Prof. Dr. Sadhan Kumar Ghosh, Professor, Mechanical Engineering, & Chief Coordinator, Centre for Sustainable Development and Resource Efficiency Management, Jadavpur University, Kolkata, Prof. Dr. India Sasmita Samanta, Pro-Vice Chancellor, KIIT Deemed to be University, Bhubaneswar, Odisha, India, Prof. Dr. Harish Hirani, Director, CSIR-CMERI, Durgapur, India, Mr. Carlos Roberto Vieira da Silva, Director, Presidente, ABRELPE, Sao Paulo/SP - Brazil & Presidente, ISWA—International Solid Waste Association, Netherlands.
Description: First Edition. | Boca Raton, FL: CRC Press, 2023. |
Series: The circular economy in sustainable solid and liquid waste management | Includes bibliographical references and index.
Identifiers: LCCN 2022005843 (print) | LCCN 2022005844 (ebook) | ISBN 9781032137254 (hardback) | ISBN 9781032139463 (paperback) | ISBN 9781003231608 (ebook)
Subjects: LCSH: Refuse and refuse disposal. | Circular economy. | Recycling (Waste, etc.)—Law and legislation.
Classification: LCC HD4482.E44 2023 (print) | LCC HD4482 (ebook) | DDC 363.72/82—dc23/eng/20220428
LC record available at https://lccn.loc.gov/2022005843
LC ebook record available at https://lccn.loc.gov/2022005844

ISBN: 9781032137254 (hbk)
ISBN: 9781032139463 (pbk)
ISBN: 9781003231608 (ebk)

DOI: 10.1201/9781003231608

Typeset in Times
by codeMantra

Contents

Section I   Legal Framework, Strategies in Waste Management and Circular Economy

Section II   Circular Economy Adoption in Waste Management

Section III   Use of Mathematical and Statistical Modelling in Setting Waste Management Strategies

Section IV   Case Studies on WM, Sanitation and Hygiene

Foreword and Introduction

The establishment of a national regulatory control program in any country with appropriate legislation, regulations, ordinances and licenses is an extremely important step for protecting human health and the environment from the mismanagement of solid wastes. Furthermore, in the absence of regulatory controls, adequate treatment and disposal facilities are not developed. Environmental standards must be rigorously enforced in order to assure the public that the solid waste systems are operated in ways that protect human health and the environment. Many countries lack the national capacity to handle and manage solid wastes, primarily due to inadequate infrastructure, including inadequate facilities, lack of trained personnel, lack of information and monitoring systems, inadequate regulatory programs and insufficient financing. To achieve effective waste management as well as circularity of wastes, effective and appropriate policies, strategies, modeling, objectives, targets, performance indicators, best practices, measurement and monitoring are very much required in a country applicable at all the levels.

Recycling and recovery of materials and energy from solid waste not only reduces the volume of waste for disposal but also conserves natural resources. However, in order for recycling to be consistent with sustainable development, it must be economically feasible. Otherwise, resources are wasted not conserved. The elements of success of a recycling operation are the same as for any successful business; new facilities employing these new technologies capable of meeting stringent regulatory standards must be established and issued operating permits. Urban circular economy is one in which cities keep resources in use for as long as possible, extract the maximum value from them whilst in use, then recover and regenerate products and materials at the end of their life. It is a more efficient and environmentally sound alternative to the traditional linear economy model of extract-produce-consume-dispose-deplete (epcd2) to an elevated echelon of achieving zero waste. The circular economy concept is being recognized as a key driver for the much-wanted green transition, simultaneously enabling greater energy and material efficiency, lower pollution and GHG emissions and job creation. According to the World Economic Forum, moving toward a circular economy is the key, and a "trillion-dollar opportunity, with huge potential for innovation, job creation, resource conservation and economic growth".

To address the above issues, this book, "*Effective Waste Management and Circular Economy: Legislative framework and strategies*" with 24 chapters distributed in 4 different sections has been presented with a correlation among Legislative framework, Waste Management, different types of modelling and Circular Economy. This book covers four specific aspects divided into the sections, viz., (i) Legal Framework, Strategies in Waste Management and Circular Economy, (ii) Circular Economy adoption in Waste Management, (iii) Use of Mathematical and Statistical Modelling in setting waste management strategies and (iv) Case Studies on WM, Sanitation and Hygiene. Nine chapters included in Section I, presented a meso-scaled study of the pollution-specific appraisal of surface water quality in an Industrial region in India, effects of legal and technological aspects on the emission from incineration plants using MSW in Italy, management of temple wastes in one of the famous temples in India, the Shree Jagannath temple at Puri, the requirements of segregation methodologies of wastes at source in Germany. The chapter in the section also presented the new strategies adopted for solid waste management in municipalities of odisha, India using micro-composting process, strategies, Innovations and Safety Considerations in Waste Monitoring and Management in Lucknow, India and presented the trajectory planning for autonomous vehicles and its implementation for effective waste management in Kolkata, India. One chapter focuses on the waste management strategies for small and medium-sized enterprises to achieve Environmental Sustainability. Section II has three chapters, the adoption of circular economy concepts and recycling initiatives in a few countries, namely, Nigeria, Bandung City in Indonesia and Kolkata Metropolitan City in India have been presented by several researchers under the section title, Circular Economy adoption in Waste Management.

Section III is entitled as, Use of Mathematical and Statistical Modelling in setting waste management strategies. The section includes six chapters that involves the use of SWM Site selection indexing, composite index to form urban score card, linear and nonlinear regression methods for equilibrium modelling, Multi-criteria decision making (MCDM) tools, namely AHP, ANN, DSS, and Gate to Gate life cycle assessment in various aspects of waste management.

Case studies on WM, sanitation and hygiene entitled Section IV, presented seven chapters. Waste management has an impact on sanitation and hygiene in the society. A few case studies from four countries, namely, sanitation and hygiene coverage in Bangladesh at Narail District in Dhaka, Urban flood induced solid waste management in some metropolitan cities in India, best practices and models on waste management in the Philippines and to curb problems evolved from plastics wastes exploring the alternatives for food packaging in place of plastics in Egypt. One chapter in the section presented an interesting meta-analysis of studies on solid waste management in the Philippines, which gives a thorough idea of interest in research on SWM in the country.

The chapters in this book will definitely be helpful for the researchers, practitioners, implementation agencies in different countries, waste management service providers and policy makers to establish effective waste management system.

3rd July 2022
Jadavpur University
Kolkata, India
Prof. Sadhan Kumar Ghosh
Prof. Sasmita Samanta
Prof. Harish Hirani
Mr. Carlos Roberto Vieira da Silva
Editors

Acknowledgements

The editors acknowledge the support of the chairman and the organizing committee of the 8th IconSWM 2018, 9th IconSWM-CE 2019 and 10th IconSWM-CE 2020, which was attended by 44 countries and the governing council of the International Society of Waste Management, Air and Water for allowing the authors to contribute the chapters for this book as well as the contributing authors.

The editors acknowledge the support and encouragement of the following organizations with gratitude.

Members in the Secretariat of the International Society of Waste Management, Air and Water; UNCRD, Japan

IPLA, Global Secretariat at ISWMAW.

Representatives who joined from UNEP, UNDP, World Bank in the IconSWM-CEs.

Researchers in the INDIA H20 Horizon 2020 research project supported by DBT, Govt of India and European Union; Ocean Plastic Turned into an Opportunity in Circular Economy (OPTOCE) – Project funded by SINTEF, Norway; Indo-Hungary Industrial research Project at mechanical engineering department, Jadavpur University supported by DST, Govt of India and Hungarian Government; and the Circular Economy adoption within small and medium-sized enterprises in India and the United Kingdom funded by Royal Academy of Engineering; UK Government under the initiative of Newton Fund.

Members in the Centre for sustainable Development and Resource Efficiency Management, Jadavpur University;

Members of Quality Management Consultants;

Members of the Editorial Board of IconSWM-ISWMAW Publication Secretariat

The editors express their gratitude to the individuals whose support and encouragement are thankfully acknowledged.

Mrs. Pranati Ghosh, Dr. Sannidhya Kumar Ghosh,

Prof. Achyuta Samanta, Mr. C. R. C. Mohanty, Prof. P. Agamuthu,

Mrs. Ritasree Chatterjee, Mr. Soumen Chatterjee,

Members of CRC Press, Taylor & Francis Group who are involved in this publication.

Editors

Dr. Sadhan Kumar Ghosh, Professor in Mechanical Engineering since 1998 and Chief Coordinator, Centre for Sustainable Development and Resource Efficiency Management at Jadavpur University, India. He also served as the Dean, Faculty of Engineering and Technology and Head, mechanical engineering. He was the Director, CBWE, Ministry of Labour and Employment, Govt. of India and Larsen and Toubro Ltd. He is a renowned personality in the field of Waste Management, Circular Economy, Green Manufacturing, Supply Chain Management, Sustainable Development, Co-processing of Hazardous and MSW in cement kiln, Plastics Waste and E-waste management and recycling, management system standards (ISO) and TQM having three patents approved. Prof Ghosh is the founder Chairman of the IconSWM; President, International Society of Waste Management, Air and Water (ISWMAW) and the chairman, Consortium of Researchers in International Collaboration (CRIC). He received several awards in India and abroad including the distinguished visiting fellowship by the Royal Academy of Engineering, United Kingdom to work on *"Energy Recovery from MSW"*. He wrote 9 books, 40 edited volumes, more than 230 national and international articles and book chapters. He is the Associate Editor of Waste Management, Journal, Elsevier and International Journal of Materials Cycle and Waste Management (IJMCWM) and Editor-in-Chief of IconSWM-ISWMAW Secretariat. His significant contribution has been able to place the name of Jadavpur University on the world map of research on waste management. He is a consultant and international expert of UNCRD/DESA, Asian Productivity Organization (APO), Japan, China Productivity Council (CPC), SACEP Sri Lanka, IGES Japan etc. His international research funding includes European Union Horizon 2020, Erasmus plus, UKIERI, Royal Society–DST, GCRF UK, Royal Academy of Engineering, Georgia Govt. etc. He is the leader of the Collaborative International Research Project on *"Global Status of Implementation of Circular Economy (2018–2022)"* by ISWMAW involving experts from 44 countries. He was the convener of ISO TC 61 WG2, member in the Indian mirror committee of ISO TC 207 and ISO TC 275. He is an expert committee member of government initiatives and was the State Level Advisory Committee Member of Plastics Waste (Management and Handling) Rules 2011, expert committee member for the Preparation of standards for RDF for utilisation set up by the Ministry of Housing and Urban Affairs (MoHUA), govt of India, Chair Elect of 12th IconSWM 2022 to be held in Nov-Dec 2022 at SVU Tirupati. He is available at: sadhankghosh9@gmail.com and www.sadhankghosh.com.

Prof. Dr. Sasmita Rani Samanta, an insightful academician loaded with high degree of incisiveness and passion for research and schooling which is intermingling with innovative pragmatically approach is the Pro Vice-Chancellor, and former Reader and Professor, KIIT Deemed to be University, Odisha, India. Her theoretical know-how has been mellowing day by day out of shouldering the assigned task for the last one and half decades in different capacities of the academic institution. She implements the waste management policy of her own institution and implemented an energy recovery process for kitchen and garden wastes. Her leadership roles in KIIT ranges from the admission process, human resource management, talent acquisition, general administration, to institute governance and competency management and has taken several steps for execution of innovative academic practices in teaching–learning methodology, evaluation and connecting the students and faculty to the academia, engaging and retaining people, bringing transparency in system leading to high motivation, ownership and commitment among the employees, increasing their productivity for the triumph of organizational goals. Prof. Samanta is a fellow member of the Indian Social Science Congress and Royal Society of Arts, United Kingdom. She has a number of publications in national and international journals. She was the Co-chairman of the 10th IconSWM-CE (International Conference on Sustainable Waste Management towards Circular Economy) in November 2019.

Prof. (Dr.) Harish Hirani, Director, CSIR-CMERI and Professor, Department of Mechanical Engineering, Indian Institute of Technology Delhi is a prominent figure in the fields of Waste Management (MSW, Bio-medical waste, wastewater treatment), Inclusive Innovation, Rural Development and Tribology. Under his leadership, CSIR-CMERI has grown as a laboratory for waste management and engineering. He is recognized for Outstanding Vision, Leadership and Contribution in the Field of Engineering Innovation and R&D Management by The Institute of Engineers (India) in 2019. He was Vice-Chairman of JEE Advanced, 2014–16, Editor of Indian Journal of Tribology (IJT) and Vice-Chairman of Delhi Chapter of Tribology Society of India (TSI). He was awarded the BOYSCAST fellowship in 2004 by the DST, Govt. of India to pursue research with Prof. Nam P. Suh at Massachusetts Institute of Technology, United States. Before joining IIT Delhi he served IIT Bombay. He has a number of IPRs to his credit. He is the author of the book entitled *"Fundamentals of Engineering Tribology with Applications"*. He has guided several Ph.D. and M. Tech. theses and published nearly 70 research papers in peer-reviewed journal. Through his innovative thinking, he could transform waste into wealth and could setup a green colony at CSIR-CMERI colony Durgapur with absolutely zero waste. He started effectively integrated (mechanized segregation of waste, biomethanation, sludge to briquette, pyrolysis, plasma gasification, ashes to bricks) waste management systems, where each system is designed to complement the others. Using such system of decentralized waste management, he could prove to eliminate Landfills and minimize societal and ecological distress.

Mr. Carlos Roberto Vieira da Silva, Attorney at Law from São Bernardo do Campo School of Law, holds a PG title in Administrative and Economic Regulation Law from Mackenzie University. He is fluent in Portuguese, English and Spanish languages. He is the President and CEO of ABRELPE since 2013 and the President of ISWA since 2020. He is a non-state partner at the Climate and Clean Air Coalition (CCAC) since 2018 with active participation in the development and implementation processes of some Brazilian Federal Laws applied to waste management, such as Law n. 11.079/2004 (Public-Private Partnership), Law n. 11.445/2007 (National Basic Sanitation Law) and Law n. 12.305/2010 (National Solid Waste Law). Carlos is the coordinator of the Panorama of Solid Waste in Brazil, annual reference document about the waste industry in the country and the Principal Director of international projects developed to assist Brazilian municipalities such as Sao Paulo, Santos, Rio de Janeiro and Curitiba. He is the member of the Steering Committee for the Global Waste Management Outlook and for the Latin America and Caribbean Regional Waste Outlook, developed by UN Environment in 2015 and 2018 and member of the Technical Research Team for the project *"Analysis of Technologies directed to Solid Waste and Treatment and Destination in Brazil, Europe, USA and Japan"*, hosted by BNDES, and concluded in 2014. He is the Principal coordinator of the project to develop a manual focusing on Best Practices for Solid Waste Planning and Financing, developed by ISWA. Carlos is the consultant on PPPs and Waste Financing projects throughout Brazilian municipalities having active participation in drawing, developing and reviewing

Solid Waste Master Plans, such as the National Solid Waste Plan in 2011–2012 and its review since 2019. He is the co-author of books related to Waste Management: *Gestão de Resíduos Sólidos: o que diz a lei*" ("Solid Waste Management: what the Law says"); "*Municipal Waste Management Systems*"; "*Reverse Logistics and E-Waste*"; "*New Perspectives for Waste Management in Brazil*"; "*Legal framework for an adequate management of used pesticide packaging*" and author of articles, researches and studies focused on the solid waste industry.

Contributors

Arlen A. Ancheta
University of Santo Tomas
Manila, Philippines

Kaveti Anithamma
JNTUA College of Engineering
JNT University
Anantapur, India

Nirjhar Bar
Department of Chemical Engineering
University of Calcutta
Kolkata, India

Dilip Kumar Behara
JNTUA College of Engineering
JNT University
Anantapur, India

Pritha Bhattacharjee
Department of Environmental Science
University of Calcutta
Kolkata, India

Ritwija Bhattacharya
Department of Environmental Science
Calcutta University
Kolkata, India

Paromita Chakraborty
School of Civil Engineering
KIIT Deemed to be University
Bhubaneswar, India

Sourya Subhra Chakraborty
Civil Engineering Department
Meghnad Saha Institute of Technology
Kolkata, India

B. Chakradhar
Ramky Enviro Services Private Limited (wholly
    owned subsidary of REEL)
Hyderabad, India

Munish K. Chandel
Environmental Science and Engineering
    Department
Indian Institute of Technology Bombay
Powai, India

Suman Chatterjee
Department of Geography
Adamas University
Kolata, India

Bitanjaya Das
School of Civil Engineering
KIIT Deemed to be University
Bhubaneswar, India

Debasish Das
Department of Jute and Fiber Technology
Institute of Jute Technology
University of Calcutta
Kolkata, India

Sudip Kumar Das
Chemical Engineering Department
University of Calcutta
Kolkata, India

Sukanya Dasgupta
School of Architecture and Planning
KIIT University
Bhubaneswar, India

Brajesh Kumar Dubey
Indian Institute of Technology Kharagpur
Kharagpur, India

Saikat Dutta
Ramky Enviro Services Private Limited (wholly
    owned subsidary of REEL)
Hyderabad, India

Isabell Eickhoff
Department Waste & Resource Management
Rostock University
Rostock, Germany

Amira Sabet El-mahrouky
Food Engineering and Packaging Department
Food Technology Research Institute
Agriculture Research Center
Cairo, Egypt

Aml Salah Elnawawy
Food Engineering and Packaging Department
Food Technology Research Institute
Agriculture Research Center
Cairo, Egypt

Bhaskar Guin
Department of Mechanical Engineering
Jadavpur University
Kolkata, India

Marwa Mohamed Helmy
Food Engineering and Packaging Department
Food Technology Research Institute
Agriculture Research Center
Cairo, Egypt

Ratnawati Kusuma Jaya
Department of Social Welfare, Faculty of Social
    and Political Sciences
University of Indonesia
Depok, Indonesia

T.S. Sasi Jyothsna
Ramky Enviro Services Private Limited (wholly
    owned subsidiary of REEL)
Hyderabad, India

Kare H. Karstensen
SINTEF
Oslo, Norway

N. Niranjan Kumar
JNTUA College of Engineering
JNT University
Anantapur, India

Amit Kundu
School of Management
Techno India Group
Kolkata, India

Sari Viciawati Machdum
Department of Social Welfare, Faculty of Social
    and Political Sciences
University of Indonesia
Depok, Indonesia

Tanmoy Majumder
School of Civil Engineering
KIIT Deemed to be University
Bhubaneswar, India

Ashanendu Mandal
Chemical Engineering Department
University of Calcutta
Kolkata, India

Francesco Di Maria
LAR5 Laboratory - Department of Engineering
University of Perugia
Perugia, Italy

Nipu Modak
Department of Mechanical Engineering
Jadavpur University
Kolkata, India

Gert Morscheck
Department Waste & Resource Management
Rostock University
Rostock, Germany

Salwa Raafat Mostafa
Chemical Engineering Department
Cairo University
Cairo, Egypt

Alokesh Mridha
Department of Environmental Science
University of Calcutta
Kolkata, India
and
Department of Jute and Fiber Technology
Institute of Jute Technology
University of Calcutta
Kolkata, India

Aniruddha Mukhopadhyay
Department of Environmental Science
University of Calcutta
Kolkata, India

Rajdeep Mullick
Department of Chemical Engineering
University of Calcutta
Kolkata, India

R.K. Nanda
School of Management
KIIT Deemed to be University
Bhubaneswar, India

Tarun Kanti Naskar
Department of Mechanical Engineering
Jadavpur University
Kolkata, India

Michael Nelles
Department Waste & Resource Management
Rostock University
Rostock, Germany
and
Deutsches Biomasseforschungszentrum gGmbH
  (DBFZ)
German Biomass Research Centre gGmbH
Leipzig, Germany

Oluwadare Joshua Oyebode
Civil and Environmental Engineering Department
Afe Babalola University
Ado-Ekiti, Nigeria

Sudha Panda
School of Architecture and Planning
KIIT University
Bhubaneswar, India

Sagarika Panigrahi
Indian Institute of Technology Kharagpur
Kharagpur, India

Lynlei L. Pintor
Science Research Specialist
ERDB College
Laguna, Philippines

Moshiur Rahman
Bangladesh Public Administration Training
  Centre
Savar, Bangladesh

Baisali Rajbansi
Department of Chemical Engineering
University of Calcutta
Kolkata, India

Tejashwi Rana
Mechanical Engineering Department
Jadavpur University
Kolkata, India

P. Rautaray
School of Management
KIIT Deemed to be University
Bhubaneswar, India

Hari Prasad Reddy
National Institute of Technology (NIT)
Warangal, India

Parimal Kumar Roy
Bangladesh Public Administration Training
  Centre
Savar, Bangladesh

Swathy Sadala
Ramky Enviro Services Private Limited (wholly
  owned subsidary of REEL)
Hyderabad, India

Ipsita Saha
Department of Mechanical Engineering
Jadavpur University
Kolkata, India

Palash Kumar Saha
SINTEF
Oslo, Norway

Kaniska Sarkar
Assistant Registrar
Jadavpur University
Kolkata, India

S.V. Satyanarayana
JNTUA College of Engineering
JNT University
Anantapur, India

Richa Sen
Department of Environmental Science
University of Calcutta
Kolkata, India

Bhupendra K. Sharma
Environmental Science and Engineering
  Department
Indian Institute of Technology Bombay
Mumbai, India

V.P. Sharma
CSIR-Indian Institute of Toxicology Research
Lucknow, India

Manal Abdel Rahman Sorour
Food Engineering and Packaging Department
Food Technology Research Institute
Agriculture Research Center
Cairo, Egypt

Jan Sprafke
Department Waste & Resource Management
Rostock University
Rostock, Germany

Saikrishna Venna
National Institute of Technology (NIT)
Warangal, India

T. Vijaya Lakshmi
Centre for Environment
Jawaharlal Nehru Technological University
    Hyderabad
Hyderabad, India

Section I

Legal Framework, Strategies in Waste Management and Circular Economy

1

Pollution-Specific Appraisal of Surface Water Quality with Limited Resource Approach: A Meso-Scaled Study of the Durgapur Industrial Region, West Bengal, India

Suman Chatterjee
Adamas University

Kaniska Sarkar
Jadavpur University

CONTENTS

1.1 Introduction

Regular environmental monitoring, spatiotemporal assessment and wastewater management are critical in managing ecological degradation and maintaining green growth in a region undergoing rapid economic growth, especially in developing nations (OECD, 2012). However, environmental monitoring is a daunting task to carry out, and a pollution-specific appraisal should be conducted with region-specific waste-management strategies. There are numerous environmental parameters (more than a hundred) that can depict the environmental status. Industrial pollution and industrial wastes are different from other environmental pollutions (Woodard, 2006, p. 1). Hence, water quality variables should be selected based on a chemical inventory approach, i.e., the nature of the operation, chemicals used and the by-product of the specific operation. Parameter's environmental significance, feasibility of monitoring and availability of resources are some other important factors for selecting the parameters (Chapman & Kimstach, 1996). The number of samples, selection of water quality variables and spatiotemporal scale

DOI: 10.1201/9781003231608-2

play important roles in determining the costs of the programme (Alilou et al., 2018). The task of sample designing becomes daunting for large-scale monitoring since the number of variables tends to increase with the expansion of spatiotemporal domain (McDonald, 2003). Besides, in developing nations due to the inaccessibility or remoteness of the sampling locations, the survey itself becomes difficult and expensive (Safarzadeh-Amiri et al., 2011). Therefore, cost-effective sampling design and optimum variable selection for large-scale monitoring programmes become necessary. The United States Environmental Protection Agency (USEPA) has described the process of determining the parameters by two methods, which are to be applicable only when prior knowledge is available for a specific area. There is no such framework that has been developed which deals with the situation where there is no prior knowledge and when the resources are scarce. In most cases, investigators or decision-makers determine those crucial decisions subjectively.

The study focuses upon the impact of industrial pollution; hence, managing wastewater is inevitable to mitigate the problems arising from surface water pollution. Although there are many conventional techniques of wastewater management, not all the techniques can be implemented in all the areas of concern indiscriminately. The strategy should be region specific and should also be based on the pollution profile of the region. There are many issues regarding implementing the conventional techniques; hence, along with developing the strategies for environmental monitoring, emphasis has also been given upon ease of implementations, sustainability, cost-effectiveness, and other factors identified during the assessment of the study area (Edokpayi et al., 2017; Naidoo & Olaniran, 2013). Hence, based on the case study of surface water pollution and causes of region-specific noncompliances in environmental regulation, we have suggested and discussed some of the region-specific wastewater treatment strategies in the recommendation part.

It is evident that pollution impact studies have been mostly undertaken as site-specific (Banerjee & Gupta, 2011; Gupta et al., 2013), e.g., assessing effluent discharge points, stack emission to avoid the cost and labour needed for a large-scale survey. However, assessing the mediums at emitting and discharge points is not sufficient to elucidate the carrying capacity of any region. Besides, the nature of pollutants also determines the spatial extent of their impact. Most of the industrial pollutants are long-range transportable (LRT) and persistent in nature, e.g., the majority of metals, metalloids and persistent organic pollutants. These pollutants influence the environment on a large spatiotemporal scale (Franklin, 2006; Peine, 1999, p. 122). Industrial emissions also lead to air pollution and acidic deposition, which alters water chemistry and disrupts the natural process of maintaining the pH of the water on a regional scale (Chapman & Kimstach, 1996). In this study, we have assessed the impact of industrial pollution on the surface water quality of the Durgapur industrial region and its surroundings in a meso-scaled study with a limited resources approach, followed by suggesting region-specific wastewater management strategies based on the findings of pollution assessment and issues related to regulation's enforcement.

The following are the objectives of the study:

1. To form a methodological approach for selecting parameters for pollution-specific (industrial) environmental monitoring programmes in the following scenarios:
   - In a data-scarce region or having no prior knowledge about the study area
   - In the context of scarcity of resources and funds
2. To conduct a meso-scaled spatio-seasonal assessment of the surface water quality in and around the Durgapur industrial region, India and
3. To explore the strategies of region-specific wastewater management.

The study is conducted on a meso-spatial scale, which typically ranges within 1–10,000 sq. km (Becker & Nemec, 1987; Huggett, 1997). The region of interest is Durgapur industrial region and its surrounding area located between 23°44'22.88″ and 23°21'08.97″ north latitudes and 87°03'51.02″ and 87°29'46.072″ east longitudes (Figure 1.2) covering an area of ~2096.11 sq. km. The administrative units (community development blocks) that come under this region are Durgapur Municipal Corporation, Faridpur Durgapur, Ondal, Pandabeswar, Jamuria, Raniganj, Kanksa, Aushgram II, Galsi I of Bardhaman District, IllamBazar (partial), Dubrajpur (partial), Khoyrasol (partial) of Birbhum District and Mejhia (partial), Barjora (partial), Gangajalghati (partial), Sonamukhi (partial) and Saltora (partial) of Bankura District of the state West Bengal, India.

1.2 Materials and Methods: Cost-Effective Monitoring Strategies

To reduce the monitoring costs, (i) first we have determined the parameters through region-specific characterisation of by-products and effluents of the industries, (ii) second, we have excluded the correlated parameters, and finally, (iii) the parameters have been prioritised using an analytical hierarchy process (AHP) with the criteria, i.e. frequency of occurrence, parameter's significance and expenses of analysis (Figure 1.1). A monitoring programme aimed at assessing the anthropogenic pollution should focus on the chemical inventory approach, where types of chemical by-products generated in a specific process are to be considered (Chapman & Kimstach, 1996). The industrial process produces a broad spectrum of by-products and is sometimes released into the environment untreated. The spectrum of substances and their impact upon specific water characteristics vary across different types of industries. First, parameters have been listed as per their occurrence from the region-specific industrial profile based on the database from various works of literature (Chapman & Kimstach, 1996; Environmental Protection Agency, 1997; USEPA, 1973; Wen, 2009).

Later, parameters denoting similar aspects of water quality have been excluded, e.g. suspended solids and turbidity, dissolved oxygen and biochemical oxygen demand (BOD) (Chilingar et al., 2012), chloride and conductivity, acidity and pH (Walton, 1989), hardness, calcium and magnesium. Furthermore, we have compared several rate lists of water quality parameter's analytical costs from several government and nongovernment laboratories, e.g. Central Instrumentation Facility, BIT Mesra, Ranchi;

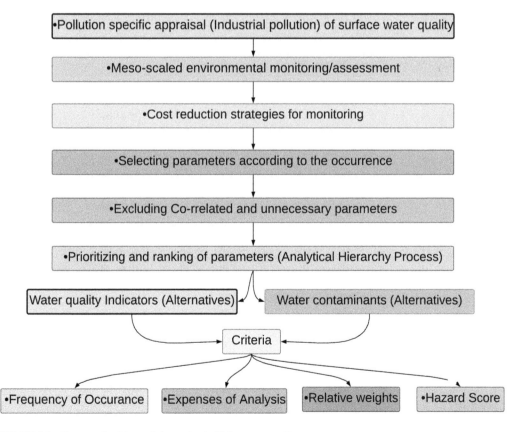

FIGURE 1.1　Framework of determining and prioritising water quality parameters.

Department of Public Health Engineering Bangladesh; Maharashtra Pollution Control Board; Pune Municipal Corporation; etc. and categorised the parameters as per their cost of testing as very high, high, moderate, low and very low (Table 1.2). Water quality parameters have been prioritised as per their frequency of occurrences, general environmental importance and expenses of analysis using the AHP technique. The Water Quality Parameters (WQPs) have been classified into two broad categories, i.e. water quality indicators (WQi) and contaminants or pollutants (Wc), and have been prioritised category-wise separately (Figure 1.1).

The AHP introduced by Saaty (1977, 1980) is a multiple criteria decision-making process that prioritises the alternatives with a pairwise comparison matrix of several criteria, yielding a ranking of parameters from which we have selected the parameters with higher priorities. The decision-maker can choose the threshold level of inconsistency. The general environmental importance is depicted as a hazard score and relative weights of the environmental importance of the contaminants and indicators, respectively. The system of Hazard Scoring (IRCH) involves scoring and ranking chemicals as per their hazard score, developed by Purdue University's Indiana Clean Manufacturing Technology and Safe Materials Institute (Indiana Clean Manufacturing Technology and Safe Materials Institute, 2001). We have used (i) frequency of occurrence and (ii) analytical cost as criteria for both WQi and Wc, and the criteria (iii) relative weight and (iv) hazard score are used for the WQi and Wc, respectively, as the environmental importance of the parameters (Table 1.1). The relative importance given to the criteria is determined subjectively by the author. For WQis, the relative weight has been considered three times more important than the frequency of occurrence and five times that of the analytical costs. Frequency of occurrence is considered three times more important than the analytical costs. Total Hazard score (IRCH) has been considered

TABLE 1.1

Showing Total Hazard Score (IRCH), Relative Weights, Frequency of Occurrence, Cost and Cost Scores of the Contaminants and Indicators

Pollutants/ Contaminants	Total Hazard Score (IRCH)	Occurrence (f)	Cost	Cost Score
Lead (Pb)	33	11	Very high	1
Mercury (Hg)	29	11	Very high	1
Cadmium (Cd)	33	11	Very high	1
Copper (Cu)	29	33	Very high	1
Nickel (Ni)	32	28	Very high	1
Iron (Fe)	7	37	Very high	1
Manganese (Mn)	21	5	Very high	1
Chromium (Cr)	33	37	Very high	2
Tin (Sn)	10	19	Very high	1
Zinc (Zn)	11	39	Very high	1
Arsenic (As)	30	5	Very high	1
Phenol (C_6H_5OH)	31	30	Moderate	3
Cyanide (Cn-)	41	29	Moderate	3
Indicators	**Relative Weights**	**Occurrence (f)**	**Cost**	**Cost Score**
BOD	0.182848	55	Very high	1
TSS	0.145117	55	Low	4
COD	0.130605	51	High	2
pH	0.122923	50	Very low	5
Oil and grease	0.116094	32	Moderate	3
Temp	0.111947	24	Very low	5
TDS	0.108838	24	Very low	5
Conductivity	0.081628	24	Very low	5

Source: Hazard score: Indiana Clean Manufacturing Technology and Safe Materials Institute (n.d.); Relative weight: Computed from the Delphi study and previous literature; Occurrence.

TABLE 1.2

Showing the Result of AHP

Parameter's Class	Parameters	Ideals	Normals	Raw	Remarks	Inconsistency Ratio Between Criteria	Inconsistency Ratio Between Criteria
Indicators	BOD	1	0.165062	0.082531	All Selected	0.03703	1.85E-14
	TSS	0.917834	0.151499	0.07575			
	pH	0.828485	0.136751	0.068376			
	COD	0.799642	0.13199	0.065995			
	Oil and grease	0.670412	0.110659	0.05533			
	Temperature	0.656994	0.108445	0.054222			
	TDS	0.644993	0.106463	0.053232			
	Conductivity	0.539987	0.089131	0.044566			
Contaminants	Cyanide (Cn-)	1	0.125347	0.062674	Selected		
	Phenol (C_6H_5OH)	0.864623	0.108378	0.054189	Selected		
	Chromium (Cr)	0.841999	0.105542	0.052771	Selected		
	Nickel (Ni)	0.690599	0.086565	0.043282	Selected		
	Copper (Cu)	0.675407	0.08466	0.04233	Selected		
	Cadmium (Cd)	0.61273	0.076804	0.038402	Selected		
	Lead (Pb)	0.61273	0.076804	0.038402	Selected		
	Mercury (Hg)	0.556416	0.069745	0.034873	Selected		
	Arsenic (As)	0.538042	0.067442	0.033721	Not selected		
	Zinc (Zn)	0.454446	0.056964	0.028482	Not selected		
	Manganese (Mn)	0.411335	0.05156	0.02578	Not selected		
	Iron (Fe)	0.387314	0.048549	0.024274	Not selected		
	Tin (Sn)	0.332193	0.04164	0.02082	Not selected		

Source: Computed in the study.

four times more important than the frequency of occurrence and three times that of the analytical cost. An analytical cost is considered two times more important than the frequency of occurrence.

Prioritisation result suggests that the BOD is the most important WQi followed by TSS, pH and COD (Table 1.2) with an inconsistency ratio of 0.03703 between criteria and 1.85E-14 between alternatives, which are much less than the 10% or 0.1. We have selected all the WQis in our study.

We have found cyanide, phenol, hexavalent chromium and nickel as the most important contaminants with an inconsistency ratio of 0.01759 between criteria and 2.26E-14 between alternatives. We have selected the first eight contaminants (Table 1.2) and excluded arsenic, manganese, zinc, iron and tin due to their lesser priority in the list.

1.3 Surface Water Pollution Assessment

Surface water samples for two seasons, summer (pre-monsoon: April to June) and winter (post-monsoon: November to January), from 34 stations (rivers, ponds, industrial waste ponds, streams, and canals) all over the study area have been collected (Figure 1.2) and analysed for spatio-seasonal aspects of physico-chemical properties of the water. Samples are preserved in proper condition and temperature, and have been analysed according to standard analytical methods of the American Public Health Association (APHA, 2017). Spatio-seasonal analysis of the surface water has been performed for the prioritised parameters.

We used appropriate interpolation approaches (chosen based on Root Mean Square Error) to simulate the spatial surface of SW pollution, for each parameter, and the water quality index (WQI), which is derived using the same parameters. We have computed WQI using the weighted arithmetic method due

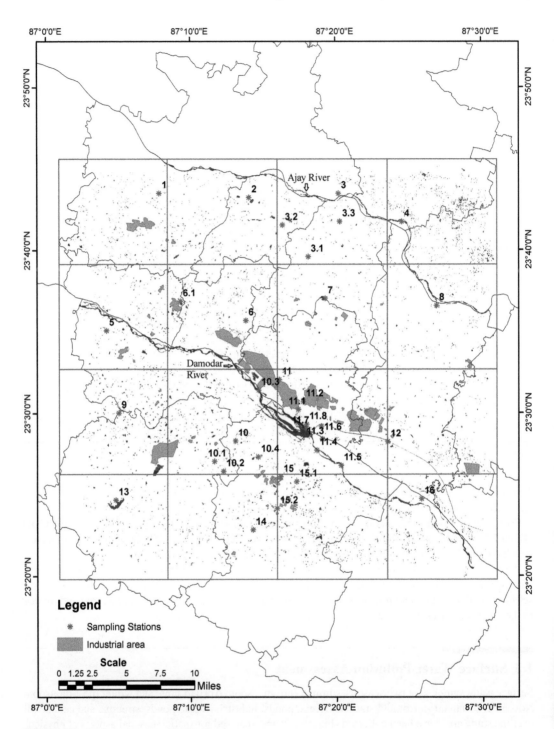

FIGURE 1.2   Surface water sampling locations.

to its simplicity and lesser data requirements (Brown et al., 1970; Tiwari & Mishra, 1985). Surface water bodies, i.e. river streams, canals, industrial waste ponds, and water bodies adjacent to both mines and industrial units, were given importance. Alarming parameters in the region have been identified, and suitable mitigation measures have been suggested.

Based on the targeted criteria in our study, i.e. water for fish culture and wildlife propagation, and water for irrigation, we have used two water quality standards: (i) Surface Water Quality Standards (as per IS: 2296) given by the Bureau of Indian Standards (Bureau of Indian Standards, 1982) and (ii) Surface Water Regulations [1989] given by the USEPA (Ryan & Council of EU communities, 1975).

1.4 Assessing Issues of Effluents Management from Firm's Perspective

Pollution abatement systems are expensive to install, operate and maintain. The government's excessively ambitious standards balloon this cost even further. It is exceedingly difficult to comply with the standards when the government does not give sufficient incentives, especially for small and medium-sized enterprises. Numerous incidents of noncompliance with the requirements have occurred in the region throughout the years, as evidenced by monthly reports (Government of West Bengal, 2018) demonstrating a high rate of noncompliance, resulting in regulatory directives for the closure of numerous units (Table 1.3). These documents may be found in the Regulatory Orders of WBPCB part of the West Bengal Pollution Control Board's official website. We noticed that the most common reasons for closure orders are the absence of an air pollution control device, malfunctioning of ETPs, discharge of ash-laden water, leaks in duct lines, improper storing and management of coal, dolochar, ESP dust, and other waste materials, and failure to meet the prescribed standard and obtain consent for hazardous substance operations. Several units have been ordered to cease operations due to regulatory breaches.

Most of the time, the high marginal cost of environmental compliance hinders businesses from investing in pollution control. As a result, they prefer to pay penalties for noncompliances when they are detected. Small businesses sometimes lack funds, fixed assets or working capital, and their lower market share inhibits them from investing in environmental protection (Priyadarshini & Gupta, 2003). Financial problems are common for small-scaled industries in pollution abatement. In addition, financial sustainability is often overlooked for the pollution abatement investments (Malik & Datta, 2005) which should be addressed at once for proper enforcement of the regulations.

TABLE 1.3

Public Complaints Lodge against Industrial Units for Pollution

Complaint Lodged Against	Memo No.	Allegation	Location
The Durgapur Projects Ltd	441-7/WPBD-Cont(416)/98 part-V	Discharge of ash-laden water to the Tamla Nullah TSS discharge standard for the ash pond overflow The wastewater generated from various activities of the plant	Tamla Nullah
M/s Durgapur Chemicals Ltd.	B-29016/04/06/IPC-I/	The unit has kept rejected salt and fly ash in open premises in the unit Source emission monitoring results indicates exceedance of parameter HCl (307 mg/NM^3 > 35 mg/NM^3)	Durgapur Industrial Area
M/s Durgapur Projects Limited	902-7/WPBD-Cont(416)/98 Part-IV	Parameters are found not to be strictly in compliance with the prescribed standards	Tamla Nullah
M/s. Durgapur Steel Plant, SAIL	190-7/WPBD-Cont(300)/98(Pt-X) Dated 30/07/2018	High emission of dust and smoke engulfing surrounding area and poor visibility on NH2	Surrounding area and poor visibility on NH2
M/S. Rajashree Iron Industries	MEMO NO.: 3618-671/WPB/ SEE(K.O.)GEN/2009 Dated 11-12-2009	The grievance was referred to as severe air pollution, especially at night	Residents of Ikhra, Bahadurpur, Dhasna villages

Source: Retrieved from http://www.wbpcb.gov.in.

TABLE 1.4

Showing Seasonal and Average Alarming SWQI Values of Water Bodies Directly Connected or Adjacent to the Industrial and Mining Units

Types of Surface water Bodies	Location	Sampling Station	Seasonal Average SWQI
Canal, streams and river	Surface water of effluent discharge point of Durgapur project limited, with Tamla Canal at Kalipur	11.2	113.9342
	Tamla Canal near Durgapur Chemicals limited	11.1	112.3907
	Tamla Canal at DPL coke oven colony near DPL sewage treatment plant	11.8	142.0949
	Damodar River near Paharpur in Barjora block, which is located on the opposite bank of the station 11.4	11.4	101.2159
	A small canal joining DVC canal and Damodar River near Sitarampur just after Durgapur barrage	11.3	154.993
	Downstream Damodar Bamandihi	11.5	106.5406
Surface water adjacent to the Industrial unit	Kajora Industrial complex, Andal	6.1	56.30897
Water bodies connected or adjacent to the mining area	Bilpahari, Sonpurbajari coal mine	2	84.19434662
	Hansdiha, Bhaluka, Sonpurbajari coal mine	3.2	71.54464

1.4.1 Result

Surface WQI values were found to be satisfactory for most of the region except the western part of the study area, i.e. the Jmuria-Ranigunj industrial and mining area, and the Downstream of Durgapur industrial area along the Damodar River. Surface water quality in winter (post-monsoon) was found to be better than that in summer (pre-monsoon). Water of a few streams, i.e. Tamla Canal (station 11.1, 11.2, 11.8), a small canal joining the DVC canal and Damodar River near Sitarampur just after Durgapur barrage (11.3), Damodar River near Bamandihi and Paharpur Barjora (11.5, 11.4), was found to be highly polluted (Table 1.4). Stations 1 and 14 during summer and Station 5 during winter are identified as outliers and can only be linked to local pollution, which is the reason for a higher value in the north-west part (Figure 1.3).

1.5 Alarming Parameters

Parameters exceeding the standard guidelines for the majority of locations or in both seasons are oil and grease, phenol, nickel and cadmium. The general slope of the region is towards the south-east; hence, due to surface run-off, it is natural to find a higher concentration of the substances in the same direction. Oil and grease is an effluent parameter, and its spatial distribution in the surface water depends upon proximity to the pollution source, the slope and surface run-off. It can be seen that the mean value of oil and grease concentration is higher during pre-monsoon (3.020942042 mg/L) than post-monsoon (1.877276744 mg/L) due to the summertime evaporation of surface water (Figure 1.4).

The study reveals that higher concentrations of phenol have accumulated along the Damodar River, mainly encompassing two centres, i.e. downstream of the Durgapur industrial belt and the Jamuria mining region. Higher mean concentrations are observed during the summer due to evaporation and lesser volume of water (Figure 1.5). In the case of nickel, it is concentrated more locally near the industrial, urban and mining region of Jamuria and Ranigunj block of Bardhaman district. The region is dominated by red-yellow ultisol soil and lower Gondwana sandstone shale, with a coal stem holding a higher concentration of Ni. For cadmium and nickel, greater concentration in surface water was found during summer due to intense evaporation of water (Pobi et al., 2019).

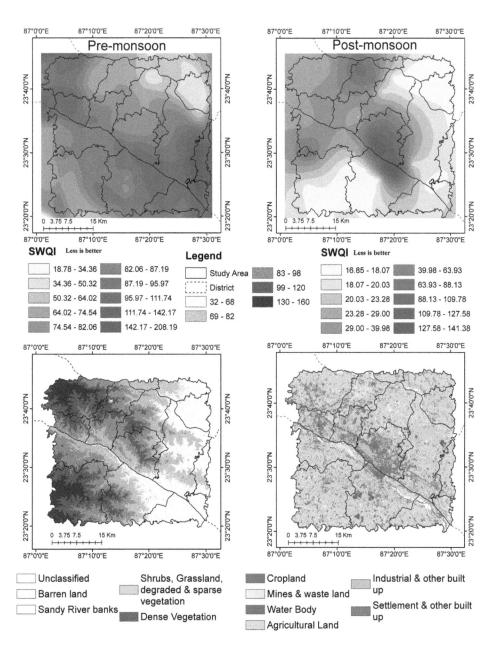

FIGURE 1.3 Spatio-seasonal variation of Surface WQI comparing with landcover and physiography of the region. (SWQI maps are computed in the study; landcover map is produced from LANDSAT OLI acquired on 04/03/2017 WRS 2 Path 139 and Row 44; physiographic map is produced from SRTM DEM 1 arc second.)

1.6 Conclusions and Recommendations

Our study aimed to reduce the cost of monitoring by systematically prioritising parameters. This systematic approach will help guide the investigator and decision-makers, especially when there is no baseline data present for a specific region. The reduction of monitoring cost becomes also necessary when we intend to conduct a large-scale study. This goal-specific parameters prioritisation has reduced the expenses without losing any important parameters that must be monitored. The approach is flexible since

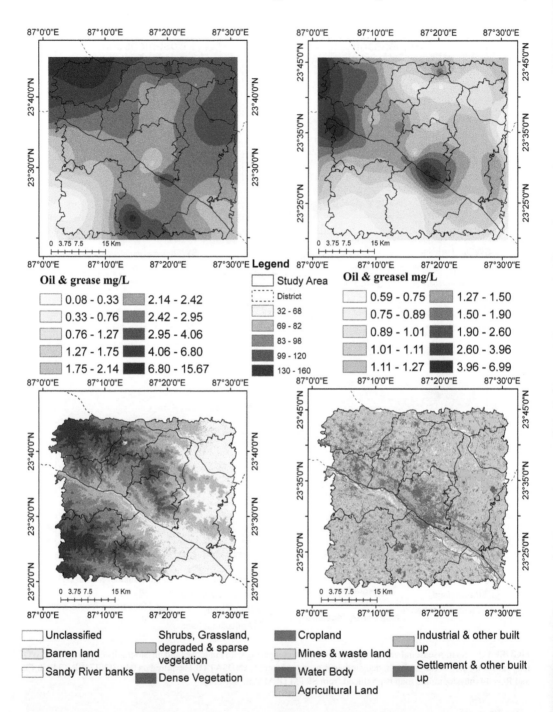

FIGURE 1.4 Spatio-seasonal variation of surface water oil and grease comparing with landcover and physiography of the region. (SWQI maps are computed in the study; landcover map is produced from LANDSAT OLI acquired on 04/03/2017 WRS 2 Path 139 and Row 44; physiographic map is produced from SRTM DEM 1 arc second.)

the importance of any factor (frequency of occurrence, importance of the parameters, and expenses) can be determined by the decision-maker.

It is evident that the western part of the study area, mostly dominated by coal mines and few industrial units, i.e. Jamuria, Ranigunj region, exhibits a serious situation of surface water quality, which indicates

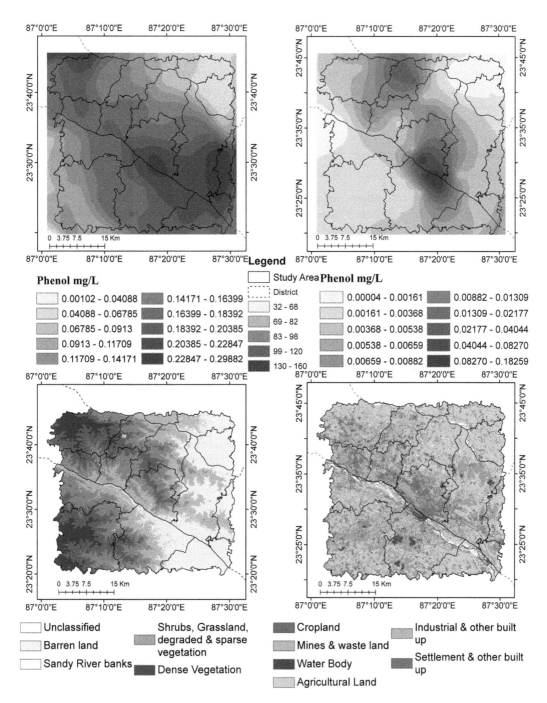

FIGURE 1.5 Spatio-seasonal variation of surface water total phenol comparing with landcover and physiography of the region. (SWQI maps are computed in the study; landcover map is produced from LANDSAT OLI acquired on 04/03/2017 WRS 2 Path 139 and Row 44; physiographic map is produced from SRTM DEM 1 arc second.)

that uncontrolled open cast coal mining also plays a significant role in the degradation of surface water quality in the region. It is concluded that surface water in the industrial vicinity area along the Damodar River is affected due to the untreated wastewater discharge coming from nearby industrial units and other local anthropogenic sources. In this study, the cost of monitoring has been significantly reduced by the exclusion of 13 parameters, i.e. iron (Fe), manganese (Mn), tin, zinc (Zn), arsenic (As), DO, chloride, turbidity, acidity, alkalinity, hardness, calcium and magnesium. By excluding these parameters, the analytical expenses have been significantly reduced. Each parameter is measured in 34 locations and in two seasons, which makes the total number of samples for analysis to be 884, which could have incurred a significant amount of costs. Because cost reduction is done in a systematic manner rather than depending just on judgement, the assessment programme gave a better result, which led to better decision-making and a more convenient and successful monitoring programme.

1.7 Recommendations

The financial constraint of medium- and small-scale enterprises, lack of awareness and resistance to technological modification in the facilities aggravated the environmental degradation in the region. One of the biggest disadvantages of traditional waste treatment/emission control methods, such as physical, chemical, and mechanical, is the high cost of installation and maintenance over time (Dubey et al., 2017). Another issue is that this technology requires highly trained technicians to operate. These two elements have become a big concern in emerging countries such as India, making it more difficult to comply with environmental standards. Phytoremediation has been proven effective in several investigations. Phytoremediation is a new pollution control technology that uses a variety of plants and microbes to reduce pollutants in the air, water, and soil while also restoring the natural environment (Garbisu et al., 2002; Lasat, 2002; Macek et al., 2000). It is a cost-effective technique compared to physical, chemical or mechanical treatment processes (Schnoor, 1997). The expenses are almost ten times cheaper than the conventional techniques (Dubey et al., 2017).

Bioremediation, biodegradation and biosorption of phenol by fungi, algae and certain plants play a role in the passive absorption of phenols, chloro- and nitro-phenols into the cell wall of plant tissue in an aqueous solution. In aerobic or anaerobic conditions, microorganisms such as bacteria and fungi break down and mineralise these chemicals to acquire energy and carbon. For example, *Bacillus stearothermophilus, Agaricus bisporus, Lentinus bisporus, Aerobic consortium* and *Acinetobacter johnsonii* can be used to break down phenol and *Pseudomonas putida* for 2-chlorophenol (Dhir, 2013).

Phenolic chemicals which are soluble in water are also taken up by the plants through their roots and leaves. Aquatic macrophytes, e.g., *Potamogeton crispus, Lemna gibba, Lemna minor*, etc., are good absorbers of phenolic compounds (Dhir, 2013). In addition, plants and microbes exude enzymes, i.e. reductases, oxidases, phosphatases and dehalogenases, that can metabolise and transform xenobiotic substances, i.e. polynitrate aromatic compounds, organochlorine insecticides, aromatic amines and phenols (Table 1.5).

1.8 Biodegradation of Oil, Grease and Fat in the Surface Water

The method of consuming the lipid through bioaugmentation and biodegradation process has increasingly come into practice. In such degradation, microbes, i.e. bacteria, break down the lipid using enzymes.

TABLE 1.5

Enzyme Induce Bioremediation of Phenol

S/N	Type of Phenol	Enzyme
1	Phenol	Phenol hydroxylase
2	Phenol	Catechol 2,3 dioxygenase
3	Phenol	Peroxidase
4	Phenol	Horse radish peroxidase
5	Methoxyphenol	Laccase

Source: Nair et al. (2008).

Oil, grease and fats are considered the slow biodegradable particulate organic matter; hence, the use of bioremediation technology became relevant compared to inorganic pollutants (Metcalf & Eddy, 2003). Bacteria and fungi that exudate lipolytic enzymes are commonly used for the conversion of complex lipids and fats into readily bioavailable long-chained fatty acid compounds in aerobic or anaerobic conditions (Tano-Debrah et al., 1999; Wakelin & Forster, 1997, pp. 37–43).

1.8.1 Oilzapper, Oilivorous-S and Oilivorous-A

An indigenous bacterial consortium has been developed by The Energy and Resource Institute (TERI) with the help of the Department of Biotechnology, Ministry of Science and Technology, Government of India. In this consortium, five different bacterial strains have been derived from various efficient bacterial cultures thriving in the natural environment from different geoclimatic regions of India. These bacterial strains feed on hydrocarbon compounds present in crude oil and oily sludge. Hence, they are capable of degrading different fractions of total petroleum hydrocarbon (TPH) of the oily waste (Saxena, 2015).

1.9 Phytoremediation and Riparian Buffer Zone

Creating a riparian buffer zone encompassing the surface water bodies, especially near the mines of Pandaveswar blocks and industrial areas of Andal and Ranigunj region, will be a very effective but low-cost management approach in this scenario. This will protect water quality from the contaminants present in the surface run-off by filtering out contaminants, maintaining a healthy aquatic ecosystem and mitigating bank erosion. The selection of plants for this approach can be combined with phytoremediation techniques which will increase the efficacy of this approach.

1.9.1 Common Effluent Treatment Plants (CETP)

One of the major issues of installing and maintaining the effluent treatment plant in the least developed countries is financial constraints. Common Effluent Treatment Plants (CETP) are a newly emerging technique, which addresses this problem by sharing the total costs between participating companies or with the government (state and central) organisations. Guidelines for establishing the CETP are discussed in the technical guidelines manual for common effluent treatment plants by the Ministry of Environment and Forest, the Government of India.

1.10 Limitations of the Study

Since industrial pollution is the foreground of the study, the study has focused on inorganic pollution and pollutants. Therefore, in this study, some of the important biological parameters, i.e. pathogens, the majority of organic hydrocarbons, herbicides, pesticides and volatile organic compounds, have been excluded. Besides being a region-specific study instead of being systematic, there can be some disagreement regarding industry-specific by-products, occurrences and general environmental importance derived in the study.

Acknowledgements

This study is supported by UGC-NET JRF fellowships by the University Grants Commission of India and the Department of Geography at Jadavpur University. The authors would also like to thank Dr Rajib Dey and Amit Kumar Bhandary of the Department of Metallurgical & Material Engineering, Jadavpur University for their cooperation.

REFERENCES

Alilou, H., Moghaddam Nia, A., Keshtkar, H., Han, D., & Bray, M. (2018). A cost-effective and efficient framework to determine water quality monitoring network locations. *Science of the Total Environment.* doi: 10.1016/j.scitotenv.2017.12.121.

APHA. (2017). *Standard Methods for the Examination of Water and Wastewater,* 23rd ed. American Public Health Association, Washington, DC. ISBN 9780875532356.

Banerjee, U. S., & Gupta, S. (2011). Statistical evaluation of hydrochemical parameters of the Damodar river water near Durgapur industrial complex, West Bengal, India. *Journal of Industrial Pollution Control,* 27(1), 51–60.

Becker, A., & Nemec, J. (1987). Macroscale hydrologie models in support to climate research. *The Influence of Climate Change and Climatic Variability on the Hydrologie Regime and Water Resources,* 431–445. http://hydrologie.org/redbooks/a168/iahs_168_0431.pdf.

Brown, R., McClelland, N., Deninger, R., & Tozer, R. (1970). A water quality index: Do we dare? Water & sewage works. A water quality index: Crashing the psychological barrier. In: W. A. Thomas (Ed.) *Indicators of Environmental Quality* (vol. 117, pp. 339–343). Springer US, Berlin, New York.

Bureau of Indian Standards. (1982). IS 2296 Tolerance limits for Inland surface water subject to pollution. Indian Standard.

Chapman, D. V., & Kimstach, V. (1996). Selection of water quality variables. In: D. Chapman (Ed.), *Water Quality Assessments,* 2nd ed. E&FN Spon. doi: 10.4324/noe0419216001.ch3.

Chilingar, G. V., Khilyuk, L. F., & Reike, H. H. (2012). Probability in petroleum and environmental engineering. *Probability in Petroleum and Environmental Engineering.* doi: 10.1016/C2013-0-15500-6.

Dhir, B. (2013). *Phytoremediation: Role of Aquatic Plants in Environmental Clean-Up.* Springer, India. doi: 10.1007/978-81-322-1307-9.

Dubey, N. K., Chandra, R., & Kumar, V. (2017). Phytoremediation: A green sustainable technology for industrial waste management. *Phytoremediation of Environmental Pollutants.* doi: 10.4324/9781315161549.

Edokpayi, J. N., Odiyo, J. O., & Durowoju, O. S. (2017). Impact of wastewater on surface water quality in developing countries: A case study of South Africa. In: H. Tutu (Ed.) *Water Quality.* InTech, London. doi: 10.5772/66561.

Environmental Protection Agency. (1997). *Wastewater Treatment Manuals.*

Franklin, J. (2006). Long-range Transport of Chemicals in the Environment. https://www.eurochlor.org/wp-content/uploads/2019/04/sd10-long_range_transport-final.pdf.

Garbisu, C., Hernández-Allica, J., Barrutia, O., Alkorta, I., & Becerril, J. M. (2002). Phytoremediation: A technology using green plants to remove contaminants from polluted areas. *Reviews on Environmental Health.* doi: 10.1515/REVEH.2002.17.3.173.

Government of West Bengal. (2018). Report of the Comptroller and Auditor General of India on Performance Audit of Pollution by Industries in West Bengal (Economic Sector) Report No. 5. https://cag.gov.in/sites/default/files/audit_report_files/Report_No_5_of_2018_Performance_Audit_of_Pollution_by_Industries_in_West_Bengal_Government_of_West_Bengal.pdf.

Gupta, S., Satpati, S., Saha, R. N., & Nayek, S. (2013). Assessment of spatial and temporal variation of pollutants along a natural channel receiving industrial wastewater. *International Journal of Environmental Engineering.* doi: 10.1504/ijee.2013.050893.

Huggett, R. J. (1997). *Environmental Change: The Evolving Ecosphere.* Routledge. doi: 10.4324/9780203440124.

Indiana Clean Manufacturing Technology and Safe Materials Institute. (2001). Indiana relative chemical hazard score: Alpha order. https://p2infohouse.org/ref/17/16325.pdf.

Lasat, M. M. (2002). Phytoextraction of toxic metals: A review of biological mechanisms soil microorganisms and metal phytoextraction. *Journal of Environmental Quality,* 31, 109–120.

Macek, T., Macková, M., & Káš, J. (2000). Exploitation of plants for the removal of organics in environmental remediation. *Biotechnology Advances.* doi: 10.1016/S0734-9750(99)00034-8.

Malik, P. K., & Datta, S. (2005). Financial sustainability of environmental investment under an empirical pollution abatement policy instrument in India: The case of wastewater treatment. *Environmental Science & Policy,* 8(1), 67–74. doi: 10.1016/j.envsci.2004.08.007.

McDonald, T. L. (2003). Review of environmental monitoring methods: Survey designs. *Environmental Monitoring and Assessment.* doi: 10.1023/A:1023954311636.

Metcalf, W., & Eddy, C. (2003). *Wastewater Engineering: Treatment and Resource Recovery*, 5th ed. McGraw-Hill Education, New York.

Naidoo, S., & Olaniran, A. (2013). Treated wastewater effluent as a source of microbial pollution of surface water resources. *International Journal of Environmental Research and Public Health*, 11(1), 249–270. doi: 10.3390/ijerph110100249.

Nair, C. I., Jayachandran, K., & Shashidhar, S. (2008). Biodegradation of phenol. *African Journal of Biotechnology*, 7(25). https://www.ajol.info/index.php/ajb/article/view/59706.

OECD. (2012). Green growth and developing countries a summary for policy makers. https://www.oecd.org/dac/50526354.pdf.

Peine, J. D. (1999). *Ecosystem Management for Sustainability : Principles and Practices*. CRC Press, Boca Raton, FL.

Pobi, K. K., Satpati, S., Dutta, S., Nayek, S., Saha, R. N., & Gupta, S. (2019). Sources evaluation and ecological risk assessment of heavy metals accumulated within a natural stream of Durgapur industrial zone, India, by using multivariate analysis and pollution indices. *Applied Water Science*. doi: 10.1007/s13201-019-0946-4.

Priyadarshini, K., & Gupta, O. K. (2003). Compliance to environmental regulations: The Indian context. *International Journal of Business and Economics*, 2(1). http://www.ijbe.org/table of content/pdf/vol2-1/vol2-1-02.pdf.

Ryan, R., & Council of EU Communities. (1975). Council directive concerning the quality required of surface water intended for the abstraction of drinking water in the Member States. *Official Journal of the European Communities*.

Saaty, T. L. (1977). A scaling method for priorities in hierarchical structures. *Journal of Mathematical Psychology*. doi: 10.1016/0022-2496(77)90033-5.

Saaty, T. L. (1980). *The Analytic Hierarchy Process: Planning, Priority Setting, Resource Allocation*. MacGraw-Hill, New York.

Safarzadeh-Amiri, A., Fowlie, P., Kazi, A. I., Siraj, S., Ahmed, S., & Akbor, A. (2011). Validation of analysis of arsenic in water samples using Wagtech Digital Arsenator. *Science of the Total Environment*. doi: 10.1016/j.scitotenv.2011.03.016.

Saxena, S. (2015). Environment and microbes. *Journal of Applied Microbiology*, pp. 55–64. Springer, India. doi: 10.1007/978-81-322-2259-0_5.

Schnoor, J. L. (1997). Phytoremediation: Technology Evaluation Report. Gwrtac E Series.

Tano-Debrah, K., Fukuyama, S., Otonari, N., Taniguchi, F., & Ogura, M. (1999). An inoculum for the aerobic treatment of wastewaters with high concentrations of fats and oils. *Bioresource Technology*. doi: 10.1016/S0960-8524(98)00181-3.

Tiwari, T. N., & Mishra, M. (1985). A preliminary assignment of water quality index to major Indian rivers. *Indian Journal of Environmental Protection*, 5, 276–279.

USEPA. (1973). *Handbook for Monitoring Industrial Wastewater*. Associated Water, Air Resources Engineers, & United States. Environmental Protection Agency. https://nepis.epa.gov/Exe/ZyPDF.cgi/20008SJY.PDF?Dockey=20008SJY.PDF.

Wakelin, N. G., & Forster, C. F. (1997). An investigation into microbial removal of fats, oils and greases. *Bioresource Technology*. doi: 10.1016/S0960-8524(96)00134-4.

Walton, N. R. G. (1989). Electrical conductivity and total dissolved solids—what is their precise relationship? *Desalination*, 72(3), 275–292. doi: 10.1016/0011-9164(89)80012-8.

Wen, X. (2009). Industrial pollution. In: Y. Chen & J. Qian (Eds.), *Point Sources of Pollution: Local Effects and Control* (vol. I, pp. 147–170). Encyclopaedia of Life Support System (EOLSS), Oxford, UK.

Woodard, F. (2006). *Industrial Waste Treatment Handbook*. Butterworth-Heinemann, Elsevier. doi: 10.1016/s0304-3894(01)00391-0.

2

Emissions from the Incineration of Municipal Solid Waste: Effects of Legal and Technological Progress

Francesco Di Maria
University of Perugia

CONTENTS

2.1 Introduction

With about 490 plants and about 96 million tonnes of waste treated, the EU is the area where incineration of waste was largely adopted (CEWEP, 2017). In early 1880s, waste incineration was adopted across Europe for managing rotting waste and for reducing the need for an area for final disposal. First incinerators were generally constituted by a combustion chamber for solid fuels (e.g. coal) and in some cases by a tiny gas treatment system generally limited to fly ash removal. These were small-sized plants largely diffused mainly in industrialized areas. No substantial differences were considered between such incinerators and other combustors. This approach was adopted for about 100 years from 1880 up to 1984. From then, something began to change. The first relevant event that introduced a gradual modification of the legislation concerning waste incineration was the work done by Kees Olie et al. (1977). In this study, Olie et al. detected the presence of polychlorinated dibenzo-para-dioxins (PCDD) and polychlorinated dibenzofurans (PCDF) in the ashes entrapped in the filters of incinerators. As a consequence of this evidence, starting from 1984, the introduction of more stringent legislation concerning technical specifications and emissions standards caused the shutdown of several old incinerators. This legislation was the consequence of the EEC directives 80/779/EEC (EEC, 1980), 82/884/EEC (EEC, 1982), 84/360/EEC (EEC, 1984) and 85/203/EEC (EEC, 1985). Of particular relevance was Article 3 of the directive 84/360/EEC, which introduced the necessity of prior authorization for specific industrial plants including incinerators and the respect of emission limits (Article 4). All these directives were referred to industrial plants, including incinerators. The transposition of these directives in the legislation of each EU Member State led to the first legal emission limits in Italy, 1990, specifically designed for incinerators (DM, 1990). The first EU directive addressing specifically waste incinerators was 2000/76/EC issued in 2000 (EC, 2000). The aim of this last directive was a further decrease in the emission levels from waste incineration (Di Maria, 2018a,b).

Concerning the impact on the environment of industrial plants, including incinerators, the Integrated Pollution Prevention and Control (IPPC) directive 96/61/EC (EC, 1996) introduced a substantial modification to the previous approach that focused only on gaseous emissions. The IPPC aimed to provide integrated prevention of pollution by introducing the following two main principles:

- Considering the whole impact that the plant can have on the environment (e.g. air, water, soil, noise, odour and landscape). In this approach, the different emission limits are a necessary condition but not sufficient for achieving high environmental protection.
- The adoption of the Best Available Techniques (BAT) (EC, 2010).

In the following, the evolution of the emission limits and the effective emissions from waste incineration were assessed with a particular focus on Italy.

2.2 Materials and Methods

The study was performed by a review of scientific papers, of official documents released by legal authorities and by an analysis of past and current legislation.

2.3 Results and Discussion

Table 2.1 represents the trend of the emission limits imposed by the legislation for waste incineration from 1990 until 2000. Limits related to the year 2000 are those imposed by the waste incineration directive 200/76/EC, the ones which are currently adopted. It is important to note that the last revision to the best reference documents and the related BAT for waste incineration that is under approval by the EC will further reduce the emission limits from 20% to 80%.

In any case, the evidence of the positive effects of both technical and emission limits implementation is reported in Figure 2.1 concerning the emission of dioxin in Italy from 1990 to 2017. Although the amount of waste incinerated increased from about 1,800,000 tonnes in 1990 to about 5,500,000 tonnes in 2017, the whole amount of dioxin emitted was drastically decreased from about 100-g I-Teq to about 7.50-g I-Teq.

Figure 2.2 shows the trend of the total emission of macro-pollutants such as CO, NOx, non-methane volatile organic compounds (NMVOCs) and SOx emitted by waste incineration from 1990 to 2017 in Italy. CO emissions were quite constant in the last year even if the amount of waste burned increased.

TABLE 2.1

Emission Limits Trend

Pollutant	Unit	1990	1997 Half hour	1997 Hourly	2000 Daily	2000 Half hour
PM	mg/Nm³	30	10		10	30
Cd	mg/Nm³	0.2	-	0.05	0.05 (half hours)	0.1 (8 hours)
Hg	mg/Nm³	0.2	-	0.05	0.05 (half hours)	0.1 (8 hours)
Pb	mg/Nm³	5	Na	Na	Na	Na
Heavy metals total	mg/Nm³	5	-	0.5	0.5 (half hour)	0.1 (half hour)
HF	mg/Nm³	2	1	4	1	4
HCl	mg/Nm³	50	20	40	10	50
Cyanate	mg/Nm³	1	Na	Na	Na	Na
SOx	mg/Nm³	300	100	200	50	200
NOx	mg/Nm³	500	200	400	200	400
TCDD-TCDF	mg/Nm³		0.0001 (8 hours)		0.0001 (8 hours)	
PCDD-PCDF	mg/Nm³	4	Na	Na	Na	Na
PCB	mg/Nm³	500	Na	Na	Na	Na
PAH	mg/Nm³	0.05	0.01 (8 hours)		Na	Na
CO	mg/Nm³	100	50	100	50	10
TOC	mg/Nm³	20	10	20	10	20

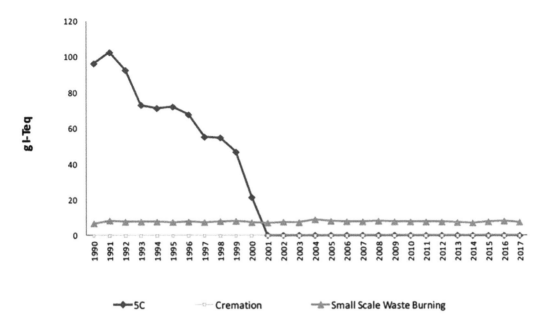

FIGURE 2.1   Dioxin emissions as g I-Teq in Italy from 1990 to 2017.

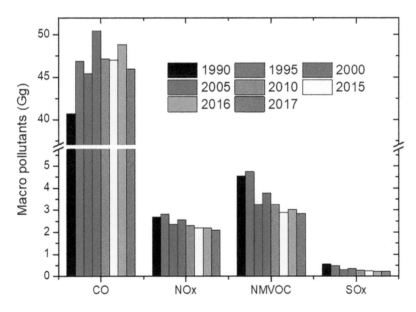

FIGURE 2.2   Total amount of CO, NOx, NMVOC and SOx emitted in Italy from 1990 to 2017.

For all the other pollutants, there was a significant reduction as the NMVOC indicated a significant progress in the combustion process. Very important is the trend of the global emission of heavy metals (Figure 2.3) for which a relevant decrease was detected during the years. In particular, some metals such as As and Hg have practically disappeared thanks to the prevention policy aimed to remove such metals from the market for the production of stuff and goods. Strong reductions have also been detected for the amount of Ni and Pb with an average reduction of >90% from 1990 to 2017.

Among the other micro-pollutants, the amount of PAH emitted was practically constant through the investigated period, whereas the emission of dioxin, HCB and PCB decreased noticeably (Figure 2.4).

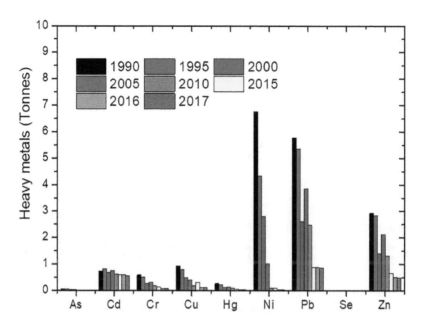

FIGURE 2.3   Total amount of main heavy metals emitted in Italy from 1990 to 2017.

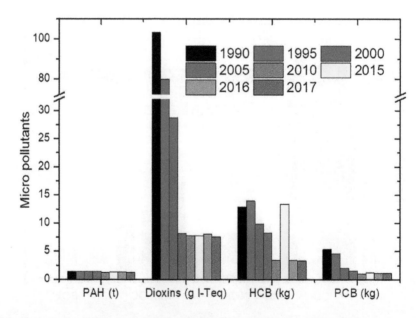

FIGURE 2.4   Total amount of PAH, dioxins, HCB and PCB emitted in Italy from 1990 to 2017.

The amount of particulate matter emitted was also quite constant from 1990 to 2017 (Figure 2.5). It is interesting to note that more than 80% of PM is represented by particle size <2.5 micron.

The concentration of pollutants in the exhaust gases at the stack was influenced by several parameters such as the waste composition and the efficiency of the combustion process. The efficiency in the removal of some pollutants resulted strongly by the gas treatment system and by the technologies adopted. Figure 2.6 shows the gas treatment system that was implemented in a new generation large-scale incinerator built in 2013. Electrostatic precipitators (EPs) were mainly used followed by a dry adsorber reactor for acid gases, fabric filters (FFs) and a selective catalytic reduction (SCR) reactor for

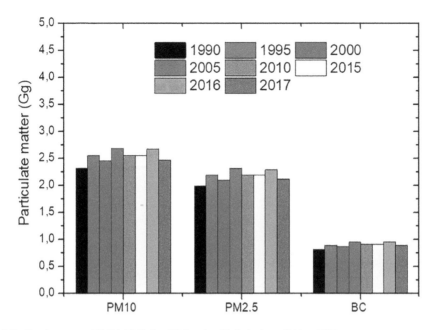

FIGURE 2.5 Total amount of PM10, PM2.5 and BC emitted in Italy from 1990 to 2017.

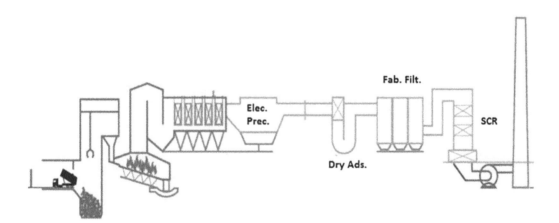

FIGURE 2.6 Flue gas treatment system with electrostatic precipitator, dry adsorption, fabric filters and SCR.

NOx removal, positioned before the stack. Figure 2.7 reports the flue gas treatment line implemented in an older small-scale incinerator operating since 2000. This line consists of a semi-dry basic reactor for acid gas removal followed by an FF. In both cases, the treatment also provides a post-combustion area inside the combustion chamber, a selective non-catalytic reduction (SNCR) reaction consisting of direct injection of urea in the combustion chamber for partial NOx removal and activated carbon injection if the flue gas treatment line is positioned before the FFs for further removal of dioxin and heavy metals.

The concentration of the pollutant at the stack of the two facilities was reported in Table 2.2. There is a possibility that the presence of SCR has a very relevant influence on the NOx concentration as on the concentration of the NH_3 as the slip factor in those systems exploits only SNCR.

Quantity of other pollutants such as acids (SOx and HCl) were generally lower in recent treatment system equipped with both EP and FF and in between a dry adsorber. CO and Hg concentrations were lower in the small-scale plant, whereas TCO was practically the same.

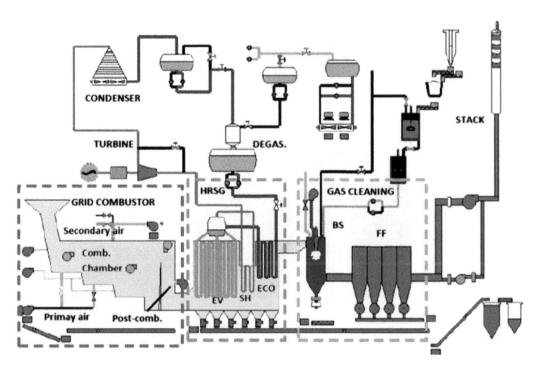

FIGURE 2.7 Flue gas treatment system with semi-dry basic reactor (BS) and fabric filters (FFs).

TABLE 2.2

Emission Levels from Two Different Flue Gas Treatment Systems (Figures 2.6 and 2.7)

Pollutant	U.M.	Value (Figure 2.6)	Value (Figure 2.7)
HCl	mg/Nm³	1.6	4.46
CO	mg/Nm³	6.5	2.28
NOx	mg/Nm³	41.0	160.3
SOx	mg/Nm³	0.83	11.0
TOC	mg/Nm³	0.37	0.32
PM	mg/Nm³	0.00	0.00
HF	mg/Nm³	0.00	0.00
NH_3	mg/Nm³	0.90	3.64
Hg	µg/Nm³	8.10	1.00

2.4 Conclusion

The progress in both legal prescriptions and gas treatment technologies had a relevant positive effect on the reduction of the amount of the main pollutant emitted by the incineration of municipal waste. For some pollutants and dioxins and furans, the reduction was >80%, whereas the percentage was less in case of the other pollutants. A relevant role due to technological progress was detected for the reduction of NOx and NH_3 and for some acid gases. Otherwise less relevant benefits were detected on the parameters detecting the efficiency of combustion that resulted very high in the full-scale plants for more than 20 years.

REFERENCES

CEWEP, 2017. Waste-to-energy plants in Europe 2017. Available at: http://www.cewep.eu/waste-to-energy-plants-in-europe-in-2017/ (accessed 02.08.2019).

Di Maria F, Bidini G, Lasagni M, Boncompagni A, 2018a. On time measurement of the efficiency of a waste-to-energy plant and evaluation of the associated uncertainty. *Applied Thermal Engineering* 129, 338–344.

Di Maria F, Sisani F, Contini S, 2018b. Are EU waste-to-energy technologies effective for exploiting the energy in bio-waste? *Applied Energy* 23, 1557–1572.

DM, 1990. Linee guida per il contenimento delle emissioni degli impianti industriali e la fissazione dei valori minimi di emissioni. Available at: https://www.provincia.pd.it/sites/default/files/paginabase/dm_12-07-1990.pdf.

EC, 1996. Council Directive 96/61/EC of 24 September 1996 concerning integrated pollution prevention and control. Available at: https://eur-lex.europa.eu/legal-content/EN/TXT/PDF/?uri=CELEX:31996L0061&from=EN.

EC, 2000. Directive 2000/76/EC of the European Parliament and of the Council of 4 December 2000 on the incineration of waste. Available at: https://eur-lex.europa.eu/legal-content/EN/TXT/PDF/?uri=CELEX:32000L0076&from=IT.

EC, 2010. Directive 2010/75/EU of the European Parliament and of the Council of 24 November 2010 on industrial emissions (integrated pollution prevention and control). Available at: https://eur-lex.europa.eu/legal-content/EN/TXT/PDF/?uri=CELEX:32010L0075&from=EN.

EEC, 1980. Council Directive of 19 October 1981 adapting, consequent upon the accession of Greece, Directive 80/779/EEC on air quality limit values and guide values for sulphur dioxide and suspended particulates. Available at: https://eur-lex.europa.eu/legal-content/EN/TXT/PDF/?uri=CELEX:31981L0857&from=IT.

EEC, 1982. Council Directive of 3 December 1982 on a limit value for lead in the air (82/884/EEC). Available at: https://eur-lex.europa.eu/legal-content/EN/TXT/PDF/?uri=CELEX:31982L0884&from=IT.

EEC, 1984. Council Directive of 28 June 1984 on the combating of air pollution from industrial plants (84/360/EEC). Available at: https://eur-lex.europa.eu/legal-content/EN/TXT/PDF/?uri=CELEX:31984L0360&from=EN.

EEC, 1985. Council Directive of 7 March 1985 on air quality standards for nitrogen dioxide (85/203/EEC). Available at: https://eur-lex.europa.eu/legal-content/EN/TXT/PDF/?uri=CELEX:31985L0203&from=EN.

Olie K, Vermeulen PI, Hutzinger O, 1977. Chlorodibenzop-dioxins and chlorodibenzofurans are trace components of fly ashe and flue gas of some municipal incinerators in the Nethernalnds. *Chemosphere* 6, 454–459.

3

Legislative Support to Curb Marine Littering through Wastes Co-processing by Promoting Circular Economy

Sadhan Kumar Ghosh
Jadavpur University

Sourya Subhra Chakraborty
Meghnad Saha Institute of Technology

Tejashwi Rana
Jadavpur University

Kare H. Karstensen and Palash Kumar Saha
SINTEF

CONTENTS

3.1 Introduction

Marine litter consists of items that have been made or used by people and deliberately discarded into the sea or rivers or on beaches; brought indirectly to the sea with rivers, sewage, storm water or winds; or accidentally lost, including material lost at sea in bad weather (UNEP, 2017).

At present, plastics have become the primary source of marine pollution responsible for about 80% of sea debris globally (International Union for Conservation of Nature, 2018). Eight million tonnes of plastics are found to be entering oceans annually. According to Jambeck et al. (2015), the total amount of plastic debris in the ocean which was 50 million tonnes in 2015 is projected to rise to 150 million tonnes by 2025. Plastic debris is now found in a variety of sizes from microns to metres in most marine environments, from coastal to pelagic habitats, from deep to superficial waters and from the most anthropogenic to the most remote regions (Bergmann et al., 2013). Being part of the bigger problem of waste management, marine litter or debris has been defined as "any persistent, manufactured or processed solid material discarded, disposed of or abandoned in the marine and coastal environment". Marine debris has been defined as a problem of global dimensions that adversely affects human beings, wildlife as well as the economic health of coastal communities to varying degrees (Oosterhuis et al., 2014; UNEP, 2005).

Sources of marine debris originate from two principal sources, namely, ocean-borne waste disposed of within the sea and terrestrial waste sourced from coastal users. Ocean-borne waste sources from fishing vessels, pleasure crafts and merchant ships, etc., whereas terrestrial waste originates mainly from recreational visitors, beachgoers and landfills. Among these two sources, terrestrial waste from coastal users has been considered most significant, which accounts for up to 80% of global marine pollution. This source of litter mainly consists of inappropriate management or illegal dumping of domestic waste that are dumped into water or carried by sewers, drains, creeks and rivers following human activities in the coastal areas. The marine environment is further impacted as the degradation period of these items is very slow. The principal reasons for terrestrial litter ending up as marine debris have been held responsible to human behaviour, actions and activities at sea (Sheavly & Register, 2007; Oosterhuis et al., 2014).

The marine environment is home to considerable biodiversity, along with tourism, which is an important and continually growing economic sector for many coastal nations. Pollutions, fisheries, deep-sea hydrocarbon exploration, eutrophication, ocean acidification and global ocean warming going along with sea-level rise as a result of rapid glacier melting and thermal expansion of seawater are noticeable examples of man-made pressures applied on the oceans with severe environmental and socio-economic consequences. Plastic is an important and ubiquitous material in our economy and daily lives. In 2018, plastic production crossed the 350 million tonnes (Mt); more than 50% produced in Asia. It is expected to double over the next 20 years (IEA Report, 2018).

It has been assessed that 10% of all plastic debris finishes up in the oceans, estimated a least 5.25 trillion plastic particles weighing 268,940 tonnes to be afloat in the sea.

Opportunities and challenges linked to plastic waste are increasingly global, and addressing them will significantly contribute to achieving the 2030 Sustainable Development Goals.

3.2 Objective

This chapter, through an extensive literature review, investigates the research questions such as the scale of the marine litter issue and its associated impacts on the environment, what solutions and legislative framework exists today and how resource recovery options such as co-processing can solve the problem of non-recyclable plastic wastes. In terms of resource efficiency, it is particularly important to prevent landfilling or dumping of plastic waste, where plastics might be converted to methane and microplastics. Any landfilling of plastic is an obvious waste of resources, which should be avoided in favour of recycling or energy recovery as the next best option.

3.3 Legislations, Regulation and Policy Instruments to Monitor Marine Littering

A growing number of countries are developing targeted laws and policies to address marine litter—from laws mandating more research (e.g., in the United States) to laws banning certain types of products (e.g., plastic bags in Bangladesh and Rwanda), to overarching frameworks to address the growing problem (e.g., in Japan and Singapore) (UNEP, 2016).

3.3.1 European Union

Marine Strategy Framework Directive (MSFD): The Marine Strategy Framework Directive (*Directive 2008/56/EC of the European Parliament and of the Council of 17 June 2008 established a framework for communal action in the ground of marine environmental plan*) is a European Directive targeted at accomplishing or maintaining Good Environmental Status in European seas. The primary goal is to reduce marine littering significantly through various steps like the introduction of a general waste policy and regulatory framework, education, awareness, communication campaigns, international cooperation, etc. EU's recent activities with regard to plastic wastes are demonstrated in Table 3.1.

Action for Marine Litter (OSPAR): In 2014, OSPAR agreed to cultivate a Regional Action Plan (RAP) for marine litter along with an execution strategy, to achieve its objective to considerably reduce amounts of both land-based and sea-based marine litter for the period 2014–2021 to levels where properties and quantities to the marine environment are left unharmed.

MARPOL Annex V(1990): Undertaken by more than 170 countries including the United States, MARPOL Annex V pursues to eradicate and reduce the amount of garbage being discharged into the sea from ships of any type whatsoever functioning in the marine environment, from merchant ships to temporary/fixed platforms to non-commercial ships like pleasure boats and yachts.

HELCOM Regional Action Plan for Marine Litter (2013): Its goal is to accomplish a noteworthy drop in marine litter by 2025 through different activities like developed and testing public monitoring methods to estimate riverine contributions of litter, including microplastics, continued training, continual consciousness-raising campaign, raising awareness and promotion of public monitoring tools, etc.

G20 Implementation Framework for Actions on Marine Plastic Litter (2019): The G20 framework is to simplify the reduction of marine plastic litter and microplastics, along with the G20 Action Plan on Marine Litter, while taking into the story the nation's suitable policies, approaches and national conditions, voluntarily. This agenda is estimated to counterpart the work of the UNEP. Table 3.2 demonstrates the significant legislations in different countries in the EU that will help in curbing marine littering.

TABLE 3.1

EU's Recent Activities with Regard to Plastic Wastes

Timeline	Activities
By 2022	In the CEAP, the Commission promised to come up with a proposal for waste prevention targets by 2022
End of 2021 but more likely in 2022	Between 75,000 and 300,000 tonnes of microplastics are released into the environment each year in the EU. A proportion of these microplastics are escaping into our environment after having been intentionally produced, supplied and used in products like cosmetics, detergents, paints, pesticides and even sports fields. In line with the EU Plastics Strategy (2018), the European Chemicals Agency (ECHA) was asked to make a proposal for **restricting the use of intentionally-added microplastics** under the Regulation on the Registration, Evaluation and Authorisation of Chemicals (REACH). Adoption of this new piece of legislation is expected soon
End of 2021	The Commission is expected to adopt the EU Ecolabel for Retail Financial Products by the end of 2021
June 2021	Legislative proposal to amend the EU's consumer law – Directive of the European Parliament and the council on consumer credits
22 February 2021	The Global Alliance on CE and Resource Efficiency (GACERE) was established by the European Commission on behalf of the European Union (EU), and by the United Nations Environment Programme (UNEP), in coordination with the United Nations Industrial Development Organization (UNIDO)
11 March 2020	European Commission adopts new circular economy action plan (CEAP), including revised legislative proposals on waste
2 July 2019	Directive on single-use plastics enters into force
28 May 2019	Commission proposal for a Directive on single-use plastics
16 January 2018	Commission adopts EU Plastics Strategy

TABLE 3.2

National Legislation in EU Countries to Curb Marine Littering

Country	Act/Legislation	Year	Targets and Activities
Austria	"Pfiat di Sackerl" ("Farewell Plastic Carrier Bag")	2016	This initiative targets to reduce the amount of plastic bags in Austria by 50% to a maximum of 25 pieces per head per year by 2019
	The "Zero (Plastic) Pellets Loss" initiative	2015	This pact includes a ten-point plan to lessen the discharge of plastic granules into the environment
	"Bewusst Kaufen" ("Buy Aware")		This act serves to increase consumer's awareness of sustainable products and provides extensive information on sustainable consumption
Bulgaria	Ordinance on Packaging and Packaging Waste	2005	This ordinance deals with packages used in the market and effective collection of the waste and reclaiming/recycling them to reduce litter in the open environment
Croatia	The Ordinance on the packaging and waste packaging (OG 88/15)	2015	This makes the packaging companies obliged to produce packaging that can be reused, recovered, and/or recycled following the best available technologies to minimize littering
	Waste management plan of the Republic of Croatia		This plan sets the establishment of a marine waste management system to be met by 2022
Cyprus	Indirect Fee System for the Collection of Ship Waste	2000	They have applied an "Indirect Fee System" applicable to all ships arriving in Cyprus as a charge that gives it the right to dispose of their waste
Denmark	1994 taxon plastics	1994	Denmark's 1994 tax on plastics includes bags and all packing materials, as well as a tax on transferring waste to a landfill or incinerating it
Estonia	Packaging Act		This Act provides the general necessities for effective management of packaging and its waste produced thereby throughout its whole life cycle
Finland	Green Deal agreement – Plastic Carrier Bag Agreement		This agreement was concluded to reduce the usage of lightweight plastic carrier bags and prevent littering
Greece	Waste Act	2011	To check the hazard and harm to human health and the environment threat by waste and waste managing, to validate the sustainable use of natural resources, to guarantee functioning waste management preventing littering
	Ministerial Decision 126635/2016	2016	Launches the monitoring programmes for the constant assessment of the environmental status of marine waters
	Joint Ministerial Decision 126856/2017	2017	Creates the institutions responsible for the monitoring of the quality of marine waters
	Joint Ministerial Decision 8600/416/2009	2009	An annual application of programmes of measures, which include cleanup activities on the beaches
Ireland	Waste Management (Environmental Levy) (plastic bag) (Amendment) (No. 2) Regulations 2007 (S.I. No. 167 of 2007)	2007	To increase the quantity of environmental levy on plastic bags
	Water Framework Directive (2000/60/EC)	2000	To decrease litter entering rivers and other water bodies by addressing pollution from surface water runoff and drainage
Italy	Law No. 221 of December 28, 2015, Provisions on the Environment	2015	To endorse the Green Economy and limit the excessive use of Natural Resources ensures compliance with necessities related to environmental influence assessments in coastal and marine regions and to reducing marine pollution

(Continued)

TABLE 3.2 (*Continued*)

National Legislation in EU Countries to Curb Marine Littering

Country	Act/Legislation	Year	Targets and Activities
Malta	Waste Management Plan for the Maltese Islands	2014	To recycle 50% of paper, plastics, metal and glass waste from family units by 2020
	Port Reception Facilities for Ship-generated Wastes and Cargo Residues Regulations (L.N. 278 of 2004)	2004	These protocols lay down provisions appropriate to port reception facilities for ship-generated waste and cargo remains to decrease the ejections of ship-generated waste into the sea, thereby improving the protection of the marine surroundings
Poland	Regulation on packaging wastes management	2005	Launches detailed rules on collecting and recycling packing wastes of paper, plastics, aluminium, glass, steel and multi-materials
Sweden	The Swedish Environmental Code (1998:808)	1998	Forbids littering and dumping of waste in unspecified areas

Remarkable Actions and Achievements

- An objective to separate 90% plastic bottles by 2029 was proposed (77% by 2025). A target to include 25% of recycled plastic in PET bottles (to produce R-PET) from 2025 and 30% in all plastic bottles as from 2030 will have a declining trend in the quantity of marine litter washed ashore and/or dumped on coastlines.
- The number of items of the top ten litter classifications was found to reduce by 30% in each provincial sea by 2020, compared with 2015, where screening methods from the technical guidelines documents on monitoring of marine litter were implemented.
- The ban in the Marine Environment Act against selected single-use products made of plastic to promote alternatives items existing on the market: cutlery, straws, plates, balloons sticks, cups, food and beverage containers made from expanded polystyrene and products made of oxo-degradable plastic.
- The EU has dedicated itself to guaranteeing the conservation of 10% of its coastal and marine areas by 2020. In 2012, 5.9% of Europe's seas had already been nominated as marine protected areas, and work is ongoing to attain 10% coverage by 2020.
- The strategies and directives towards mitigating plastic waste and marine litter have been linked to other strategies such as European Green Deal, circular economy action plan, European industrial strategy and biodiversity strategy.

3.3.2 In Other Countries

Many countries in Asia, Africa and countries like the United States have constructed their own legislation, regulations and policy instrument to monitor marine littering. The policy instruments described below are divided into the following groups: "International/Global", "Regional" and "National". Significant legislations and policy instruments at Global, Regional and National levels have been listed here in Table 3.3.

3.4 Integrated Waste Management Options

An estimated 9.3 billion tonnes of virgin plastics were produced globally up to 2019. Out of this, 6.3 billion tonnes have already ended up being plastic waste, of which only 9% were recycled, 12% incinerated and 79% dumped. More than 5 billion tonnes of plastic waste are today accumulated in dumpsites/landfills around the world. Plastic in dumpsites and landfills will eventually break down and be released into groundwater and rivers and constitute a continuous source of microplastics to our oceans. If current production and waste management trends continue, roughly 12 billion tonnes of plastic waste will be in landfills or in the natural environment by 2050 (Geyer et al., 2017).

TABLE 3.3

Significant Global, Regional and National Level Policy Instruments in Non-EU Countries

Policy Instrument	Details	Significant Target and Activities
International		*Thailand* has drafted Plastic Debris Management Plan (2017–2021) which targets to increase plastic waste recycling to at least 60%.
International Agreement on the Prevention of Marine Pollution by Dumping Wastes and Other Matter (London Convention, 1972)	The London Convention fosters the control of all sources of marine pollution and takes possible steps to combat marine pollution. Under these instruments, disposal at sea of persistent plastic and other synthetic materials (such as netting and ropes) has been forbidden (UNEP, 2019)	*Indonesia's* Plan of Action on Marine Plastic Debris (2017–2025) focusing on five pillars including improving enhancing funding mechanisms, behaviour changes through raising awareness , reformation of policy etc.
United Nations Convention on the Law of the Sea (UNCLOS, 1982)	UNCLOS deals with "Protection and preservation of the marine environment" and requires states to take all restrictions and activities required to prevent, reduce and control pollution of the marine environment	*Viet Nam* has launched its "National Action Plan for Management of Marine Plastic Litter by 2030" in 2020. (Truong et al., 2020). In *China* the National Development and Reform Commission has taken policy instruments in 2019, targeting prohibition of plastic bags across all cities and towns by 2022 and aiming restriction the use of single-use plastic items by 30%.
Regional		
The ASEAN Framework of Action on Marine Debris: The Framework of Action on Marine Debris (2017)	The Framework was accomplished to act on the recommendations from the ASEAN Conference on Reducing Marine Debris in ASEAN Region, consisting of actions and suggested activities for combating marine debris	*Japan* has taken initiative to reduce the 25% of total single-use plastics by 2030, dealing with marine pollution through source segregation.
The Coordinating Body on the Seas of East Asia (COBSEA) Regional Action Plan on Marine Litter (2019)	COBSEA brings together nine countries of East Asia to the development and protection of the marine environment and coastal areas of the region, for the health and wellbeing of present and future generations	The *Sri Lankan* government has prohibited the production, usage, since 2017, and import of plastic products. Government of **India** has taken initiative for the prohibition of single-use plastic products gradually by2030 which can be further affective in combating marine littering problem in India (Raha et al., 2021).
Bangkok Declaration on Combating marine debris in ASEAN region (2019)	This declaration was designated in Southeast Asian Nations (ASEAN), focusing on sustainably using the oceans, seas and marine resources for sustainable development, specifically to attain Sustainable Development Goal 14 to prevent and significantly reduce marine pollution of all kinds	

National Level policy

Singapore utilizes both legislative controls and administrative measures to monitor marine pollution from land-based sources

In **Viet Nam,** the Law on Natural Resources and Environment of Sea and Islands was adopted in 2015 (Law No. 82/2015/QH13), dealing with regulations on control of marine pollution, oil-spill response, toxic chemicals and dumping at sea.

Presidential Decree No. 979 was declared as national policy in the **Philippines,** in the year 1976, focusing on the prevention and control of the pollution of seas by waste dumping.

The Government of **Indonesia** implemented a number of laws and regulations in order to mitigate marine litter, which include Regulation No. 19 of 1999 on Control of Marine Pollution, Government Regulation No. 21 of 2010 on Marine Environment Protection (Sea-Circular, Indonesia).

(Continued)

TABLE 3.3 (*Continued*)

Significant Global, Regional and National Level Policy Instruments in Non-EU Countries

Policy Instrument	Details	Significant Target and Activities
In **China**, The fundamental legislation in the marine environmental protection area is the Marine Environment Protection Law which provides an overall regulation on pollution control, ecosystem protection and resources conservation (Chen et al., 2012).		
In 2005, The National Diet of **Japan** passed Comprehensive National Land Development Act, focusing on the utilization and protection of sea areas including Exclusive Economic Zone (EEZ) and the continental shelf as targets for territory planning (Chu, 2008).		
In **India,** Environment Protection Act (1986) contributes an important role to control land-based marine plastic pollution. Continental Shelf, Territorial Waters, Exclusive Economic Zone, and other Maritime Zones Act (1976) to restrict and protect the marine environment (Raha et al., 2021).		

Recycling is the preferred option, but not all plastic waste is suitable for recycling. From a technical aspect, it is challenging to recycle plastic that consists of several types of polymers, as you need to separate them. From an economic perspective, the recycling sector suffers greatly from low oil prices. The main component of plastic is oil; hence the low oil prices lead to low prices of virgin plastic. Studies have also shown that most of the plastic that ends up in the oceans is of low quality that is hard to recycle.

Energy recovery from wastes and plastics in Municipal Solid Waste Incinerators with Waste to Energy (WtE) normally involves the generation of electricity in steam turbines, but the conversion efficiency to electricity is poor and will not recover the construction costs (Mutz et al., 2017). WtE-plants are expensive to build and operate; they represent an additional emission source and produce large amounts of residues (fly ash, bottom ash, etc.) that need to be treated/landfilled. Incineration of wet wastes in the rainy season is another challenge, which causes difficult burning conditions and results in elevated emissions.

3.4.1 Co-processing in Cement Kilns

Countries with cement industry may to a certain degree forego building expensive Waste-to-Energy plants/incinerators. Cement kilns are already in operation and may increase the waste treatment capacity significantly if integrated into the waste management strategy. They are usually cost-efficient and do not produce any residues that need disposal.

A preheater cement kiln possesses many inherent features which makes it ideal for waste treatment: high temperatures, long residence time, surplus oxygen during and after combustion, good turbulence and mixing conditions, thermal inertia, counter currently dry scrubbing of the exit gas by alkaline raw material (neutralises all acid gases like hydrogen chloride), fixation of the traces of heavy metals in the clinker structure, no production of by-products and efficient recovery of energy and raw material components in the waste. Figure 3.1 shows different feeding points in a kiln for co-processing of different types of waste in a cement kiln.

Cement kilns have proven to be effective means of recovering value from waste materials, and co-processing in cement kilns is now an integral component in the spectrum of viable options for treating several waste categories, practised in developed countries for the last four decades.

Replacing parts of coal with non-recyclable plastic wastes may represent a win–win opportunity – preventing the plastic from ending up in the ocean, reducing the need for large amounts of fossil coal and indirectly reducing greenhouse gas emissions by avoiding building new incinerators or landfills. What is considered waste in one sector becomes a resource in another. This concept represents a circular economy in practice and incorporates waste treatment with existing industrial production, which is also referred to as Incineration and Landfilling in the internationally accepted waste management hierarchy.

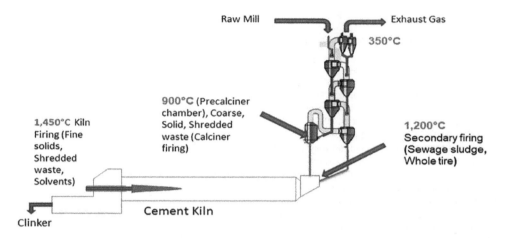

FIGURE 3.1 Co-processing of different types of waste in cement kiln showing different feeding points in a kiln. (Modified from Geocycle India's scheme.)

3.4.2 Global Status of Co-processing Technology in Cement Kiln

The United States accomplishes an average TSR (Total Substitution Rate) of 15% cement production in 2014, while Europe has an average TSR of 40%. All the European countries have different TSRs for cement production. It varies from less than 5% to more than 80%. For example, the current TSR is 70% in Norway, 65% in Germany, and 80% in Austria (ECOFYS, 2017; Ghosh et al, 2022). Table 3.4 demonstrates the current status, target and potential of different countries.

TABLE 3.4

Current Status and Target of Co-processing in Different Countries

Sl.	Country	Status of Thermal Substitution (%)	Target or Potential for Thermal Substitution (%)	References
	European countries	40.2 in 2014	-	ECOFYS (2017)
	Austria	80 in 2017	-	Sarc (2018)
	Belgium	53 in 2016	60-65 by 2020	Beer et al. (2017) and ECOFYS (2017)
	Bulgaria	21 in 2016	40	Beer et al. (2017) and ECOFYS (2017)
	Czech Republic	62 in 2014	80	ECOFYS (2017)
	France	37.3 in 2014	50	ECOFYS (2017)
	Germany	65 in 2014	80 by 2020	ECOFYS (2017)
	Greece	7 in 2014	20	ECOFYS (2017)
	Hungary	40 in 2014	60-70	ECOFYS (2017)
	Ireland	29 in 2016	80 by next 10 to 20 years from 2014	Beer et al. (2017) and ECOFYS (2017)
	Italy	13.3 in 2014	40 by 2025	ECOFYS (2017)
	Norway	70 in 2016	-	SINTEF (2017)
	Poland	52 in 2014	65	ECOFYS (2017)
	Portugal	28.4 in 2013	40	ECOFYS (2017)
	Spain	23.1 in 2014	35-50	ECOFYS (2017)
	Sweden	48 in 2014	60	ECOFYS (2017)
	United kingdom	44 in 2014	50 by 2020	ECOFYS (2017)
	United States	15 in 2014	-	WBCSD CSI GNR (2014)

(Continued)

TABLE 3.4 (*Continued*)

Current Status and Target of Co-processing in Different Countries

Sl.	Country	Status of Thermal Substitution (%)	Target or Potential for Thermal Substitution (%)	References
	Japan	-	-	-
	Australia	-	-	-
	Brazil	19 in 2014	-	WBCSD CSI GNR (2014)
	South Africa	-	-	-
	China	8		Karstensen et al. (2020)
	India	2.5 in 2016 (estimated)	25 by 2025	CII (2016)

3.4.3 Legislative Framework on Co-processing

Legislation and actions taken for the implementation of co-processing in different countries are demonstrated in Table 3.5.

3.5 Discussion and Analysis

- The awareness of the environmental, economic, social, public safety and individual health risks posed by marine litter has significantly increased. There is also the establishment of numerous and diverse sets of regulatory measures at local, national, regional, and international levels. Implementation of legislation and regulation to monitor marine littering has resulted in the reduction of at least a small portion of marine litter. Legislations like OSPAR, MARPOL and HELCOM have set a target to reduce marine litter by 2025 with a significant amount.

- Several initiatives, like the prohibition on taxes and ban on plastic usage, whether full or partial, have been implemented at national and regional levels, yet they are insufficient. Multiple authorities, producers and retailers are lagging, and the industry remains actively fighting some of the legislative propositions.

- Too often the way plastics are currently produced, used and discarded fails to capture the economic benefits of a more "circular" approach and harms the environment.

- Once the plastic has entered the ocean, it is difficult to remove; therefore, the problem should be tackled at its source. Reducing the plastic load by 50% in the 10 top-ranked rivers would reduce the total river-based load to the sea by 45% (Best, 2017). Jambeck et al. (2015) estimate that to achieve a 75% reduction in the mass of mismanaged plastic waste, waste management would have to be improved by 85% in the 35 top-ranked countries.

- There is a lack of actual data on the quantity and extent of marine litter at local, national, regional and global levels, as only such thorough information will allow the suitable improvement of adequate and well-organized regulations; most information are estimates. However, efforts are being made in this direction.

- Co-processing has been considered here to be an effective and efficient method to address the issue of non-recyclable plastic wastes. Several countries have issued policies and guidelines supporting co-processing in cement kilns in various stages of implementation currently, especially in developing countries.

- Many of the EU countries like Austria, Norway, Germany, Czech Republic and the Netherlands have TSR of more than 60%, while countries like Greece, Italy, the United Kingdom and Bulgaria are still facing problems to achieve EU average TSR. The co-processing rate in Greece is on average 7%, which is the lowest among the entire EU countries (compared to EU average 41%) and the reason behind this is the limited availability of suitable waste fuels and the lengthy permitting process for their use in cement kilns.

TABLE 3.5

Legislation and Actions Taken for Implementation of Co-processing in Different Countries

Country	Legislation	Remarkable Actions and Achievements
European countries	Landfill Directive (1999/31/EC) Waste framework directive (2008/98/EC) Integrated Pollution Prevention and Control (IPPC) Waste Incineration Directive (WID) (2000/76/EC)	Co-processing in **France** was first regulated under the French Waste Law, issued in 1975 and the facility of Pre-processing of the wastes started in 1985 Countries like **Austria, Norway, Germany, Czech Republic & the Netherlands** having TSR of more than 60% and their target is to achieve TSR of more than 80% by 2025 (ECOFYS, 2017; Beer et al., 2017). The co-processing rate in **Greece** is on average 7% which is the lowest among the entire EU countries (compared to EU average 41%) and the reason behind this is the limited availability of suitable waste fuels and lengthy permitting process for their use in cement kilns (ECOFYS, 2017). The national waste management plans of **Italy** have a strong position for co-processing. However, permits for co-processing are very often and only small volumes of waste are allowed to be co-processed (ISPRA, 2015).
United states	1970 Clean Air Act.	The US EPA established regulations to monitor air emissions (i.e., PM, NOx, and SO$_2$) from cement kilns through the agency's fence-line monitoring programme.
Japan	The Law for the Promotion of Utilization of Recyclable Resources (the Recycling Law) in 1991. Waste Disposal and Public Cleansing Law (the Waste Disposal Law), 1991.	In 1990s Japanese Cement Company, Taiheiyo has produced a specifically labelled co-processed cement product so-called eco-cement. In 1997, The Japanese Ministry of Economy, Trade and Industry and the Ministry of the Environment established the eco-town concept originated through a subsidy system (Global Environmental Center Foundation 2009).
Brazil	National Regulatory Act No. 264/99 National Regulatory Act 316/02	National Regulatory Act No. 264/99 establishes technical and operational criteria, emissions limits, and requirements of pre-permit testing for co-processing in cement kilns. National Regulatory Act 316/02 for Licensing of Incineration/Co-incineration establishes limits for emissions of dioxins and furans (0.5 nanograms per cubic nanometer [ng/Nm³]) from cement kiln co-processing.
South Africa	National Policy on the Thermal Treatment of General and Hazardous Waste (the South Africa National Policy), 2009	National Policy relies on the EU Incineration Directive 2000/76/EC and other international policies, including guidelines for co-processing by WBCSD and Holcim, as models.
China	The Development Policy of Cement Industry (NDRC, 2006) Technical Policy on Pollution Prevention and Control for Co-processing Solid Waste in Cement Kilns, 2016 Industry Green Development Plan (2016–2020) Planning for the Innovative Capacity Development of Industrial Technology (2016–2020)	By the end of 2018, there had been 57 Cement Kiln Co-processing production lines for MSW treatment distributed in 16 provinces. The top three provinces with the highest treatment capacity are Guizhou (2613 t/d), Guangxi (2500 t/d) and Anhui (1900 t/d). Green Manufacturing Specific Action Plan, 2016 launched pilot projects for co-processing MSW in cement kilns. Circular Economy Development Strategies and Near-Term Actions Encourage cement kilns to co-process solid wastes for resource utilization and circulation. In 2019, China had a total of 1681 new dry cement production lines; among these only 160 (<10%) and 57 (about 3%) were co-processing solid waste and MSW, respectively. China could not meet the 2020 targets of 15% of new dry cement clinker production lines which would have capable of co-processing solid waste, and the waste utilized as AFR to reach above 20% of total fuels and materials used for cement production due to COVID Pandemic. Ten co-processing projects with a total capacity of 1.7 Mt/yr have been reported the end of the 2021 mostly based in Guangxi and Yunnan provinces, process municipal solid waste, urban sludge and industrial waste.

(Continued)

TABLE 3.5 (*Continued*)

Legislation and Actions Taken for Implementation of Co-processing in Different Countries

Country	Legislation	Remarkable Actions and Achievements
India	Hazardous Waste Management Rules 2016 Solid Waste Management Rules 2016	The Ministry released Guidelines for Co-processing of Plastic Waste in Cement Kilns (As per Rule Plastic Waste Management Rules, 2016) in May 2017. The guideline included the collection, segregation, transportation and co-processing of plastic waste in cement kiln. In July 2017, CPCB released guidelines for pre-processing and co-processing of hazardous and other wastes in cement plant. The guidelines have been revised for complying with the conditions as specified in the Hazardous and Other Wastes (Management and trans-boundary Movement) Rules, 2016. The estimated TSR of the **Indian** cement plant is around 2.5 % and a target is set to achieve TSR of 25 % is by 2025 (CII, 2016).

A SWOT (Strength, Weakness, Opportunity and Threat) analysis reveals the important drivers and limitations of co-processing in cement kilns.

3.5.1 SWOT Analysis on Implemented Legislations on Co-processing of Waste in Cement Kiln

A SWOT analysis is done (Figure 3.2) based on information collected on the legislations implemented and actions taken in different countries in (Section 3.3.2 and Table 3.2).

3.5.2 Proposed Model for Prevention of Marine Littering through Legislations and Regulations

In terms of resource efficiency, it is particularly important to prevent landfilling or dumping of plastic waste, where plastics might be converted to methane and microplastics. Any landfilling of plastic is an obvious waste of resources which should be avoided in favour of recycling or of energy recovery as the next best option. As not all plastic waste can be recycled, we need to find additional solutions to avoid the plastics that strangle us and our planet! The non-recyclable plastic waste represents the largest share of marine litter which could be cost-efficiently used as alternative fuels in the local energy-intensive industries, such as the cement industry.

International action is key to addressing the most significant sources of plastic litter in the oceans, i.e. insufficient waste management in developing countries and emerging economies, especially connected to major world river basins, dumpsites, landfills and industrial hotspots.

Leaders around the world recognize the importance of working to solve the problems, as we have seen through different international initiatives on marine litter, like the UN Global Partnership on Marine Litter and the action plans put forward by the G7 and G20. The Basel Convention has been addressing the issue of marine plastic litter and microplastics and the Stockholm Convention on Persistent Organic Pollutants (POPs) is relevant for plastic in which it restricts the use of commercial flame retardants such as Penta and Octa Bromo Diphenyl Ether (BDE).

The following model (Figure 3.3) has been developed as a result of the present study that shows the value chain of plastic with a focus on different waste treatment options. The model suggests that with proper legislation on marine litter and co-processing, it is possible to manage land-based non-recyclable plastic wastes on a large scale, which would otherwise leak into water bodies and ultimately to oceans.

3.6 Conclusion

In this chapter, an attempt has been made to review the existing rules and regulations and their enforcement at national, regional and global levels to curb marine litter and integrated waste management option, such as co-processing in cement kilns. A SWOT analysis has been conducted to understand the

Strength	Weakness
➢ Promote CE & support several SDGs ➢ Higher in waste management hierarchy compared to landfilling and incineration ➢ High energy and material recovery ➢ Reduce GHG emissions ➢ Improves overall waste management in the country ➢ Comparatively smaller investments as cement kilns already exists & operates 24x7 throughout the year	• Lack of regulatory framework • No incentives or level playing field compared to other technologies, for example Incineration • Poor competence and lack of capacity in developing countries • Unwillingness to invest in waste pre-treatment facilities • Lack of a business case - polluters pay principle not applied
Opportunities	**Threats**
❖ Material and fuel substitution- less dependence on non-renewable/fossil sources ❖ Cost competitiveness for cement industry ❖ Large scale solution for plastic wastes ❖ Opportunity to earn tipping fee for industrial wastes- polluters pay principle ❖ As integrated waste management solution, help achieve government's goals/targets on waste and GHG emissions reductions	▪ Waste availability at reasonable distance to the cement plants ▪ Limiting constituents in wastes affecting process, quality & emissions ▪ Wastes affecting lower production in sold out market situation ▪ Not implementing BAT & proper management routines ▪ Erratic unsustainable waste supplies affecting payback period on investments.

FIGURE 3.2   SWOT analysis on plastic waste co-processing in cement kilns.

current drivers and limitations of co-processing and a model has been proposed showing the linkage of different waste treatment options and legislative framework for preventing marine litter.

Policies and laws need to address not only the removal of litter but are generally more successful when they govern the production, use and disposal of products that would otherwise become marine litter. To this end, using a circular economy approach to prevent the generation of waste products can reduce the overall production of marine litter (UNEP, 2016).

Billions of tonnes of plastic wastes are accumulated in dumpsites/landfills or mismanaged around the world. Everyone is talking about recycling, but what do we do with all the plastic waste that is non-recyclable and slowly released into the ocean and the food chain? The potential of integrating the energy-intensive cement industry with plastic waste management is enormous. However, it is important to show that it works locally (Karstensen et al., 2020).

Acknowledgements

The authors gratefully acknowledge the funding support of the "Ocean Plastic Turned into an Opportunity in Circular Economy" – OPTOCE,[1] academic collaboration research project by SINTEF, Norway at the Department of Mechanical Engineering, Jadavpur University. The authors also acknowledge the research support by the International Society of Waste Management, Air and Water (ISWMAW), Dalmia Cement Plant, Ultra Tech Cement Plant, Kolkata Municipal Corporation, Howrah Municipal Corporation, Baidyabati Municipality, Konnagar Municipality, Champdani Municipality, Rishra Municipality, Sherampore Municipality, Uttarpara Municipality, Sonarpur-Rajpur Municipality, Budge Budge Municipality, Maheshtala Municipality in West Bengal and Muzaffarpur Municipality in Bihar, India and all others who supported this research.

---

[1] The regional project Ocean Plastic Turned into an Opportunity in Circular Economy – OPTOCE - will investigate how the involvement of Energy Intensive Industries, like cement manufacturing, can increase the treatment capacity for Non-Recyclable Plastic Wastes in China, India, Myanmar, Thailand and Vietnam and thereby contribute to reduce the release of plastics to the Sea.

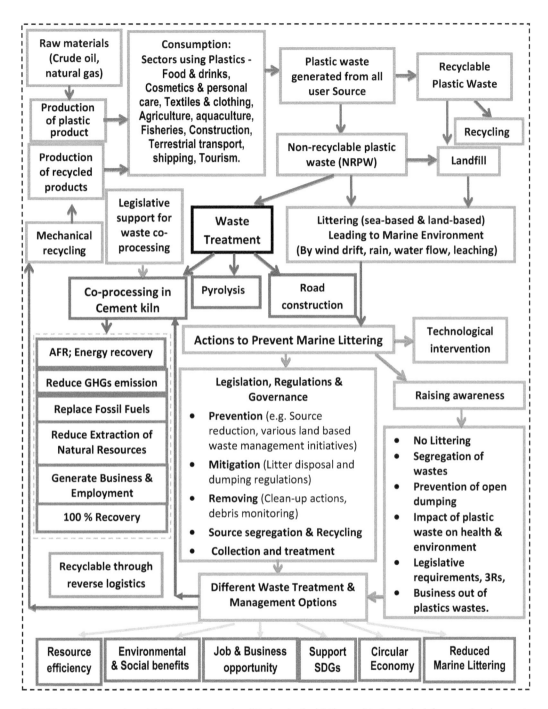

FIGURE 3.3 Proposed model: Preventing marine littering by legislative and technological framework using waste co-processing.

BIBLIOGRAPHY

Agamuthu, P., Mehran, S.B., Norkhairah, A., Norkhairiyah, A (2019) Marine debris: A review of impacts and global initiatives. *Waste Management & Research*, Vol. 37(10), 987–1002. doi: 10.1177/0734242X19845041.

ASEAN (2019) Regional action plan for combating marine debris in the ASEAN member states. ISBN 978-623-6945-33-9.

Assembleia da República. Available at: https://info.portaldasfinancas.gov.pt/NR/rdonlyres/06E2525B1E76-454C-AB73-75539519258F/0/Lei_82D_2014.pdf.

Beer, J., Cihlar, J., Hensing, I., Zabeti, M. (2017) Status and prospects of co-processing of waste in EU cement plants. ECOFYS, April 26 2017.

Bergmann, M., Gutow, L., Klages, M. (2013) *Marine Anthropogenic Litter.* doi: 10.1007/978-3-319-16510-3.

Best, S. (2017) Shocking report reveals that 95% of plastic polluting the world's oceans comes from just TEN rivers including the Ganges and Niger. *Daily Mail* (Mail Online). http://www.dailymail.co.uk/sciencetech/article-4970214/95-plastic-oceans-comes-just-TEN-rivers.html, Accessed on 06 October 2018.

CEMBUREAU (2009) Co-processing of alternative fuels and raw material in the European cement industry, Sustainable cement production.

CII (2016) Promoting alternative fuel and raw material usage in Indian cement industry, approach paper for achieving 25% thermal substitution rate in Indian cement industry by 2025.

Costa, J.P.D., Mouneyrac, C., Costa, M., Duarte, A.C., Rocha-Santos, T. (2020) The role of legislation, regulatory initiatives and guidelines on the control of plastic pollution. *Frontiers in Environmental Science*, 8. doi: 10.3389/fenvs.2020.00104.

Crawford, C.B., Quinn, B. (2017) 3- Plastic production, waste and legislation. *Microplastics Pollution*, 2017, 39–56. doi: 10.1016/b978-0-12-809406-8.00003-7.

Directive 2008/98/EC of the European Parliament and of the Council. (19 November 2008) Waste and repealing certain directives, Official Journal of the European Union.

Enveco Environmental Economics Consultancy. (2012) Marine litter in Sweden: A study for the Economic and Social Analysis of the Initial Assessment of the Marine Strategy Framework Directive, Havs-och vattenmyndighetens rapport 2012: 3.

Geyer, R., Jambeck, J.R., Law, K.L. (2017) Production, use, and fate of all plastics ever made. *Science Advances*, 3, e1700782.

Guidelines on Usage of Refuse Derived Fuel in Various Industries (July 2018) Prepared by Expert Committee Constituted by Ministry of Housing and Urban Affairs (MoHUA) & Central Public Health and Environmental Engineering Organisation (CPHEEO).

Guimaraes, A.G., Vaz-Fernandes, P., Rosario Ramos, M., Martinho, A.P. (2018) Co-processing of hazardous waste: The perception of workers regarding sustainability and health issues in a Brazilian cement company. *Journal of Cleaner Production*, 186(1), 313–324. doi: 10.1016/j.jclepro.2018.03.092.

Hardesty, B.D., Wilcox, C. (2017) A risk framework for tackling marine debris. *Analytical Methods*, 9(9). doi: 10.1039/c6ay02934e.

Hasanbeigi, A., Lu, H., Williams, C., Price, L. (2012) International best practices for pre-processing and co-processing municipal solid waste and sewage sludge in the cement industry.

Hermawan, S. (2019) Law and Economic approach to reduce marine plastic litter in Indonesia. *Proceedings of the 3rd International Conference on Globalization of Law and Local Wisdom (ICGLOW 2019)*. doi: 10.2991/icglow-19.2019.56.

IEA Report (2018) The future of petrochemicals: Towards more sustainable plastics and fertilisers.

International Union for Conservation of Nature (2018) National marine plastic litter policies in EU member states: An overview.

ISPRA (2015) Rapporti rifiuti urbani 2015. Available at: http://www.isprambiente.gov.it/it/pubblicazioni/rapporti/rapporto-rifiuti-urbani-edizione-2015.

Jambeck, J.R., Geyer, R., Wilcox, C., Siegler, T.R., Perryman, M., Andrady, A., Narayan, R., Law, K.L. (2015). Plastic waste inputs from land into the ocean. *Science*, 347(6223), 768–771.

Jang, Y.C., Ranatunga, R.R.M.K.P., Mok, J.Y., Kim, K.S., Hong, S.Y., Choi, Y.R., Gunasekara, A.J.M. (2018) Composition and abundance of marine debris stranded on the beaches of Sri Lanka: Results from the first island-wide survey. *Marine Pollution Bulletin* 128, 126–131. doi: 10.1016/j.marpolbul.2018.01.018.

Kao, J.C., Marine Policy and Environmental Management in Taiwan (2011) UMI U567021.

Karstensen, K.H., Saha, P.K., Vigerust, E., Paulsen, A.A., Engelsen, C.J., Ahmadi, M. (2020). Asia's plastic potential. *International Cement Review*. ISSN 0959-6038.

Kosajan, V., Wen, Z., Zheng, K., Fei, F., Wang, Z., Tian, H. (2021) Municipal solid waste (MSW) co-processing in cement kiln to relieve China's MSW treatment capacity pressure. doi: 10.1016/j.resconrec.2020.105384.

Lachmann, F., Almroth, B.C., Baumann, H., Broström, G., Corvellec, H., Gipperth, L, Hassellov, M., Karlsson, T., Nilsson, P. (2017). Marine plastic litter on small Island Developing States (SIDS) Impacts and measures, Swedish Institute for the Marine Environment.

Lam, C.-S., Ramanathan, S., Carbery, M., Gray, K., Vanka, K.S., Maurin, C., Bush, R., Palanisami, T. (2018) A comprehensive analysis of plastics and microplastic legislation worldwide. *Water, Air, and Soil Pollution*, 229, 345. doi: 10.1007/s11270-018-4002-z.

Llorca, M., Álvarez-Muñoz, D., Ábalos, M., Rodríguez-Mozaz, S., Santos, L.H.M.L.M., León, V.M., Campillo, J.A., Martínez-Gómez, C., Abad, E., Farré, M. (2020) Microplastics in mediterranean coastal area: Toxicity and impact for the environment and human health. *Trends in Environmental Analytical Chemistry*, 27, e00090. doi: 10.1016/j.teac.2020.e00090.

Marazzi, L., Loiselle, S., Anderson, L.G., Rocliffe, S., Winton, D.J., Aschonitis, V.G. (2020) Consumer-based actions to reduce plastic pollution in rivers: A multi-criteria decision analysis approach. *PLoS One*, 15(8). doi: 10.1371/journal.pone.0236410.

MoEFCC (2016a) Hazardous and other wastes (management and trans boundary movement) rules, published in the Gazette of India, extraordinary, Part III, Section-3, sub-section (i), Government of India, Ministry of Environment, Forests and Climate Change, New Delhi.

MoEFCC (2016b) Solid waste management rules, published in the Gazette of India, extraordinary, Part II, Section-3, Sub-section (ii), Government of India, Ministry of Environment, Forests and Climate Change, New Delhi.

Morseletto, P. (2020) A new framework for policy evaluation: Targets, Marine Litter, Italy and the marine strategy framework directive. *Marine Policy*, 117, 103956. doi: 10.1016/j.marpol.2020.103956.

Mutz, D., Hengevoss, D., Hugi, C., Gross, T. (2017) Waste-to-Energy Options in Municipal Solid Waste Management. A Guide for Decision Makers in Developing and Emerging Countries. Deutsche Gesellschaft für Internationale Zusammenarbeit (GIZ), Germany.

Nguyen, V.T., BeiPing, C. (2020) Plastic marine debris in Vietnam: Source, impacts and management. *Environmental Policy and Law*. doi: 10.3233/EPL-200201.

Oosterhuis, F., Papyrakis, E., Boteler, B. (2014) Economic instruments and marine litter control. *Ocean & Coastal Management*, 102, 47e54. doi: 10.1016/j.ocecoaman.2014.08.005.

Raha, U., Kumar, B.R., Sarkar, S.K. (2021) Policy framework for mitigating land-based marine plastic pollution in the Gangetic Delta region of bay of Bengal: A review. *Journal of Cleaner Production*, 278, 123409. doi: 10.1016/j.jclepro.2020.123409.

REPORT (2013) Marine litter study to support the establishment of an initial quantitative headline reduction target, European Commission DG Environment.

REPORT (2017) Cost effective and cost-benefit analysis of new measures put forward as part of Malta's Marine strategy framework directive programme of measures.

Ghosh, S.K., Parlikar, U.V., Karstensen, K.H. (2022) *Sustainable Management of Waste through Coprocessing*, Springer Nature, ISBN: 978-981-16-6073-3.

Saha, P.K., Karstensen, K.H. (2017) Co-processing of Alternative Fuels & Resources in Indian Cement Industry Baseline and Potential.

Sarc, R., Seidler, I.M., Kandlbauer, L., Lorber, K.E., Pomberger, R. (2019) Design, quality and quality assurance of solid recovered fuels for the substitution of fossil feedstock in the cement industry.

Sea Circular Annual Report (2020) United Nations Environment Programme; Coordinating Body on the Seas of East Asia. https://wedocs.unep.org/20.500.11822/36243.

Sheavly, S.B., Register, K.M. (2007) Marine debris & plastics: Environmental concerns, sources, impacts and solutions. *Journal of Polymers and the Environment*, 15, 301–305. doi: 10.1007/s10924-007-0074-3.

Sheridan, H., Johnson, K., Capper, A. (2020) Analysis of international, European and Scot's law governing Marine Litter and Integration of Policy within Regional Marine Plans. *Ocean and Coastal Management*, 187, 105119. doi: 10.1016/j.ocecoaman.2020.105119.

UNEP (2013) Regional plan for the marine litter management in the Mediterranean.

UNEP (2016) *Marine Litter Legislation: A Toolkit for Policymakers*. ISBN: 978-92-807-3594-9.

Vigerust, E., Paulsen, A.A., Engelsen, C.J., Ahmadi, M. (2019). Ocean plastic: An opportunity in the circular economy? *International Cement Review*. ISSN 0959-6038.

WBCSD CSI (2014) Global cement database on CO and energy information "Getting the Numbers Right" (GNR). http://www.wbcsdcement.org/pdf/GNR%20dox.pdf.

4

Waste Segregation at the Source in Germany: A Key Component of Sustainable Waste Management Systems

Michael Nelles
Rostock University
German Biomass Research Centre gGmbH

Jan Sprafke, Isabell Eickhoff, and Gert Morscheck
Rostock University

CONTENTS

4.1 Development, Status and Prospects of the Waste Management System

Municipal waste includes waste from private households (domestic waste, bulky waste, bio-waste, separately collected recyclables, such as glass, paper, packaging and metals), as well as commercial waste resembling household waste (waste from doctors' offices, administrative buildings, schools and kindergartens). Furthermore, municipal waste also includes market waste, road sweepings and litter, waste from public places, park waste and waste from water-management measures (sewage sludge). Table 4.1 shows the waste generation balance in 2018 [1].

Germany implemented separate collections for waste, household and other kinds of waste, more than 20 years ago. Diverse environmental damage, lack of landfill space and the use of finite resources led, in the early 1990s, to a rethink in waste management. Today, climate change and energy demand are important arguments for the separate collection and utilisation of all kinds of waste.

The German Closed Cycle Management Act is aimed at turning waste management into resource management [2]. The realisation that waste can be a useful source of raw materials and energy is not new; metals, glass, organic waste and textiles have been collected before and put to new use. The waste management policy, which has been adopted in Germany over the past 30 years, is based on closed cycles and assigns disposal responsibilities to the manufacturers and distributors of products. This has made people even more aware of the necessity to separate waste, leading to the introduction of new disposal technologies and increased recycling capacities.

Today, 14% of the raw materials used by German industry are recovered waste, thus leading to a reduction in the extraction levels and related environmental impact [3]. Modern closed cycle management

DOI: 10.1201/9781003231608-5

TABLE 4.1

Waste Generation Balance in Germany in 2018 [1]

Types of Waste	Total Quantity of Waste Generated	Waste Deposited in Waste Treatment Plants					Recovery Rate	Recycling Rate
		Disposal Operations			Recovery Operations			
		Landfilling	Thermal Disposal	Treatment for Disposal	Energy Recovery	recycling		
	1,000 tonnes						%	
Total	417,197	71,201	3,558	3,962	48,021	290,456	81	69
Of Which								
Municipal wastes	50,260	109	179	603	15,646	33,723	98	67
Wastes resulting from mining and treatment of mineral resources	28,846	27,916	0	67	14	849	3	3
Construction and demolition wastes	228,120	25,908	91	994	1,351	199,776	88	88
Secondary wastes	54,885	5,004	342	769	18,518	30,253	89	55
Remaining wastes (in particular of manufacturing and other economic activities)	55,086	12,265	2,945	1,529	12,492	25,855	70	48

contributes, with a share of ~20%, to achieving the German Kyoto targets for the reduction of climate-relevant emissions.

Closed cycle management not only contributes to environmental protection but also pays economically. The waste management industry has become an extensive and powerful economic sector in Germany: almost 290,000 people have been employed in ~11,000 companies which generate an annual turnover of ~76 billion euros [4]. A total of 15,500 installations contribute to resource efficiency by recycling and recovery procedures. High recycling rates of ~66% for municipal waste, 48% for commercial waste and 88% for construction and demolition waste speak for themselves [3].

The legal foundations for proper waste management were provided far back in history. With the Prussian Local Tax Act of 1893, municipal finances were reorganised and the prerequisite for the establishment of municipal cleaning facilities was created. Municipalities were henceforth entitled to levy charges for waste disposal. Later, in 1935, the German municipal authorities established the general principles of collection and use of the waste collection system. This ensured the collection of all waste and forbids all illegal disposal routes. In the mid-1960s, cities and municipalities were finally identified as waste disposal authorities and were thus responsible for waste disposal. During the same period, the first bulletins were drafted on the issues of waste disposal, which were the guidelines for dealing with waste.

The obligatory self-administration tasks include those municipal affairs that must be carried out by the municipalities on the basis of a federal or state law. Within the framework of self-administration, the municipality is thus obliged to ensure the energy and water supply and waste disposal of the citizens. The way in which the municipality fulfils this task is its own affair. The municipality can not only run its own waste management company but can also obtain this service from another provider. However, the responsibility for waste disposal always remains with the municipality!

The most important tasks of self-administration are, for example:
- Supply citizens with water, electricity, district heating and gas
- Sewerage and waste management
- Planning of the municipal area by indicating residential, commercial and other areas (land use plans).

The municipalities need income in order to fulfil their tasks. The municipalities' main sources of income include fees and contributions, taxes, federal and state financial contributions, from the sale of building land or from real estate and loans.

Fees are levied when a particular municipal service is used, e.g. waste disposal. Fees in the waste sector cover costs.

However, waste management in Germany is constantly changing, due to new political and legal requirements as well as technical and organisational developments, and has developed into a large and powerful economic sector. Thus, modern waste management is the result of a long development process. Figure 4.1 summarises this development along a time axis as per the Federal Ministry for the Environment, Nature Conservation and Nuclear Safety (BMU), 2017 [5].

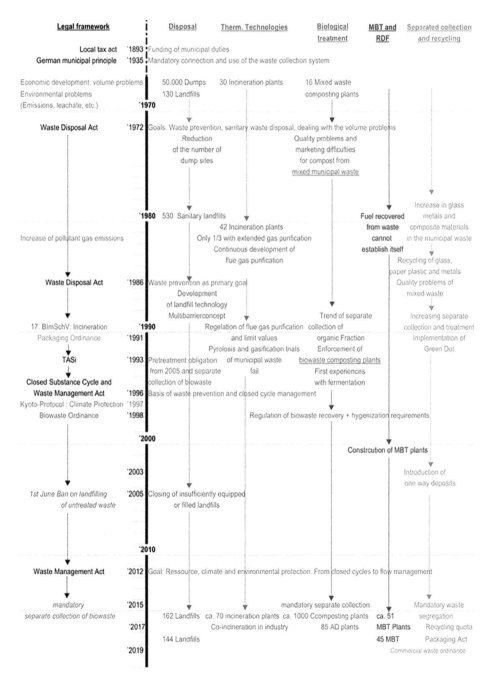

FIGURE 4.1 Summary of the legal (a) and technical (b) development of waste management in Germany along the time axis.

4.2 From Waste Disposal to Recycling

Until the late 1960s, waste had mostly been deposited in one of the ~50,000 uncontrolled landfills. Only about 37% of the municipal waste was treated and deposited in one of the ~130 sanitary landfills, 16 composting plants and 30 incineration plants [1]. Back then, the technologies were not yet mature and caused secondary environmental problems: groundwater pollution by leachate and gas emissions, due to the degradation of biogenic waste, pollutant emissions from waste incineration and quality problems of mixed waste compost.

In addition, economic growth in the early 1970s led to an increase in industrial production and private consumption, as well as disposable packaging and products. During this time, waste disposal had, on the one hand, to deal with the waste masses and, on the other hand, to develop an orderly waste disposal, preventing any risk to human and animal health. The first constitutional legal framework, the 1972 Waste Disposal Act, should address both problems.

With this law, the number of the original 50,000 uncontrolled landfills had been considerably reduced, while at the same time, a continuous improvement of landfill technology could be achieved by the end of the 1980s. In the same period, the number of waste incineration plants increased, but still with insufficient flue gas cleaning. Although all the plants had dust removing facilities, only one-third of the plants had an extended flue gas purification system. As a result of this, and by the increasing share of chemical products in domestic waste, the quantities of emitted pollutants continued to rise.

The technologies for the treatment of mixed waste are not yet environmentally compatible. The first attempts to produce solid recovered fuel (SRF) are discontinued as there are major problems regarding emissions. Also, the composting of the municipal waste could not be implemented. As a result of the above-mentioned change in the household waste composition, with increasing proportions of metals and composite materials, the heavy metal loads increased. Because of the low quality, the compost was not accepted to be used by farmers.

The first attempts to recover recyclables from municipal waste failed. The reasons for that were, amongst others, the inadequate product quality and the poor efficiency of the sorting systems. From mid-1980s onwards, individual municipal waste streams began to be separately collected (glass and paper).

While good progress has been made in the proper disposal of waste, only poor results were achieved in the management of waste quantities. For this reason, the 1986 Waste Act declared the prevention of waste as a superior objective, before recovery and disposal. Waste prevention includes low-waste technologies, the recycling of products and its recyclable construction, as well as the increase in its service life.

Due to the quality problems of the recovery technologies, a rethinking of the waste disposal routes took place at the beginning of the 1990s and separate collections became more important. The collection of the biogenic fraction, separated from the rest of the municipal waste, was clearly increasing. Only from this separately collected bio-waste could an economic product be produced. Composting of separately collected bio-wastes thus gains importance, as well as the anaerobic treatment of this fraction as an addition to composting which was developing. Therefore, at the beginning of the 1990s, fewer than 10 plants for mixed municipal waste composting and about 80 plants for bio-waste composting were operated. At the same time, recycling, which mainly concentrated on glass, paper, plastic and metal, has proven that a separate collection of valuable substances is the prerequisite for the production of high-quality secondary raw materials.

With the 17th Federal Emission Control Ordinance of 1990, existing waste incineration plants have to be upgraded with a sophisticated exhaust gas purification system or shut down, and the prescribed limit values have to be met. Despite this regulation, the incineration of waste was still disapproved by the population, as new pollutants such as dioxins became of concern. The 17th Federal Emission Control Ordinance is constantly revised; the exhaust gas purification systems are further upgraded and the quality of emissions from waste and waste co-incineration plants is greatly improved. Up to the 2000s, however, not only the number of plants but also the average plant throughput increased. In the area of thermal treatment of municipal waste, tests with degasification and gasification technologies have been conducted, which were, however, not accepted.

With the 1991 Packaging Ordinance, users and distributors are obliged to take back and recycle their product packaging, in accord with the principle of "product responsibility". The dual system (symbol "Green Dot") was founded to fulfil this duty.

The multi-barrier concept was implemented in landfilling technology and thus liquid and gaseous emissions were minimised. However, it is easy to see that, above all, the gaseous emissions are not completely prevented. In order to minimise the negative environmental impact of waste deposition, the pre-treatment of waste became obligatory in 2005 with the Technical Instructions for Waste Management.

In 1996, the Waste Act was amended, forming the Closed Substance Cycle and Waste Management Act. In addition to waste prevention, the focus is on the recycling industry. In addition, the quality requirements for the recycling of separately collected biogenic waste are regulated by the 1998 Biological Waste Ordinance.

4.3 Climate and Resource Protection Are Gaining Importance

With the Kyoto Protocol in 1997, the issues of climate and resource protection are becoming more important. Thanks to the strict legal requirements, waste management is able to contribute significantly to climate protection. The methane formation in landfills was avoided, above all, by the ban on the deposition of untreated municipal waste, which had been issued in June 2005. Also, the increased material and energy recovery from waste contributes to climate and resource protection.

At the beginning of the 2000s, mechanical–biological waste treatment has been developed, in addition to thermal waste treatment, in order to meet the pre-treatment requirements. The 30th version of the Federal Emission Control Ordinance hereby regulates the technical standards to be met, with regard to emissions. At first, the focus is on the input of the biogenic fraction; however, later on, the recovery of the high calorific fraction is becoming more and more interesting. By the energetic recovery of the waste, fossil fuels can be replaced.

In 1990, waste management still contributed to about 38 million tonnes of CO_2-equivalents, while in 2006, it was able to save about 18 million tonnes of CO_2 equivalents. From 1990 to 2006, waste management reduced its annual emissions of climate-damaging gases by ~56 million tonnes [6].

Furthermore, by 2005, most insufficiently equipped or filled landfills were shut down and the so-called "one-way deposit" was introduced.

In Germany, waste segregation is the answer to tackle several environmental problems arising from waste management. Since the biogenic fraction is mainly responsible for climate-relevant emissions, the segregated collection and treatment of this fraction has significantly reduced this environmental impact. The separate collection of bio-waste is also a precondition for the production of high-quality composts and, therefore, for the recirculation of organic matter and nutrients. Furthermore, the segregated collection of biogenic waste and its recycling reduces the amount and changes the composition of residual waste, reducing its water content. In this light, Figure 4.2 summarises the different benefits achieved by the segregated collection of biogenic waste as per the Federal Ministry for the Environment, Nature Conservation and Nuclear Safety (BMU), 2017 [5].

4.4 High Recycling Rates – On the Path to Material Flow Management

The current version of the Waste Management Act of 2012 [2] provides the path of the waste and closed cycle management to a resource-efficient economy. The objective of waste management is to conserve natural resources and manage waste in an environmentally sound manner, so that a sustainable improvement in environmental and climate protection, as well as resource efficiency, is achieved. Waste is regarded as a valuable raw material, whose effective use saves natural resources. While waste prevention leads to a reduction in raw material consumption and environmental pollution, waste recovery focuses on the recycling of raw materials and energy into the economic cycle.

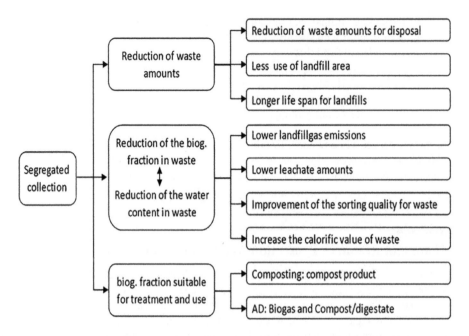

FIGURE 4.2 Importance of segregated collection of biogenic waste for climate and resource protection.

The key to the Waste Management Act is the implementation of the five-stage waste hierarchy: waste prevention, reuse, recycling, other utilisation (mainly energetic) and finally, waste disposal. The best option, with regard to environmental protection, has always been a priority, even if technical, economic and social aspects are considered. This ensures a consistent focus on waste prevention and recycling.

Further emission reduction can also be achieved in the field of climate protection, for example, by the increase in capacity in mechanical–biological pre-treatment.

The Waste Management Act continues the proven division of tasks between private and public waste management companies. According to the "polluter pays" principle, commercial producers and the owners of waste are responsible for the disposal of their wastes.

According to the principle of public services, municipalities are responsible for the disposal of waste from private households and other areas of origin.

In 2017, 68 waste incineration plants were in operation in Germany (capacity ~20 million tonnes) and 32 substitute fuel incineration plants (capacity ~5 million tonnes). In 2017, 45 bio-mechanical waste treatment plants (capacity 5 million tonnes) treated about 4.5 million tonnes of waste [3].

4.5 Separate Collection, Treatment and Utilisation

4.5.1 Legal Background and Importance for Resources and Climate Protection

For all municipal waste fractions, there are legal requirements as to how they are to be collected, transported, recycled or treated.

The implementation of the hierarchy in the steps of preventing and reducing waste generation, Reuse and Preparation for reuse (giving the products a second life before they become waste), Recycle (material recycling and composting), Recovery (high-quality thermal recycling) and Disposal (landfilling, incineration and other finalist solutions) is already prescribed by law. The determination of the priority of a type of utilisation (reuse, recycling and recovery – including energy) is regulated with the defined heating value criterion of 11 MJ/kg (§ 8 KrWG).

In addition, recycling is promoted and secured by the introduction of the nationwide mandatory seg-regated collection for bio-waste (§ 11KrWG) as well as of paper, metal, plastic and glass waste (§ 14 KrWG). By 2020, a recycling rate of at least 65% has to be reached for municipal waste (§ 14 KrWG). Further, there is a general prohibition on mixing hazardous waste with other waste streams in the future (§ 9 KrWG). The separate collection of bio-waste and recyclables is to make the high resource potential of this waste material more efficient.

After the ban on the landfilling of untreated municipal waste, the only possibility for reducing the annual emissions of climate-damaging gases in Germany is to increase the use of material and energy by increasing the efficiency of waste treatment plants and, in particular, by increasing the material recycling of waste streams [4].

4.5.2 German Concept for Waste Separation

In Germany, it is common to divide waste into different groups instead of throwing everything into the same bin. The reason is that only separated garbage is reusable. Therefore, waste is not only separated in residential houses and dormitories but also on the whole campus area and in other public places, such as the railway station.

Various container systems and vehicles are used for waste collection and transport, depending upon the type of waste involved, whereby a distinction is made between systematic and system-less waste collection.

System-less collection of household waste has, for the most part, given way to the use of a broad range of container systems whose main purposes are to allow for source separation of various types of waste. The containers are placed either in very close proximity to households (pick-up system) or at central locations (drop-off system). Residual waste is deposited in grey containers, to which end, the following elements are used: wheeled bins that can accommodate 120–140 L of waste, and garbage bags and 1.1 cubic metre containers that are used in settings such as large apartment buildings.

In the 1990s, Germany established a second waste collection system besides the household waste system, so-called dual systems, which allow for separate collection recovery and disposal of product packaging in close proximity to households. Dual-system use is governed by the Waste Management Act and the Packaging Act. Table 4.2 provides the used transport and secondary packaging collected in 2017 [1].

Recycling management in Germany is based on (waste avoidance) and waste recycling. Approximately 66% of domestic waste is already recycled. In the future, only material recycling will count! But Germany must significantly increase the recycling quotas for all types of plastics in the future.

TABLE 4.2

Collected Packaging in 2017 [1]

Types of Packaging	From Commercial and Industrial End-Users	From Households
	in 1,000 Tonnes	
Glass	225.2	1887.4
Paper and board	3107.6	5675.6
Ferrous metals	49.8	271.4
Aluminium	3.4	61.2
Other scrap metal, metal composites	18.5	
Plastics	329.1	133.8
Wood	495.8	1302.5
Composites; lightweight packaging	69.5	2753.2
Other materials	569.0	
Textiles		168.5
Packaging for hazardous filling goods	12.5	
Total	**4880.3**	**12235.6**

The Circular Economy Package of the European Union [7] stipulates a further increase in recycling for all member states. Member states will have to meet the following targets as they increase the reuse and recycling of municipal waste (Table 4.3).

	By 2025	By 2030	By 2035
Municipal waste	55%	60%	65%

Member states will set up, by 1 January 2025, separate collections of textiles and hazardous waste from households. Member states will ensure that by 31 December 2023, bio-waste is either collected separately or recycled at the source (e.g. home composting). This is in addition to the separate collection, which already exists for paper and cardboard, glass, metals and plastic.

The Circular Economy Package defines specific recycling targets.

For Germany, the Packaging Act (VerpackG) sets stricter goals [8]. The targets for the recycling of packaging will be increased from 1 January 2019 and then again from 1 January 2022. Packaging systems are required to achieve the following minimum annual averages for their contractually agreed quantities of packaging with a view to preparing for recycling and reuse (Table 4.4) [8].

The Packaging Act provides for a target reuse rate of 70% for beverage packaging. In addition, the mandatory deposit was extended to beverage packaging containing carbonated fruit and vegetable nectars and beverages with a share of milk products of more than 50%.

The Packaging Act places much higher demands on the material recycling of plastics in particular. As a result, sorting plants have to be retrofitted and the use chain for plastics for recycling has to be redesigned. At present, for example, the plastics processing industry's demand for plastics from sorting plants is too low. The market for plastic waste is too small! This is why plastic waste is exported to Asia. This export is now controlled and reduced to zero.

Another challenge is posed by the Commercial Waste Ordinance [9]. This ordinance also requires the separation of recyclable waste from trade, commerce and industry. Universities are also obliged to improve the separation of recyclable waste.

TABLE 4.3

Specific Recycling Targets According to the EU Circular Economy Package

	By 2025 (%)	By 2030 (%)
All packaging	65	70
Plastic	50	55
Wood	25	30
Ferrous metals	70	80
Aluminium	50	60
Glass	70	75
Paper and cardboard	75	85

TABLE 4.4

Specific Recycling Targets According to the German Packaging Act

Material	Previously (%)	Starting 2019 (%)	Starting 2022 (%)
Glass	75	80	90
Paper and cardboard	70	85	90
Ferrous metals	70	80	90
Aluminium	60	80	90
Beverage carton packaging	60	75	80
Other composite packaging	60	55	70
Plastics (material recycling)	36	58.5	63

The implementation of the Commercial Waste Ordinance currently represents one of the major challenges for German waste management. The Federal Government is endeavouring to reduce the use of plastics as single-use packaging. Politically, the best way is controversial.

Germany also wants to reduce exports of plastic waste. Waste that cannot be recycled in the target countries must not be exported.

More and more Asian countries are rightly refusing to accept waste from industrialised countries.

Hope for improvement, however, is promised by a resolution passed in May. In the Basel Convention, several waste exporters, including Germany, committed themselves to stricter rules for trade in plastic waste.

Only if the goods are cleaned, sorted and recyclable can they be traded freely; otherwise, the authorities must give their consent. The EU even wants to go one step further. By 2021, the export of poorly recyclable waste to developing countries will be completely banned.

4.6 Conclusion

In recent decades, Germany has developed a fairly good system for the separate collection of recyclable waste. A comprehensive body of legislation supports this system. A combination of waste charges and financial contributions under the Extended Producer Responsibility (EPR) finances separate collection and recycling without state financial support.

Waste producers are offered a comprehensive range of different waste collection containers and disposal options. But too many citizens do not separate their waste well enough. Partly out of ignorance, partly intentionally!

The recycling capacities in Germany and the European Union are still too small. That is another reason why recyclable waste is exported. Unfortunately, this also leads to illegal exports.

There are not enough possibilities for recycling plastic waste; too much plastic is thermally recycled. At present, recycling too often involves a lack of a chain of demand for secondary plastics.

Overall, a lot has been achieved in the last 30 years, but much still needs to be improved.

LITERATURE

1. Statistisches Bundesamt (Destatis), Abfallwirtschaft 2018, 2020. http://www.destatis.de/.
2. The Waste Management Act (KrWG - ClosedSubstance Cycle Act) - Gesetz zur Förderung der Kreislaufwirtschaft und Sicherung der umweltverträglichen Bewirtschaftung von Abfällen (Kreislaufwirtschaftsgesetz - KrWG), 24.02.2012, BGBl. I S. 212, 20. Juli 2017, BGBl. I S. 2808.
3. Federal Ministry for the Environment, Nature Conservation and Nuclear Safety (BMU), 2019, Waste Management in Germany 2018, http://www.bmu.de/en/publications.
4. BDE - Bundesverband der Deutschen Entsorgungs-, Wasser und Rohstoffwirtschaft e. V., 2018: Statusbericht der deutschen Kreislaufwirtschaft - Einblicke und Aussichten; https://www.bvse.de/.
5. Federal Ministry for the Environment, Nature Conservation and Nuclear Safety (BMU), 2017, Recycling and Recovery of the biogenic fractions from municipal solid waste in the PR of China, https://www.retech-germany.net/.
6. German Environment Agency (UBA), 2010. Klimaschutzpotenziale der Abfallwirtschaft; Umweltbundesamt, UBA Text 06/2010, https://www.ifeu.de/abfallwirtschaft/.
7. EC - European Commission, 2019. The European Commission Circular economy package, https://ec.europa.eu/environment/circular-economy/.
8. Gesetz über das Inverkehrbringen, die Rücknahme und die hochwertige Verwertung von Verpackungen (Verpackungsgesetz - VerpackG), 2017, BGBl. I S. 2234.
9. Verordnung über die Bewirtschaftung von gewerblichen Siedlungsabfällen und von bestimmten Bau- und Abbruchabfällen (Gewerbeabfallverordnung - GewAbfV), 2017, (BGBl. I S. 896), Artikel 2 Absatz 3 des Gesetzes vom 5. Juli 2017 (BGBl. I S. 2234).

5

A Study on the Legal and Administrative Support for Proper Management of World-Famous Shri Jagannath Temple at Puri

Sasmita Samanta
KIIT Deemed to be University

R.K. Nanda and P. Rautaray
KIIT Deemed to be University

CONTENTS

DOI: 10.1201/9781003231608-6

5.1 Introduction

The holy city of Puri, the abode of Lord Shri Jagannath, situated in the East, acquired increasing distinction as one of the Dhams, others being Badrinath in the North, Dwarka in the West, and Rameswaram in the South. It was recognized by Adi Shankaracharya as one of the renowned places of pilgrimage in the eighth century AD. The city has assumed the holy status over a period of time with the visit of holy saints like Shri Ramanuja, Shri Vishnuswamy, Shri Nimbarka and Sri Madhvacharya. The place has been visited by the famous saints of various religions, namely, Nanak, Kabir and Shri Chaitanya. Many saints have visited Puri and were so attracted by the Deities and the other places in Puri that they established monasteries and stayed for a long period, which have subsequently been continued by their disciples.

Puri town is situated on the sacred bank of the Bay of Bengal in Odisha, India, which is otherwise called Mahodadhi, the 7-km beach that attracts pilgrims from far and wide to assemble here for rituals as well as to enjoy the natural beauty. Puri has 3-km grand road that is treated as very sacred, where the annual car festival is organized.

5.2 Review of Literature

Shree Jagannath Temple, which has gained importance from the 12 century AD, is shrouded in mystery as to the origin of the Lord. As per Skanda Puran, Utkal Khanda, King Indradyumna of Malwa, had installed the four Deities in the Purusottam Kshetra, but History is silent about the subsequent reforms taken up by the kings in the temple. However, descriptions have been made in the Patalkhanda of Padma Puran and Vishnu Puran about the existence of the temple of Purusottam in the said ancient sacred place. According to Kapila Samhita, Lord Brahma accepted the prayers of King Indradyumna and came down to the earth and installed the divine idols. This has been confirmed in Niladri Mahodaya that Lord Brahma chanted Vedic hymns for the Deities while installing the idols. Madala Panji, the chronicle of the history of Shree Jagannath Temple, states that the old temple was built by King Jajati Keshari. The present temple was started by King Chodaganga Dev and completed by King Anangabhima dev in 1230 AD.[2]

The guidelines for running the temple date back to Skanda Puran, where Lord Shree Jagannath has given direction about the various festival to be observed each year with a detailed description of each ritual. During that time, no administrative guidelines were given except for directing the King to organize the festivals. However, with the passage of time, the kings felt the need for sharing the responsibility of the work. The concept of Chhatisha Nijog started from the time of Jajati Keshari when the service to the Lord was divided into 36 categories and responsibilities were fixed. Thereafter, many changes were made by the kings, but finally the Government intervened and formed a Managing Committee by making the King of Puri the Chairman and keeping members from the Government as well as Servitors and other public figures. Elaborate administrative guidelines were notified in the Manual.[3] Ministry of Environment and Forests (MoEF), Government of India promulgated solid waste management rules in 2,000, mandating the municipal authorities to implement the approved system of waste management. Finally, the case of solid waste management came into force when the Puri Municipal Corporation took the responsibility of handling the waste of the temple, which was simultaneously taken up by Puri Konark Development Agency and other voluntary organizations.

5.3 Objectives

The chapter aims to discuss various laws and resolutions for the successful running of the world-famous temple and the surroundings especially the Puri town, which is always full of tourists who keep coming around the year. It also aims to find out the reasons for the importance given to protect the environment from environmental pollution and to find out the source of garbage inside the temple and in the holy city and the problem posed by the excessive generation of solid waste and the role played by the Government as well as Non-Governmental Organizations.

5.4 Methodology

A. Collection of information on various laws, rules and regulations for proper management of the temple affairs including waste management.

B. Collection of various rules, enactments and rules for the holy city of Puri.

C. Analysis of the sources of generation of waste material inside the temple and Puri.

D. Contacting the temple authorities, Municipal Authorities, NGOs as well as executives of Puri Konark Development Authority (PKDA) to review the action taken by them towards handling the solid waste.

E. Reviewing the progress made by the agencies and suggesting for the betterment of the Temple and the holy city.

5.5 The Study

The main attraction of the holy city is the world-famous Shree Jagannath Temple where Lord Shri Jagannath is worshipped as the presiding Deity. Centuries of myth, legends and history are blended into a grand composite culture, which is centred around Lord Shri Jagannath, the Lord of the Universe, one of the most revered and ancient of the Deities of the Hindu pantheon. Legendary sources suggest that Shri Jagannath was originally worshipped by the tribals. In course of time, the cult of Lord Jagannath took an Aryanized form, and various major faiths like Shaivism, Vaishnavism, Jainism and Buddhism were assimilated into the concept of Jagannath as an all-pervasive and all-inclusive philosophy, symbolizing unity in diversity. The Jagannath culture preaches love, integrity and peaceful co-existence among the human community. The characteristics of culture are generosity, endurance and coordination. The temples of Lord Shri Jagannath have been built and are being built all over the world because the Lord symbolizes love and humanism. Although there are temples throughout the world, Puri, the original abode of Lord Jagannath, assumes importance as the original abode of Lord Jagannath. The importance attached to Puri and Lord Shri Jagannath temple is varied. The main reasons among them are as follows:

1. The place is defined as Purushottam – Puri has a reference to many scriptures as the most sacred shrine in India. The sacred shrines, according to tradition, existed even during the great deluge and also at the beginning of the creation. Its greatness is unparalleled, its importance is unique and its sanctity is unquestionable.

2. Puri is considered one of the seven salvation centres of the Hindu faith the other six being Ayodhya, Mathura, Varanasi, Kanchi, Avantika, and Dwaraka.

3. Named after Lord Jagannath as Jagannath Puri the city is known as Niladri, Shree Kshetra, Sankha Kshetra and even Martya Baikuntha or heaven on earth. Therefore, the city assumes importance as the holiest of the holies (Figure 5.1).

The Jagannath temple is built on an elevated ground of about 20 ft above the level of surrounding area. It is 215 ft high, the imposing sight of which can be seen from kilometres away. The construction of the present edifice dates back to the era of Anangabhimadeva III (1078–1147) of Ganga Dynasty. The main shrine of four distinct buildings (i) Vimana, or the great Temple; (ii) Jagamohan or the Hall of Audience; (iii) Nata Mandap or the Dancing Hall; and (iv) Bhogamandap or the Refractory.

The images of Lord Jagannath, Lord Balabhadra, Goddess Subhadra and Lord Sudarshan are installed in the inner Sanctuary of the temple on a raised platform called the Ratnavedi or the jewelled platform. The miniature images of Goddess Laxmi, Goddess Saraswati and Lord Madhab are placed on the Ratnavedi along with the main Deitesans are worshipped. The raised platform is made of stone, which is 16-ft long, 13-ft wide and 4-ft high.

FIGURE 5.1   From left to right, Lord Balabhadra, Goddess Subhadra, Lord Jagannath and Lord Sudarshan pillar shape.

The temple has two big enclosures. The inner walls are known as Kurma Prachira measuring 400×276 ft and the outer wall Meghnad Prachir measuring 665×644 ft with a height varying from 20 to 24 ft. The whole temple complex is on 10.7 acres.

The temple has four gates at the eastern, southern, western and northern mid-points known as Lion's Gate, Horse's Gate, Tiger's Gate and Elephant's Gate, respectively.

Inside the temple, there are around 60 small and medium-size temples prominent among them are Hanuman, Patitpaban, Kashi Vishwanath, Ramachandra, Nrusingha, Bimala, Mangala MahaLaxmi, Bhubaneswari, Sasthi, Savitri, Gayatri, Navagraha, Sun and Ganesh.

On the other hand, the place assumes greater importance as this forms a part of the golden triangle; the other two places are Konark and Bhubaneswar. The distance from one place to other is within 60 km. Therefore, there is flow of tourist round the year.

Puri has many other places of tourist interest apart from the temple. They are as follows:

A. Astasambhu (eight famous Shiv Temples) Markandeshwar, Kapal Mochan, Lokanath, Jameswar, Nilakantheswar, Bileswar, Isaneswar and Pataleshwar

B. Asta Chandi (eight Goddess symbolizing Shakti, the power) Bata Mangala, Alamchandi, Ardhasini, Marchika, Narayani, Shyamakali, Dakshina Kali and Ramachandi

C. Pancha Tirtha (five Holy Tanks) Markandeswar Tank, Indradyumana Tank, Sweta Ganga, Mahodadhi and Rohini Kunda, Kalpabata and Pancha Shiva.

In addition to this, there are other places of tourist interest inside Puri and the adjacent area. Prominent among them are Baseli Thakurani Temple, Tota Gopinath Temple, Swargadwar, Gundicha Temple, Sunara Gauranga, Ugrasen Temple, Rameswar temple, etc.[4]

Puri is also famous for religious monasteries belonging to various sects centring around the temple of Shree Jagannath. Most of the monasteries are connected with the worship of Lord Jagannath and have typical sevas to perform in the temple. There are around 50 such monasteries and ashrams belonging to different sects active in the town of Puri. Some important monasteries and ashrams are Languli Baba Ashram, Satalahadi Matha, Radhakanta Matha, Guru Nanak Ashram, Jatia Baba Ashram and Ginarbant Ashram (Figure 5.2).

Puri is always full of tourists due to the observance of religious festivals that are spread around the year. Few of the events that can give the impression of crowd movement to Puri are ChandanYatra, Snana Purnima, Car festival (which draws around 10 lakh crowd in a span of 10 days), Holi, Diwali and birth-day celebration of Lord Ganesh, Baladeva, Nrusingha, Sri Ram, Sri Krishna, Hanuman and Goddess Saraswati. It also celebrates the marriage function of Lord Jagannath and Goddess Lakshmi.

In addition to these festivals, the Deities are decorated with different attires to remember various occasions in the past. These are called Beshas, which are performed on 24 occasions, drawing heavy crowd to witness the Deities.

FIGURE 5.2   The famous car festival of Lord Jagannath, which attracts around ten lakh devotees to Puri.

As per the Government Report, the tourist inflow to Odisha was 1,41,11,243 in the year 2017 out of which Puri had alone 1,25,85,158 visitors (Statistical Bulletin, Department of Tourism, the Government of Odisha 2017). With the concentration of more tourists, Shree Jagannath Temple and Puri deserve special attention.

Major Issues in Puri

Lack of field drains to the drainage system

Insufficient discharge of surplus floodwater from the catchment area

Encroachment of drainage path by the inhabitants

Insufficient vent size provided by in the National Highway WHICH is not adequate to discharge the surplus rainwater

In the case of Lord Shri Jagannath Temple, the problems relating to waste generation and management are analysed below.

5.6 Sources of Waste from Shree Jagannath Temple

a. **Flowers**: The Deities are adorned with flowers everyday as per rituals. There are specified places inside the temple – one near Koili Baikuntha and the other in Neelachal Upaban for preparing garlands and special type of ornamental decoration from flowers. At present, 14 ladies are engaged for this work from 7 am to 9 pm everyday. There are six flower-growing farms maintained by the temple authorities. Apart from these flowers, every day devotees give flower in these places for making garlands. The floral decoration used for Lord Jagannath, Lord Balabhadra, Lord Sudarshan and Goddess Subhadra are changed minimum four times a day, out of which the night decoration is famous as Bada Singar when the Deities are heavily decorated. Apart from the main Deities, the other Deities surrounding the temple are florally decorated three times a day. The main Deities among them are Goddess Mangala, Goddess Vimala, Goddess Maha Laxmi, Lord Patitpaban, Lord Satyanarayan, Lord Ganesh and Lord Hanuman.

b. **Tulsi (Basil) Leaves:** Tulsi (Basil leaves) garland is another heavy consuming item in the temple as Lord Jagannath is fond of Tulsi. Devotees are interested to offer Tulsi garland to the Deities sitting in the Ratna Bedi. Lots of sellers of Tulsi leaves garland are seen inside the temple.

c. **Residue of the Vegetables from the Temple Kitchen:** It is said that the kitchen of Lord Jagannath is the biggest in the world. It has the capacity to feed one lakh devotees per day. On average, cooking for 10,000 people is done in the kitchen that increases to 30,000 on festive occasions. The residue of the vegetable and the coconut shell which are generated on daily basis forms the waste from the kitchen.

d. **The Earthen Pot for Cooking and Food Leftover Food Items:** The cooking in the kitchen is made in earthen pots that are used once only. Therefore, everyday new earthen pots are used and in the earlier days, earthen pots were disposed of in the dump yard. Further to this, the residual food item, leftover by the devotees after eating, the plantain leaves, lemon and unsold Mahaprasad also form a part of the dumping material.

e. **The Earthen Lamp:** Everyday thousands of devotees purchase earthen lamps filled with ghee and light the lamp to offer their prayer to the God and Goddesses inside the temple as a mark of respect to them. And those earthen lamps become redundant once they are burnt. These lamps accumulate inside the temple on a daily basis, which are dumped inside the yard on the south side.

5.7　Management of Waste Inside Shree Jagannath Temple

Shree Jagannath Temple generates 15–17 tonnes of waste every day.

The temple authorities have tied up with ITC, which has erected 3 compost plants of 1.5 MT each. The agency has engaged two employees who have been working from dawn to dusk. Whatever is adjusted in the compost plants remains inside the temple premises. The excess solid waste is dumped outside the temple dump yard on the north and south side on a daily basis, which is then collected by Puri Municipal Corporation daily.

5.8　Work Done by Various Agencies on Waste Management in Puri

a. **Puri Municipality in Collaboration with JUSCO**

Puri Municipality has joined hands with Jamshedpur Utilities and Services Company Ltd (JUSCO) to make the city clean. A fully mechanized road sweeping machine, five dumpers placers, pickups, tractor with hydraulic trolleys and other related equipment are deployed by JUSCO for pick-up of the city, thereby diverting waste compost.

b. **Puri Municipality in Collaboration with Feedback Foundation**

Feedback Foundation is providing technical support to Puri Municipality for an end-to-end solution as per Solid Waste Management Rules of 2016 for decentralizing solid waste management in Puri. Under the scheme of intervention and mass mobilization of people for segregation of waste, two Micro-compossping Centres have been installed to avoid dumping and garbage.

c. **Puri Municipality in Collaboration with HCCB**

Municipal Corporation with Hindustan Coca Cola Beverage (HCCB) has joined hands to ensure the car festival plastic-free. Puri generates 100 tonnes of solid waste every day and a lot of it is plastic waste. Therefore, the agency in collaboration with Municipality Officers educates the people on the benefits of separately keeping the solid waste for processing.

5.9 Bio Composting – Solid Waste Management

The civic body Puri Municipality was formed in 1984. In 1998, under the support of Indo-Norwegian Development Corporation and with the active participation of the Government of Odisha, a bio-compost plant was set up in the town.

The above works are being done on regular basis by various organization, the major being done by Puri Municipality and Puri Konark Development Agency.

However, for the smooth running of the temple and the betterment of the holy city of Puri, lots of rules, regulations and acts have been framed.

5.10 Shri Jagannath Temple Act 1954 (Odisha Act 11 of 1955)

This is known as Shree Jagannath Temple Act 1954. It is the first complete and historic act prepared by state and the act is still in force. This act was approved by the President on 15 October 1955. This Act was notified vide notification no. 5923, dated 2 November 1955 and available in Orissa Gazette, dated 4 November 1956; part X, no. 44. At the beginning of this Act, it is stated that it is "an act to provide for better administration and governance of Shree Jagannath Temple at Puri and its endowments."

The following statements were also there depicting its important need and objectives for the smooth management and administration of the temple.

1. The ancient Temple of Lord Jagannath of Puri has since its beginning been an institution of sole national significance in which millions of Hindu devotees from different regions have kept their faith and belief and have regarded it as the essence of their practice and ethnicity.
2. After the British invasion, the superintendence, control and management of the Temple have been the direct responsibility of successive rulers, the Government and its officers and of the public exchequer.
3. By Regulation IV of 1809 of the Governor-General in Council on 28 April 1809 and thereafter by other laws and regulations and in accordance of arrangement entered into with the Raja of Khurda, later known as the Raja of Puri, entrusted hereditarily with the management of the Temple and its affairs as the superintendent. They will be subject to the control and supervision of the ruling power.
4. In view of serious irregularities, the Government had to interfere on various occasions in the past.
5. The management under the Superintendent has deteriorated and a situation has compelled it expedient to restructure the scheme of management of the affairs of the temple and offer better administration in restraint of all previous laws, having regard to the ancient customs and usages and the traditional nitis and rituals Orissa contained in the Record of Rights arranged under the Puri Shree Jagannath Act XIV of 1952 Temple (Administration) Act, 1952, in the method hereinafter appearing.
6. It is hereby enacted by the Legislature of the State of Orissa in the sixth year of the Republic of India.

5.10.1 Orissa Ancient Monuments Preservation Act 1956

This Act, known as the "1956 Act", was provided for the protection of ancient monuments and control over-excavation and protection of acquirement in cases of ancient monuments and objects of archaeological, historical or artistic interest in the state.

5.10.2 Coastal Regulation Zone

The Government of India has declared the coastal stretches of seas, bays, estuaries, creeks, rivers and backwaters, which are influenced by the tidal action (on the landward side) up to 500 m from High Tide

Line (HTL) and the land between the Low Tide Line (LTL) and the HTL as Coastal Regulation Zone Under Clause (d) of sub-rule (3) of Rule 5 of the Environment (Protection) Rules, 1991.

Under the Act, the Regulations for CRZ II are

a. Buildings shall be allowed neither on the seaward side of the existing road nor on the seaward side of the proposed road.
b. Reconstruction of the authorized building to be permitted by the specified authority.
c. The design and construction of buildings shall be consistent with the rules and regulations made by the authority.

5.10.3 Orissa Irrigation Act 1959

The Section 4 (6) of the act defines "Government water source" as any water source formed naturally or otherwise by collection or deposit of water at a fixed place. In Section 4(17), it further defines "Water course" as any channel or pipe not maintained at the cost of the State Government which is supplied with water from an irrigation work and includes all secondary works connected with any such channel or pipe, except the sluice or outlet through which water is supplied from an irrigation work to such channel or pipe.[5]

5.10.4 Sweet Water Zone Notification

The water supply management of Puri town was given to Public Health Engineering Organization (PHEO). Previously it was with Puri Municipality up to 1882. Puri town gets water supply from two well fields, Chakratirtha and Baliapanda. These two places are located at two ends of the town. These well fields were assigned by PHEO as Puri Water Works in order to protect the said field and guard the aquifers from the pollution by urban waste. Puri does not have any other source of ground water. The notification was published in 24 May 2000 for the safety of the ground water resource by Housing and Urban Development Department and Revenue Department to restrain unauthorized construction activities either by encroachments or without permitted plan in these zones, to control Puri Municipality from further sale/lease/transfer of these revenue lands and to reinstate lands already allotted/leased out.[6]

5.10.5 Tourism Policy

This policy letter was submitted to the Department of Tourism and Culture (Tourism), the Government of Odisha on 17 April 2013 by the expert committee for approval and subsequent implementation. Among other things, the broad outline of the policy is stated as follows:

The tourism sector has been recognized worldwide as a major industry with the potential of being a creator of economic growth. Recognizing the above, the Odisha Tourism Policy, 2013 introduces tourism as a profitable economic sector – as a sector generating 78 jobs per million rupee investment as opposed to the 45 jobs produced through any other manufacturing sector.[7]

5.10.6 Puri Konark Development Authority

PKDA was constituted under Odisha Development Authorities Act, 1982. Following are the functions of the PKDA.

Preparation of growth plans, undertaking works pertaining to different Town Planning Schemes and public amenities.

a. Regulating the development and utilization of lands, including private land.
b. Working as a coordinating agency between various Government Departments and other agencies associated with development activities.
c. The planning area consists of Puri Municipal limits, Konark NAC area and the adjacent rural areas, including 131 numbers of revenue villages, with a population of about 4.96 lakhs.

d. "The main objective of the authority is to make certain planned and systematic development of the areas under its jurisdiction, with the preparation of development plans, undertaking works pertaining to different Town Planning Schemes and public amenities. The PKDA area consists of three administrative and planning jurisdictions,[8] which are (Table 5.1).

5.10.7 Puri Municipality

Puri Municipality is the first city in Odisha to have arrangements for disposal of waste in a scientific manner. Puri Municipality is able to recover a portion of its expenditure from the royalty it gets from the sale of composts. "Puri Municipality limits extends over an area of 16.84 sq km, whereas the Master Plan Area extends over 63.578 sq km. The municipal area stretches along the sea-shore measuring about 6.59 km. The municipality has 32 wards extending over 34 mouzas. The master plan area has 16 more villages in addition to the municipal area. The master plan area stretches over 20.35 km along the seashore (Table 5.2)."[9]

5.10.8 Public Health Engineering Organization

Public Health Engineering Division, Puri is under the administrative control of the Engineer-in-Chief, PH Odisha. The Division belongs to Housing & Urban Development Department, the Government of Odisha and is headed by one Executive Engineer (PH). This division consists of three PH subdivisions like Puri, Nimapada and Nayagarh. It is responsible for planning, designing, construction, operation and maintenance of water supply system and management of water schemes, including their transportation and distribution.[10]

TABLE 5.1

Waste Storage Facilities in Puri Municipal Corporation

Sl. No.	Component	PM Area
1	Dust bins	60
2	Open collection points	7
3	Push carts/wheel barrows	480
4	Conservancy staff	477
5	Barrow wheel	480
6	Excavators	1

Source: Puri Municipality.

TABLE 5.2

Population Forecasting by Puri Konark Development Authority 2031

Block	Decadal Growth Rate (%)	Population		
		2001	2021	2031
Puri Town	30	1,57,837	2,65,374	3,44,098
Puri Sadar Block	32	14,880	25,473	33,328
Block A	19	17,238	24,414	29,054
Konark Block	34	16,979	40,040	61,487
Block B	20	6,872	8,827	10,005
PKDA (Total)	29	2,13,806	3,64,127	4,77,972

Source: Census data, CEPT analysis.

5.10.9 Odisha Water Supply and Sewerage Board (OWSSB)

OWSSB is responsible for the construction and preservation of major stormwater drains within Puri town limits. The board is also responsible for the construction and provision of sewerage network within the town.

5.10.10 Odisha State Archaeology

Odisha State Archaeology is headed by Superintendent of Archaeology, assisted by Curators, Archaeological Engineers and Conservation Assistants to looks after maintenance and protection of ancient monuments. It undertakes the programme for conservation and safeguarding of ancient monuments and Buddhist heritage of the State.[11]

5.10.11 Nabakalebara Projects

Nabakalebara is an important festival of the state, which was celebrated during 2015 in Puri in the month of July. For the festival, ~6 million people visited the city. The Government of Odisha implemented several infrastructure projects for up gradation of existing facilities, which were undertaken through multiple departments. These projects have been called Nabakalebara 2015 projects, most of which have been completed before July 2015. Several of the projects have been undertaken to make stronger the existing infrastructure of the city and hence catering to improved experience of visitors and improved quality of life for the local community. Without the help of these works, no crowd control policy and traffic management policy could be possible. Finally, the authorities proved that the project was successfully completed without any casualty.[12]

5.11 AMRUT Scheme

Atal Mission for Rejuvenation and Urban Transformation (AMRUT) is a scheme of Ministry of Urban Development, the Government of India where the main components of AMRUT consists of capacity building, reform implementation, water supply, sewerage and seepage management, storm water drainage, urban transport and development of green spaces and parks.[13]

5.11.1 Urban Transport

1. Ferry vessels for inland waterways.
2. Footpaths/walkways, sidewalks, foot over bridges and facilities for non-motorized transport.
3. Multi-level parking.
4. Bus Rapid Transit System (BRTS).

5.11.2 Water Supply

1. Water supply systems including augmentation of existing water supply, water treatment plants and universal metering.
2. Rehabilitation of old water supply system, including treatment plants.
3. Rejuvenation of water bodies specifically for drinking water supply and recharging of groundwater.
4. Special water supply arrangements for difficult areas.

5.11.3 Sewerage

- Decentralized, networked underground sewerage systems, including augmentation of existing sewerage systems and sewage treatment plants.
- Rehabilitation of old sewerage systems and treatment plants.
- Recycling of water for beneficial purposes and reuse of wastewater.

5.11.4 Seepage

1. Sludge Management-cleaning, transportation and treatment in a cost-effective manner.
2. Mechanical and biological cleaning of sewers and septic tanks and recovery of operational cost in full.

5.11.5 Pilgrimage Rejuvenation and Spiritual Augmentation Drive (PRASAD)

The Union Ministry of Tourism provides Central Financial Assistance (CFA) to State Governments/Union Territory Administrations, including the places of religious significance, for various tourism projects subject to availability of funds, inter-se priority, liquidation of pending utilization certificates and adherence to the scheme guidelines. For development of tourism infrastructure in the country, the Ministry of Tourism has introduced two new schemes in 2014–2015, i.e. PRASAD – Pilgrimage Rejuvenation and Spiritual Augmentation Drive. Under PRASAD 12 cities have been recognized for development initially, namely Amritsar, Ajmer, Dwaraka, Mathura, Varanasi, Gaya, Puri, Amaravati, Kanchipuram, Vellankanni, Kedarnath and Guwahati. Under Swadesh Darshan Scheme, Spiritual Circuit has been identified as theme circuit for development. During the year 2015–2016 under Budget Estimates Rs. 100.00 crore have been allocated for PRASAD to develop tourist destinations of global standards.[14]

The projects sanctioned under PRASAD scheme for Puri are of 50.00 crore.

It is to be concluded that apart from the constitutional law support and support of the Human Rights Act, the different Acts and Rules discussed in this chapter are regarded as highly indispensable. If any of these Acts will be removed or withdrawn, no temple will function properly. In the case of Shree Jagannath Temple, the SHREE JAGANNATH TEMPLE ACT 1954 (ODISHA ACT 11 OF 1955) is the principal Act on the basis of which the Record of Rights are prepared in which all the important rules are framed for the smooth management of the temple.

5.12 Conclusion

Realizing the importance of Lord Shree Jagannath temple and the ever-increasing flow of traffic to the holy city of Puri, both the Government of India and Government of Odisha have started many new schemes for attraction of tourists and tourist management. Of late, they have started giving importance to the environment to control air, water and noise pollution and to handle the waste material being generated in the temple and the ever-growing city of Puri. It is heartening that many of the Non-Government organization and corporate houses have come forward to help in the implementation of various types of projects concerning waste management and prevention of pollution to protect the environment.

It is the prime duty of both Central and State Governments not only to implement solid waste management projects but also to encourage more and more outside agencies, including corporate houses, to participate in various types of projects to create awareness about the environment among the general public to make Puri a model city under Swachh Bharat Project.

REFERENCES

1. Shree Jagannath Temple Act, 1955.
2. Shree Jagannath the Supreme Lord of the Universe Dr Bhaskar Mishra and Sri Susant Mishra, 2015.
3. The shrines of Shree Jagannath an overview Dr Bhaskar Mishra and Susant Das, 2015.
4. Shree Jagannath, MahimohanTripathy, 2018.
5. Orissa Irrigation Act, 1959.
6. Sweet Water Zone Notification, 2000.
7. Tourism Policy of Govt of Odisha, 2013.
8. PKDA Act of Govt of Odisha, 1982.
9. Puri Municipality Rules, Puri, Govt of Odisha, 1982.
10. Health and Family Welfare Policy Guidelines, Govt of Odisha.
11. Orissa State Archeology Act, Govt of Odisha, 2019.

12. Nabakalebar Project Document, Government of Odisha, 2015.
13. AMRUT Scheme, Govt of India, 2015.
14. PRASAD Scheme, Gove of India, 2015.

6

Gate-to-Gate Life Cycle Assessment of Hydrothermal Carbonization Process for Food and Yard Waste

Saikrishna Venna
National Institute of Technology (NIT)

Sagarika Panigrahi
Indian Institute of Technology Kharagpur

Hari Prasad Reddy
National Institute of Technology (NIT)

Brajesh Kumar Dubey
Indian Institute of Technology Kharagpur

CONTENTS

6.1 Introduction

Universally, food waste discarded is not only produced from the consumer's end but also from the production, processing, and retailing (Thacker, 2018). As per the United Nations Food and Agriculture Organization (FAO), the food waste produced costs around $750 billion of the total world's economy (Thacker, 2018). Furthermore, the food waste generated is about one-third (1.3 billion tons) of the total global food production (Thacker, 2018). The waste generated includes 20% of meat products, 20% of dairy products, 30% of cereals, 35% of seafood, and 45% of vegetables and fruits (Ross, 2015).

The clean India journal reported that India process 67 million tons of food waste annually (Thacker, 2018). The waste diverted to open landfill site is 70%–75% of waste generated resulting in 3.3 Gigatons of carbon footprint, leading to global warming (Saqib et al., 2018). The worldwide topmost producers of GHG emissions are US and China. The food waste produced per person in North America and Europe is 95–115 kg per year, and 6–11 kg in Africa and southern parts of Asia (Ross, 2015). The waste generation per kg per year across the different parts of the world is shown in Figure 6.1 (magnet.co.uk, 2018) as per statista.com.

Waste conversion to energy is a previous technology to use leftover material as a resource. This process decreases resource depletion and environmental impacts as well (Kalyani & Pandey, 2014). There are methods like anaerobic digestion (AD), pyrolysis, and gasification to transform the waste into energy (Kalyani & Pandey, 2014). Although the methods generate energy from food waste, there are also some environmental disadvantages. AD includes long duration, release gases like nitrogen and hydrogen sulfide (Phuong et al., 2014). The incineration includes the usage of a high amount of heat and releases toxic gases like dioxins (Phuong et al., 2014). Pyrolysis and gasification involve high temperatures (Phuong et al., 2014). But the difficulties involved to perform these processes took a path to choose the easiest and most efficient process called "hydrothermal carbonization" (HTC) (Phuong et al., 2014).

HTC is a promising route for the efficient yield of biomass through a variety of hydrolysis, dehydration, decarboxylation, condensation–polymerization, and aromatization processes (Kent Hoekman et al., 2018; Sharma and Dubey, 2020a). The process will take place in an aqueous phase under high temperatures (180°C–350°C) and autogenous pressure, oxygen-limited environment (Ramke et al., 2009; Sharma et al., 2020c). Since the water temperature is above 100°C, the reaction pressure also must be elevated (more than 1 atm) to maintain the water in a liquid form (Wyczalkowska & Sengers 2002). The process uses water as a solvent and catalyst to break the complex material in an easy way (Ramke et al., 2009). The outputs of the process are gaseous, water, and a carbon-rich, high-energy-density solid material known as "Hydrochar" (Ramke et al., 2009). Hydrothermal carbonization of solid waste is also advantageous: offering low-cost, low temperature, and environmentally friendly production of novel carbon materials from natural precursors without the need to use toxic chemicals (Titirici et al., 2010). Carbon-based materials have attracted considerable interest in many energy-related applications, such as biofuels (Sharma and Dubey, 2020b), soil amendments, energy storage in supercapacitors, Li-ion batteries, catalysis/electro-catalysis, adsorption, and gas separation and storage (Chemistry, 2019). The abundant chemical and thermal stability, processability, and the possibility of tuning their textural and structural characteristics fulfil the requirements for specific applications (Jain et al., 2016). Hydrochar

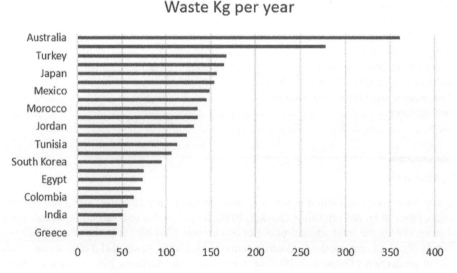

FIGURE 6.1 Waste generation, kg per year.

prepared by HTC contains amounts of surface functional groups. The surface functional groups affect faradic reactions and wettability between the electrolyte solution, and the carbon surface leads to higher electrical conductivity (Zhao et al., 2010).

The product from HTC of organic waste has market value and has the potential of creating local employment, thus bringing socioeconomic benefits (Sharma et al., 2021). The resource recovery approach using HTC is economically and environmentally friendly (Lin et al., 2017). Even though the technology is in the early stage of development, literature concerning it does suggest it has the potential of being future technology. Despite numerous sustainable applications, the integration of HTC with other treatment processes (like AD for process waste treatment) would help in realizing benefits associated with the circular economy approach of the treatment process (Sharma et al., 2020a; Mittapalli et al., 2021). Apart from these, the energy usage during the HTC process depends upon the operating conditions of the process. Therefore, detail process intensification and integration are necessary to improve the environmental and economic advantages of the process (Sharma et al., 2020b).

The supremacy of HTC over other thermal and bio-based energy recovery techniques includes eco-friendly, simple and fast process, low activation energy and power consumption requirement, low operating temperatures, and no predrying is required (Funke & Ziegler, 2020; Libra et al., 2011; Liu & Balasubramanian, 2014; Lu et al., 2014; Lucian & Fiori, 2017; Pham et al., 2015; Titirici & Antonietti, 2010). Even though various studies concluded HTC to be an eco-friendly process, some LCA studies reported that HTC process water (a by-product of HTC treatment) can impose a significant footprint (Berge et al., 2015; Saqib et al., 2019) at a large-scale application of HTC which is not sustainable and destroys its circular economy approach. Therefore, a gate-to-gate life cycle assessment (LCA) of HTC of MSW's using Simapro software was performed to evaluate the environmental impacts generated from the process for better process improvement.

6.2 Materials and Methods

The present study aims to display the environmental impacts caused by the HTC process and to ensure the efficient use of emissions associated with the process. LCA is chosen as the key tool used to find out the impacts.

6.2.1 Life Cycle Assessment

LCA contains a set of methods like a goal and scope definition, inventory analysis, impact assessment, and an interpretation phase (Standard, 2006). LCA is an iterative process and the methodology adopted for analysis is as per the ISO 14040 (2006) and ISO 14044 (2006) standards. A gate-to-gate LCA is performed using universally very successful LCA software Simapro® 9.0.0.35 with database support from an Eco invent v3 database, which is available in Simapro. The variant used in the study is a gate-to-gate analysis. The detailed information about the four stages of LCA is discussed in further steps. The accuracy of the results depends on the specified goal of the study and the input to output related to the functional unit chosen (Standard, 2006).

6.2.1.1 Goal and Scope Definition

The evaluated chosen system boundary's objective is to produce hydrochar from the food waste and yard waste as feedstocks by an HTC process. The major objective is to compare the negative impacts like global warming and soluble chemical oxygen demand (SCOD) from the HTC process of food waste and yard waste through a gate-to-gate analysis. The compared results were used to help the process development by analysing their environmental performance discussed in conclusions. The system boundary contains the input materials such as raw waste, de-ionized water, and electrical energy, whereas the output materials are process water, hydrochar, and gas.

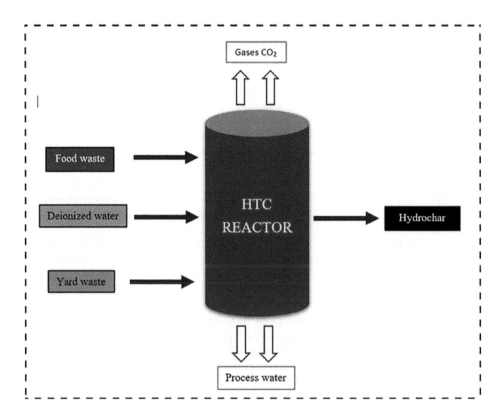

FIGURE 6.2 System boundary of study area.

6.2.1.2 Functional Unit

The functional unit adopted is 1 MJ of hydrochar production. The produced hydrochar is used as solid fuel, soil amendment, and adsorbent and as an energy storage device (Chemistry, 2019). A gate-to-gate assessment is performed for the system.

6.2.1.3 System Boundary

The system boundary includes the consumption of materials, energy for the process, and release of pollutants associated with air, water from the process, and products associated. The upstream processes that are out of the system boundary are waste collection and transport of raw food and yard wastes, rice and yard cultivation, transportation of the wastes to a lab, machinery used for grinding, and the usage stage of hydrochar due to nonavailability of the appropriate data. The analysers used to find the properties of raw wastes and hydrochar after the production. The food waste is collected from the hostel mess and the yard waste is collected from the curbsides. The food waste is mixed with some vegetables and it is an oil-free waste. The curbside yard waste consists of stems, sticks, roots that are segregated from the leaves, and the transformation of these individual wastes is processed by HTC reactor at specific operating temperature and residence time (Figure 6.2).

6.3 Process Description

The hydrochar is made through a process called HTC through a high-pressure reactor. The reactor parts are assembled, and the carbonization is done in an oxygen-free condition with water as a catalyst and solvent. First, the quantities of the inputs were taken. The inputs include food waste, de-ionized water, yard waste, and the quantities are shown in Table 6.1.

TABLE 6.1

Quantities of the Inputs

Name	Quantity
De-ionized water (input)	400 mL
Food waste (FW-input)	25 g
Yard waste (YW-input)	25 g
Hydrochar FW (output)	10 g
Hydrochar YW (output)	15 g
Yield = (I/O) * 100	40%
Yield = (I/O) * 100	60%

The whole individual processes were operated at 250°C for 2 hours to bring the temperature to 250°C electricity is used as a source. The overall time taken to complete the individual process is almost 6 hours. Before starting the process, an inlet and outlet water pipes are connected to the reactor to maintain the room temperature in the outer portion of the reactor.

The total process time is divided into three parts: (i) time taken to obtain the required temperature is 2 hours, and (ii) time taken to run the actual process is 2 hours—during this stage the pressure develops to maintain the water in a liquid state. The end product of this stage is a slurry type of material and (iii) the time taken to cool down the reactor is 2 hours. After the reactor has reached zero pressure from elevated pressure, the gas valve is opened to release the generated gas during which the process reactor parts are disassembled and the slurry from the end product is separated into solid and liquid products. The solid product is called hydrochar and the liquid product is called process water.

6.3.1 Data Collection

The yard waste is collected from campus discarded green cuttings; food waste was collected from the hostel mess.

6.3.2 Proximate and Ultimate Analysis

The properties of hydrochar and raw food like moisture content, volatile matter, the ash content in percentages are examined using proximate analyses according to ASTM standards. The sample was first oven dried as per ASTM-E871 for 24 hours at 103°C ± 2°C. Then later it was placed in a desiccator to remove the moisture that is adsorbed on the crucible and for cooling purposes. The same sample was placed in muffle as per ASTM-E872 at 950°C for 7 minutes to get volatile matter, and then again placed in a desiccator to serve the same purpose. The leftover sample was placed in muffle as per ASTM-E873 at 575°C for 5 hours to get ash percentage. The fixed carbon is calculated by negotiating the above percentages from a total of 100. The C, H, N, S and O percentages are examined by the ultimate analyser, and the high heating values in MJ are founded by a bomb calorimeter. Before performing these two tests, the sample should be oven-dried properly to achieve constant weight. The properties are shown in Table 6.2.

6.3.3 Calculations

$$\text{Fuel ratio: Fixed carbon/volatile matter} \tag{6.1}$$

$$\text{Hydrochar yield: } \left(\text{Hydrochar weight/feedstock weight}\right) \times 100 \tag{6.2}$$

$$\text{Energy densification (ED):} \left(\text{HHV of Hydrochar/HHV of feedstock}\right) \tag{6.3}$$

$$\text{Energy yield: Hydrochar yield} \times \text{Energy densification} \tag{6.4}$$

TABLE 6.2

Physical and Chemical Characterization of Feedstock and Hydrochar Synthesized

Name	VS = %TS[a]	Ash[b] (%)	FC[c] (%)	C (%)	H (%)	N (%)	O (%)	ED[d]	Fuel Ratio[e]	HHV[f] (MJ/kg)
Food waste	91.2	4.2	3	49.3	8.1	1.7	40.7	-	0.03	21
Yard waste	84.6	8.4	6.8	44.2	5.3	0.6	49.8	-	0.08	15.3
Hydrochar FW	70.8	8	21.2	63.6	6.6	1.3	28.3	49.2	0.29	25.9
Hydrochar YW	68.9	6.2	24.9	50.3	5.1	0.4	43.0	76.5	0.36	19.6

[a] Volatile solids.
[b] Ash content.
[c] Fixed carbon.
[d] Energy densification.
[e] Fuel ratio.
[f] High heating value.

6.3.4 Inventory Data

TABLE 6.3

Inventory Data Collected from the System Boundary-Quantification of Input, Output, and Emission

Name	Input (Raw) (g)	Distilled Water (mL)	Output(Hydrochar) (g)	Process Water Soluble COD (mg/L)	Mass of Carbon in Solid Char (g)	Mass of Carbon in Gas (g)
Food waste	25	400	10	18,400	18.5	0.25
Yard waste	25	400	15	16,528	19	0.125

6.3.5 Advantages of HTC over Pyrolysis

- The HTC process is very advantageous for wet feedstocks than pyrolysis that will decrease the fuel cost for drying (Titirici et al., 2010) (Table 6.3).
- The formation of polar functional groups will take place on the surface of hydrochar than aromatization groups which can be easily modulated for specific applications (Kent Hoekman et al., 2018).
- The maximum amount of carbon is converted into solid through the HTC process rather than in gas through pyrolysis (Ramke et al., 2009).

6.4 Impact Categories

The assessment method chosen for this analysis is Impact 2002+ (Jolliet et al., 2017) because it analyzes the four major categories—eco-toxicity, climate change, human toxicity, and resource usage. The major objective for choosing this assessment method is it gives a real mode implementation of both damage and midpoint assessments. To showcase, the characterization and impact results, Simapro software is used with the eco-invent database. This software assessed the impacts according to the inventory data given to the software. The impact assessments and damage assessments are shown in Figures 6.3 and 6.4. The major analysed impact in this study is global warming.

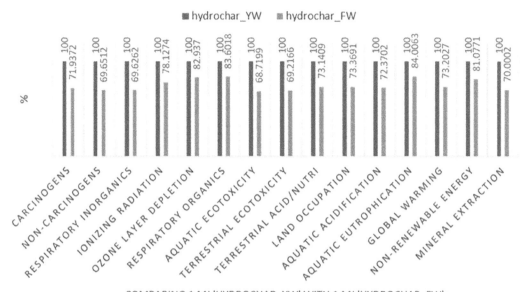

COMPARING 1 MJ 'HYDROCHAR_YW' WITH 1 MJ 'HYDROCHAR_FW';
METHOD: IMPACT 2002+ V2.15 / IMPACT 2002+ / CHARACTERIZATION / EXCLUDING
INFRASTRUCTURE PROCESSES / EXCLUDING LONG-TERM EMISSIONS

FIGURE 6.3 Impact characterization results of Yard waste hydrochar (YW) and Food waste hydrochar (FW).

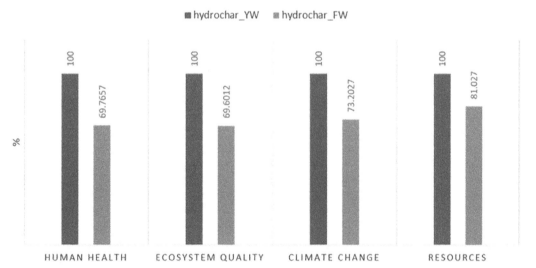

COMPARING 1 MJ 'HYDROCHAR_YW' WITH 1 MJ 'HYDROCHAR_FW';
METHOD: IMPACT 2002+ V2.15 / IMPACT 2002+ / DAMAGE ASSESSMENT / EXCLUDING
INFRASTRUCTURE PROCESSES / EXCLUDING LONG-TERM EMISSIONS

FIGURE 6.4 Damage assessment results of yard waste hydrochar (YW) and food waste hydrochar (FW).

6.5 Impact Assessment Results

The objective of the study concluded that the soluble COD value of food waste (18,400 mg/L) is more compared to the soluble COD of yard waste (16,528 mg/L). The proper analytical calculation is not done to find the global warming comparison. Instead of an analytical procedure, the global warming comparison is done by Simapro software. Along with the global warming potential (GWP) values of YW and FW, the other impact categories are also analysed using Simapro software. The impact categories that are analysed include carcinogens, non-carcinogens, respiratory inorganics, ionizing radiation, ozone layer depletion, respiratory organics, aquatic toxicity, terrestrial eco-toxicity, terrestrial nitrification, land occupation, aquatic acidification, aquatic eutrophication, global warming, non-renewable energy, mineral extraction (Jolliet et al., 2017), as shown in Figure 6.3. The output from the Simapro showed that the GWP potential is more for FW than YW. This might be due to more carbon fraction released through a gas form in FW than YW.

The damage assessment was also analysed using Simapro to compare the damage that occurred to human health, ecosystem quality, climate change, and resources. The assessment method used impacts 2002+ which is a combination of IMPACT 2002 (Mahath et al., 2019), Eco-indicator 99 (Mahath et al., 2019), CML (Berge et al., 2015), and IPCC. The results are shown in Figure 6.4.

6.6 Applications of HTC Secondary Products

After the completion of the HTC process, there is a release of gaseous products. The first diving factor to consider while studying the LCA of biofuel is the emissions of greenhouse gases (mainly CO_2). However, in many LCA studies, the concept of carbon balance ignores this effect as the combustion of biofuels is replacing the carbon in the atmosphere as it is consumed during the growing stage (Kent Hoekman et al., 2018). Berge et al. (2015) also mentioned the same in their LCA study (Table 6.4). But the process water that has a complex mixture of water-soluble organic solids has high potential for recovery and alternate applications. There are several paths to treating this to make the entire HTC process eco-friendly (Table 6.5). One of the treatment techniques involves recycling the HTC process water. This technique not only decreases total water consumption and reduces expensive treatment and disposal, but it also shows opportunities for enhanced heat recovery, which decreases the energy requirements (Kambo and Dutta, 2015). Stemann et al. (2013) also recycled the process water of hydrothermally treated poplar wood through a batch-type HTC reactor. For every recycling, process water from the last phase was utilized to carbonize a newer poplar feedstock. It is observed that the Total Organic Carbon (TOC) and organic acids concentrations increased after each recycling step and reached maximum limits after five to eight steps of recycling. It was also observed that this recycling process slightly increased the mass, energy content, and dewaterability of the hydrochar product, presumably because some of the components from the process water became incorporated into the hydrochar.

The other attractive way to treat HTC process water is AD. Due to their high levels of TOC and chemical oxygen demand (COD), HTC process waters are potentially good candidates for AD treatment. The resulting methane product can be captured and used in a variety of ways, thereby improving the overall economics of the operation. The investigated AD treatment of process water from HTC of corn silage (Weiner et al., 2016). The feedstock had a TOC level of about 16 g/L and a COD level of 41 g/L. While initial results showed good removal efficiency for TOC and COD, along with expected methane production, there were concerns about the lack of certain trace elements that promote AD and the presence of some inhibiting organic species.

TABLE 6.4

Damage Assessment Results of the System Boundary Chosen

Damage Category	Unit	Hydrochar_FW[a]	Hydrochar_YW[b]
Resources	MJ primary	0.021	0.017
Climate change	Kg CO_2 eq	0.00136	0.000998
Eco system quality	PDF*m²*yr[c]	0.000339	0.000236
Human health	DALY[d]	1.33E-9	9.26E-10

[a] Food waste.
[b] Yard waste.
[c] Potentially disappeared action*m²*year.
[d] Disability adjusted life years.

TABLE 6.5

Impact Characterization Results of the System Boundary Chosen

Impact Category	Unit	Hydrochar_FW[a]	Hydrochar_YW[b]
Mineral extraction	MJ surplus	9.49E-5	6.64E-5
Non-renewable energy	MJ primary	0.0209	0.0169
Global warming potential	MJ CO_2 eq	0.00136	0.000998
Aquatic eutrophication	Kg PO_4 p-lim	1.89E-5	1.59E-5
Aquatic acidification	Kg SO_2 eq	5.28E-6	3.28E-6
Land occupation	M^2org.arable	1.39E-5	1.02E-5
Terrestrial acidification	Kg SO_2 eq	1.75E-5	1.28E-5
Aquatic eco-toxicity	Kg TEG soil	0.0377	0.0261
Terrestrial eco-toxicity	Kg TEG water	0.156	0.107
Respiratory organics	Kg C_2H_4 eq	2.93E-7	2.45E-7
Ozone layer depletion	Kg CFC-11 eq	8.63E-11	7.16E-11
Ionizing radiation	Kg C-14 eq	0.0041	0.0032
Respiratory inorganics	Kg $PM_{2.5}$ eq	1.68E-6	1.17E-6
Non-carcinogens	Kg C_2H_3Cl eq	2.72E-5	1.9E-5
Carcinogens	Kg C_2H_3Cl eq	2.57E-5	1.85E-5

[a] Food waste.
[b] Yard waste.

6.7 Conclusions and Future Implications

Although the comparative study is not done for pyrolysis and HTC, as per advantages are concerned, waste transformation through HTC is more beneficial. The environmental impacts of individual HTC processes are analysed, but the economic benefits should be thoroughly studied using life cycle costing. In the future, the appropriate operating conditions should be maintained to get an efficient yield but a catalyst is required to accelerate the temperature to the required temperature. The supremacy of HTC over other thermal and bio-based energy recovery techniques includes eco-friendly, simple, and fast process, low activation energy and power consumption requirement, low operating temperatures, and no predrying requirement. However, the desirable hydrochars quality and yield completely rely upon the HTC process parameters. In addition, to understand the proper material flow, there should undergo a detailed study on mass and water balance. Although hydrochar usage is increased in a wide variety of applications, in-depth research should be performed on reaction kinetic data including reaction pathways and mass transfer which are important parameters for process optimization and design of kinetic reaction during the HTC process. The gas containing CO_2 and CH_4 can be captured through proper carbon sequestration studies. The process water also contains higher levels of soluble organic products having an opportunity to recover, and possible applications are also reviewed to make the HTC process an eco-friendly technology. Also, despite HTC showing, enormous potential to treat organic waste, vast literature which is still in infancy calls for a more rigorous study focusing on total LCA. Therefore, any number of LCA studies should be performed to showcase HTC as an environmentally sustainable process.

Acknowledgement

The authors are thankful to all the staff of the environmental engineering laboratory, Indian Institute of Technology Kharagpur, India for their technical support during proximate and elemental analysis.

REFERENCES

Berge, N.D., Li, L., Flora, J.R.V., Ro, K.S., 2015. Assessing the environmental impact of energy production from hydrochar generated via hydrothermal carbonization of food wastes. *Waste Manage.* 43, 203–217. doi: 10.1016/j.wasman.2015.04.029.

Chemistry, E., 2019. Minireview of potential applications of hydrochar derived from hydrothermal carbonization of biomass. *J. Ind. Eng. Chem.* 57, 15–23.

Funke, A., Ziegler, F., 2020. Hydrothermal carbonization of biomass: A summary and discussion of chemical mechanisms for process engineering. *Biofuels Bioprod. Biorefin.* 4, 2–3.

Jain, A., Balasubramanian, R., Srinivasan, M.P., 2016. Hydrothermal conversion of biomass waste to activated carbon with high porosity: A review. *Chem. Eng. J.* 283, 789–805. doi: 10.1016/j.cej.2015.08.014.

Jolliet, O., Margni, M., Charles, R., Humbert, S., Payet, J., Rebitzer, G., 2017. Milk volume, vitamin D supplementation, outdoor activity and season are important predictors of vitamin D serum level in preschoolers: A target kids! *Study. Paediatr. Child Health (Oxford).* 8, 324–330. doi: 10.1093/pch/16.suppl_a.47ab.

Kalyani, K., Pandey, K.K., 2014. Waste to energy status in India: A short review. *Renew. Sustain. Energy Rev.* 31, 113–120. doi: 10.1016/j.rser.2013.11.020.

Kambo, H.S., Dutta, A., 2015. A comparative review of biochar and hydrochar in terms of production, physicochemical properties and applications. *Renew. Sustain. Energy Rev.* 45, 359–378.

Kent Hoekman, S., Leland, A., Felix, L., 2018. Hydrothermal Carbonization (HTC) of biomass for energy applications. *Biomass Preprocess. Pretreat. Prod. Biofuels* 196–254. doi: 10.1201/9781315153735-8.

Libra, J.A., Ro, K.S., Kammann, C., Funke, A., Berge, N.D., Neubauer, Y., Titirici, M.-M., Führer, C., Bens, O., Kern, J., Emmerich, K.-H., 2011. Hydrothermal carbonization of biomass residuals: A comparative review of the chemistry, processes and applications of wet and dry pyrolysis. *Biofuels* 2, 71–106. doi: 10.4155/bfs.10.81.

Lin, Y., Ma, X., Peng, X., Yu, Z., 2017. Hydrothermal carbonization of typical components of municipal solid waste for deriving hydrochars and their combustion behavior. *Bioresour. Technol.* 243, 539–547. doi: 10.1016/j.biortech.2017.06.117.

Liu, Z., Balasubramanian, R., 2014. Upgrading of waste biomass by hydrothermal carbonization (HTC) and low temperature pyrolysis (LTP): A comparative evaluation. *Appl. Energy* 114, 857–864. doi: 10.1016/j.apenergy.2013.06.027.

Lu, X., Flora, J.R.V., Berge, N.D., 2014. Influence of process water quality on hydrothermal carbonization of cellulose. *Bioresour. Technol.* 154, 229–239. doi: 10.1016/j.biortech.2013.11.069.

Lucian, M., Fiori, L., 2017. Hydrothermal carbonization of waste biomass: Process design, modeling, energy efficiency and cost analysis. *Energies* 10. doi: 10.3390/en10020211.

magnet.co.uk, 2018. Per capita food waste of selected countries worldwide in 2017 (in kilograms per year). *Statista* 2017–2020.

Mahath, C.S., Mophin Kani, K., Dubey, B., 2019. Gate-to-gate environmental impacts of dairy processing products in Thiruvananthapuram, India. *Resour. Conserv. Recycl.* 141, 40–53. doi: 10.1016/j.resconrec.2018.09.023.

Mittapalli, S., Sharma, H.B., Dubey, B.K., 2021. Hydrothermal carbonization of anaerobic granular sludge and co-pelletization of hydrochar with yard waste. *Bioresour. Technol. Reports* 14, 100691.

Pham, T.P.T., Kaushik, R., Parshetti, G.K., Mahmood, R., Balasubramanian, R., 2015. Food waste-to-energy conversion technologies: Current status and future directions. *Waste Manag.* 38, 399–408. doi: 10.1016/j.wasman.2014.12.004.

Phuong, T., Pham, T., Kaushik, R., Parshetti, G.K., Mahmood, R., Balasubramanian, R., 2014. Food waste-to-energy conversion technologies : Current status and future directions. *Waste Manage.* 38, 399–408.

Ramke, H.-G., Blöhse, D., Lehmann, H.-J., Fettig, J., Höxter, S.T., 2009. Hydrothermal carbonization of organic waste. *Sardinia 2009: Twelfth International Waste Management and Landfill Symposium.*

Ross, C., 2015. Which countries waste the most food? *World Econ. Forum* 6–9.

Saqib, N.U., Baroutian, S., Sarmah, A.K., 2018. Physicochemical, structural and combustion characterization of food waste hydrochar obtained by hydrothermal carbonization. *Bioresour. Technol.* 266, 357–363.

Saqib, N.U., Sharma, H.B., Baroutian, S., Dubey, B., Sarmah, A.K., 2019. Valorisation of food waste via hydrothermal carbonisation and techno-economic feasibility assessment. *Sci. Total Environ.* 690, 261–276. doi: 10.1016/j.scitotenv.2019.06.484.

Sharma, H.B., Dubey, B.K., 2020a. Co-hydrothermal carbonization of food waste with yard waste for solid biofuel production: Hydrochar characterization and its pelletization. *Waste Manage.* 118, 521–533. doi: 10.1016/j.wasman.2020.09.009.

Sharma, H.B., Dubey, B.K., 2020b. Binderless fuel pellets from hydrothermal carbonization of municipal yard waste: Effect of severity factor on the hydrochar pellets properties. *J. Clean. Prod.* 277, 124295. doi: 10.1016/j.jclepro.2020.124295.

Sharma, H.B., Panigrahi, S., Sarmah, A.K., Dubey, B.K., 2020a. Downstream augmentation of hydrothermal carbonization with anaerobic digestion for integrated biogas and hydrochar production from the organic fraction of municipal solid waste: A circular economy concept. *Sci. Total Environ.* 706, 135907. doi: 10.1016/j.scitotenv.2019.135907.

Sharma, H.B., Panigrahi, S., Sarmah, A.K., Dubey, B.K., 2020b. Downstream augmentation of hydrothermal carbonization with anaerobic digestion for integrated biogas and hydrochar production from the organic fraction of municipal solid waste : A circular economy concept. *Sci. Total Environ.* 706, 135907. doi: 10.1016/j.scitotenv.2019.135907.

Sharma, H.B., Sarmah, A.K., Dubey, B., 2020c. Hydrothermal carbonization of renewable waste biomass for solid biofuel production: A discussion on process mechanism, the influence of process parameters, environmental performance and fuel properties of hydrochar. *Renew. Sustain. Energy Rev.* doi: 10.1016/j.rser.2020.109761.

Sharma, H.B., Panigrahi, S., Dubey, B.K., 2021. Food waste hydrothermal carbonization: Study on the effects of reaction severities, pelletization and framework development using approaches of the circular economy. *Bioresour. Technol.* 333, 125187.

Standard, I., 2006. Environmental management - life cycle assessment - principles and framework. *Int. Organ. Stand.* 3, 20. doi: 10.1016/j.ecolind.2011.01.007.

Stemann, J., Erlach, B., Ziegler, F., 2013. Hydrothermal carbonisation of empty palm oil fruit bunches: Laboratory trials, plant simulation, carbon avoidance, and economic feasibility. *Waste Biomass Valorization* 4, 441–454. doi: 10.1007/s12649-012-9190-y.

Thacker, H., 2018. Food wastage in India. *CSR J.* 1–3.

Titirici, M.M., Antonietti, M., 2010. Chemistry and materials options of sustainable carbon materials made by hydrothermal carbonization. *Chem. Soc. Rev.* 39, 103–116. doi: 10.1039/b819318p.

Titirici, M.-M.M., Neubauer, Y., Funke, A., Berge, N.D., Libra, J.A., Bens, O., Kammann, C., Kern, J., Fühner, C., Emmerich, K.-H.H., Ro, K.S., Kammann, C., Funke, A., Berge, N.D., Neubauer, Y., Titirici, M.-M.M., Fühner, C., Bens, O., Kern, J., Emmerich, K.-H.H., 2010. Hydrothermal carbonization of biomass residuals: A comparative review of the chemistry, processes and applications of wet and dry pyrolysis. *Biofuels* 2, 71–106. doi: 10.4155/bfs.10.81.

Weiner, B., Wedwitschka, H., Poerschmann, J., Kopinke, F., 2016. Utilization of organosolv waste waters as liquid phase for hydrothermal carbonization of Chaff 6–11. *ACS Sustainable Chem. Eng.* doi: 10.1021/acssuschemeng.6b01665.

Wyczalkowska, A.K., Sengers, J.V., 2002. Thermodynamic properties of sulfurhexafluoride in the critical region. *J. Chem. Phys.* 111, 1551–1560. doi: 10.1063/1.479414.

Zhao, L., Fan, L.Z., Zhou, M.Q., Guan, H., Qiao, S., Antonietti, M., Titirici, M.M., 2010. Nitrogen-containing hydrothermal carbons with superior performance in supercapacitors. *Adv. Mater.* 22, 5202–5206. doi: 10.1002/adma.201002647.

7

Waste Monitoring and Management: Strategies, Innovations and Safety Considerations

V.P. Sharma

CSIR-Indian Institute of Toxicology Research

CONTENTS

7.1 Introduction

The waste-management concept involves cradle to grave approach through well-defined characterization, quantification, segregation, storasge, transport and treatment of waste. It involves the concept of 3 R's – reduce, recycle and reuse and aims at avoiding the generation of waste or recovering. The exponential growth of plastics is mainly due to their unique characteristics and comparatively low cost. Plastics have high strength-to-weight ratio and may be comfortably shaped into a variety of forms, color shapes and sizes with impermeability to liquids and resistance to physical or chemical degradation. These salient properties have led to the substitution of traditional materials by plastics in various applications, viz. textile, packaging of consumer goods, transport, construction, automobile defense or space-like strategic sectors. The versatility extends to protection or preservation of foodstuffs and reduction of food waste. Polymeric products are used to a great extent in infrastructure developments where their impermeability and durability may contribute to savings of water in urban areas [1–12].

All beneficiaries, entrepreneurs and environmentalists connected with economic growth and supporting health-care activities must ethically and legally propagate or disseminate the safety aspects for adequate management of waste. We need to produce environment-friendly products or devices to accomplish safe disposal. The WHO has reinforced that part of the taxes may be assigned to meet the budget for creation, support and maintenance of an efficient health-care waste-management system that may include intelligent or novel methods/devices to reduce the bulk and toxicity of health-care waste. Stakeholders and NGOs may undertake programmers and activities that contribute to this incentive.

7.2 Waste-Management Regulations and Innovations in Treatment

The biomedical wastes are generated during the diagnosis, treatment, or immunization of human or animal research activities pertaining thereto or in the production, medical treatment or testing of biological. The basic waste-management rules of 1998 were improvised in 2000, 2003, 2011 and 2016 [1,3]. Recently, the Ministry of Environment, Forest and Climate Change (MOEFCC) made modifications in March 2016 and have amended the BMW rules but still we experience need for strict implementation at varied levels throughout the country. It mainly focuses on segregations, transportation and disposal with appropriate scientific techniques. We need to have more municipal and biomedical or hazardous waste treatment facilities taking advantage of cost-effective and updated novel techniques. Common incinerators may emit toxic air pollutants, and ash is also potentially hazardous. Nowadays advanced incinerators, plasma-based treatment solutions and chemical-based treatment strategies are being used with few limitations. Continual improvements are being planned.

The dioxins are the by-product of waste combustion, which may be generated during incinerator operation. In addition, a group of 75 chemicals may coexist along with another group of toxins called furans. These toxicants and aerosols are potential chemicals to accumulate in fatty tissues and traverse through the food chain. The combustion of PVC-based medical devices is the largest dioxin producer in the environment and a cause of concern. Few countries, viz. the Philippines and Denmark, have banned or restricted the construction of incinerators. Thus, it is necessary to see the accepted regulation of the correct type of incinerators and temperatures with scrubbers, etc.

Several technologies are available for the management and treatment of municipal solid waste, but choosing the appropriate one depends on the nature and local conditions. Selecting the appropriate technology also helps to reduce greenhouse gas emissions and contributes to mitigating climate change. The opportunity to reduce emissions may be offered by the Clean Development Mechanism (CDM).

7.3 Potential Risks to Human Health and Environment

The pollution due to polymeric products poses several potential risks to human health. The presence of plastic in food or subsequent consumption by seafood lovers has led to concerns related to chemical bio-accumulation in the food chain. Investigations on plastics research have detected microplastic contamination in tap water, bottled water and salts across a number of countries. The two reasons are related to the longevity of plastics leading to accumulation in the environment and fragmentation into smaller microplastics and nanoplastics. The magnitudes of the damages have great uncertainty. The precise impact on marine and terrestrial ecosystems may emerge in the forthcoming years, but few environmental effects of plastics pollution are visible [8–14].

7.4 Phthalates, Bisphenol A and Other Contaminants

Phthalates are used in the fabrication of toys, vinyl flooring, wall covering, detergents, lubricating oils, food packaging, pharmaceuticals, blood bags and tubing and personal care products, viz. nail polish, hair sprays, aftershave lotions, soaps, shampoos, perfumes, etc. They are colorless, odorless, oily liquids and occupational workers may be exposed through ingestion, inhalation, dermal routes, etc. Phthalates are readily absorbed into the human body and are converted quickly to their respective metabolites. Unlike some chemicals, they tend to pass out of the body quickly in urine and feces. The release can occur during all the stages of the product lifecycle – from production, through use, to disposal. Children of 3–5 years are more vulnerable to phthalates due to their development phase, smaller body size and sensitivity. Young children have hand-to-mouth tendency for every item they get to hold and are consequently exposed to higher levels by sucking on products made of phthalate-containing polymeric products. The frequent use of chemicals with endocrine active properties in unit products and contamination of soil, water and food sources by persistent chemical pollutants result in ubiquitous exposures.

Wildlife observations and animal pharmacological medicine studies reveal adverse effects of EDC (endocrine disrupting chemical) on health.

Phthalates are a group of chemicals used in hundreds of products, such as toys, vinyl flooring and wall covering, detergents, lubricating oils, food packaging, pharmaceuticals, blood bags and tubing and personal care products, viz. nail polish, hair sprays, aftershave lotions, soaps, shampoos, perfumes, etc. Plastic as a material may contain a variety of chemicals, some potentially hazardous. Phthalates, for instance, are a group of chemicals produced in large volumes and are commonly used as plasticizers in plastics manufacturing. Phthalates are readily absorbed into the human body and are converted quickly to their respective metabolites. Unlike some chemicals, they tend to pass out of the body quickly in urine and feces. The ever-present use of phthalates as an additive to PVC (polyvinyl chloride) products to make them flexible and to personal care products to make fragrances last longer in the past 50 years has resulted in widespread general population exposure. Phthalates are readily absorbed into the human body and are converted quickly to their respective metabolites. Unlike some chemicals, they tend to pass out of the body quickly in urine and feces. Phthalates can interact with each other and increase the exposure effect.

Bisphenol A (BPA) is a monomer primarily used in the production of polycarbonate plastic and epoxy resins. BPA is an endocrine disruptor and experimentation carried out in animal studies by several workers to help us to indicate that BPA may adversely affect humans. Few epidemiological studies have examined the relationship between BPA and health effects in humans. The sporadic gestational BPA exposure studies indicate neurodevelopment and behavioral problems in early life. We antecedently reported that prenatal BPA exposure could have an effect on kid behavior and otherwise among boys and women aged 3–5. Prenatal exposure to BPA and phthalates and behavioral problems in children at preschool age has been studied and reported in environmental health perspectives in 2018. Bisphenol F (BPF) is apparently the main BPA replacement that is used increasingly. BPF has been detected in canned food, thermal paper receipts and soft drinks. In the present experiment, we did both in vitro and in vivo studies to evaluate the effect of low- and high-dose BPF exposures on testosterone concentration, oxidative stress, and antioxidants activity in reproductive tissues of male rats. BPA or its analogs are another endocrine-disrupting chemical and experimentation carried out in animal studies by several workers to help us to indicate that BPA may adversely affect humans. Few epidemiological studies have examined the relationship between BPA and health effects in humans. The sporadic gestational BPA exposure studies indicate neurodevelopment and behavioral problems in early life.

Thus, EDCs are synthetic or natural compounds that interfere with endogenous endocrine action. The frequent use of chemicals with endocrine active properties in household products and contamination of soil, water, and food sources by persistent chemical pollutants result in ubiquitous exposures. Wildlife observations and animal toxicological studies reveal adverse effects of EDCs on reproductive health. In humans, a growing number of epidemiological studies report an association with altered pubertal timing and progression. While these data are primarily reported in females, this review will focus on the small number of studies performed in males that report an association of polychlorinated biphenyls with earlier sexual maturity rating and confirm the subtle effects of lead, dioxins and endosulfan on delaying pubertal onset and progression in boys. Studies have also demonstrated that exposure may affect pubertal testosterone production without having a noticeable effect on sexual maturity rating. A limitation to understand the effects of EDCs in humans is the potential for confounding due to the long temporal lag from early-life exposures to adult outcomes. Indirect additives like inks and adhesives used during manufacturing may also pose toxicity to store-in food, pharma products or water under simulated test conditions. The ISO/CFR/FAO/IS guideline must be referred for safety aspects. The storage, transportation and labeling, disposal norms as per FSSAI regulations, need to be adhered to for plastic items. On international platforms or export of food items, we must take special care of packaging and other interrelated aspects [5–14].

7.5 Identification of Phthalates

Phthalates may be identified on labeled by a three or four-letter acronym that defines their chemical structures. However, the common packaging rarely states "contains phthalates" and there are a multitude of phthalate compounds, e.g. DMP, DEP, DPP, DBP and DEHP. Phthalates with a higher molecular weight are very slightly soluble in water; phthalates with a lower molecular weight are reasonably soluble in water. Globally, in silico method, in vitro mechanistic, in vivo preclinical and limited clinical or epidemiological human studies showed that over a dozen phthalates and their metabolites ingested passively by a man from the final setting, foods, drinks, breathing air and routine household products cause various dysfunctions. EDCs are hormonally active substances that may perturb puberty by acting both peripherally on target organs, such as adipose tissue or adrenal glands, and/or centrally on the hypothalamic–pituitary–gonadal (HPG) axis. Evidence of associations between exposures to EDCs and altered time of life temporal arrangement makes it affordable to support their relationship. Potential impacts on human health require restricted use in selected applications and a need for closer monitoring of potential sources of human exposure. Although the presence of phthalates in a variety of plastics has been recognized, the influence of plastic recycling on phthalate content has been hypothesized but not well documented. Different statistical analysis of the analytical results suggested that phthalates were potentially added in the later stages of plastic product manufacturing (labeling, gluing, etc.) and were not removed. Plastics debris, especially microplastics, has been found worldwide in all marine compartments. Much analysis has been distributed on adsorbate pollutants on plastic items and hydrophobic organic compounds (HOC) related to microplastics. Few studies reveal that leaching occurs from inferior quality products, which are fabricated not as per regulatory guidelines and thus not certified and recommended for usage.

7.6 Implications on Marine Organisms and Biodiversity

Plastic pollution may be an international threat to marine ecosystems. Plastic litter may migrate or leach a variety of substances into marine environments. In order to understand the effects on photosynthetic bacteria or the base of the marine food web studies have been undertaken on marine *Prochlorococcus*. The presence of plastic baggage on coastal dunes worldwide is well documented. Plastic bags contain additives that during rainfall events may leach out from bags into a sand dune and be absorbed by the seeds and roots of plants. Single-use plastics (SUPs), viz. plastic bags, microbeads, cutlery, straws and polystyrene are substantial sources of plastic marine pollution, yet preventable via legislative and non-legislative interventions. Several international legislative procedures are formulated to handle plastic marine pollution from plastic baggage and microbeads. These have been initiated on public interest litigations or through active environmentalists. Marine organisms may serve as bioindicators. The increasing level of marine pollution due to plastic debris/litter is a globally recognized threat that needs attention and effective actions for remediation, monitoring and mitigation. Marine organisms function as bioindicators of plastic pollution and facilitate in correlating data/info that will higher integrate the spatial and temporal presence of plastic rubble within the ocean. With sustained presence and frequent migrations, numerous marine species may ingest plastics. Sometimes it is difficult to identify and establish quantitative correlations of ingested plastics within well-defined spatio-temporal patterns and biodiversity. To monitor the amount of plastic waste in the vicinity, and also the prospect of it coming into the organic phenomenon, the bioindicator species square measure is used throughout Northern European Seas. The use of standardized indicator species may help to ensure coherence in the reporting of marine litter ingestion trends throughout the Mediterranean Sea.

Internationally, the plastics recycling rates remain low and continue to be an economically marginal activity. The present recycling rates are anticipated to be 14%–18%. Twenty-four percent of plastic waste is either incinerated or 58%–62% of plastic is disposed of in a landfill or the natural environment. The recycling rates for common industrial metals – steel, aluminum, copper, etc. – and paper are thought to exceed 50% [11].

7.7 Paradigm Shift in Innovative Designing or Plastic Products for Environmental Sustainability and Minimization of Adverse Implications

Nowadays, several approaches are readily available for addressing the environmental side effects of rapidly growing use of plastic production and disposal. Changes in product design with the use of alternative materials may reduce production and usage. Changes in design and shifting toward biobased or biodegradable plastics may reduce the adverse environmental impacts of plastics more directly by reducing their environmental footprint. The improved waste-management systems focus on facilitating waste collection and recycling appropriately before it initiates problems in the natural environment [4,11].

7.8 Interventions for Better Quality of Life

The remediation or global clean-up activities may allow the removal of plastics from the natural environment with the introduction of multiple stream collection systems allowing separated collection of recyclables, creation of incentives for better product and plastics design. The support to R&D for improved plastic management systems and the sustainable design of plastics are required in close partnership with industry. The introduction of more ambitious targets of recycling rate and harmonization of the methods used to calculate these rates may be adopted. Landfill and incineration fees may be adopted as policy instruments. The appropriate waste disposal practices are the requirement of the present generation to safeguard health and improve sanitation. The SDG goals must be always considered for sustainable growth and improving the quality of life for the future generation. We must leave behind a healthy environment with enough resources for the children.

REFERENCES

1. ASTM D6954 (United States) Standard Guide for Exposing and Testing Plastics that Degrade in the Environment by a Combination of Oxidation and Biodegradation.
2. Bernardo, C., C. Simões, and L. Pinto (2016) Environmental and economic life cycle analysis of plastic waste management options. *China Journal of Renewable and Sustainable Energy*, 10(123), 140001–140002. doi: 10.1063/1.4958429.
3. ISO 15270 (2008) Plastics — Guidelines for the recovery and recycling of plastics waste.
4. Klaassen, C. (2017) *Casarett and Doulls Toxicology: The Science of Poisons*, 9th Edition. McGraw Hill Education, New York.
5. HPRC (2015) Environmental Impacts of Recycling Compared to Other Waste Disposal Methods. https://docs.wixstatic.com/ugd/49d7a0_6bb3ebb481ec49ceafef92f0b0ba010d.pdf (accessed on 28 March 2018).
6. Jambeck, J. et al. (2015) Marine pollution: Plastic waste inputs from land into the ocean. *Science* (New York, N.Y.), 347(6223), 768–771. doi: 10.1126/science.1260352.
7. Koelmans, A. et al. (2017) Risks of plastic debris: Unravelling fact, opinion, perception, and belief. *Environmental Science & Technology*, 51(20), 11513–11519. doi: 10.1021/acs.est.7b02219.
8. Kosuth, M. et al. (2018) Synthetic Polymer Contamination in Global Drinking Water. https://orbmedia.org/stories/invisibles_final_report/multimedia (accessed on 25 April 2018).
9. Mason, S., V. Welch, and J. Neratko (2018) Synthetic Polymer Contamination in Bottled Water. https://orbmedia.org/sites/default/files/FinalBottledWaterReport.pdf (accessed on 25 April 2018).
10. OECD (2018) Review of Secondary Plastics Markets. https://one.oecd.org/document/ENV/EPOC/WPRPW(2017)4/REV1/en/pdf (accessed on 28 March 2018).
11. OECD (2018) Improving Plastics Management: Trends, Policy Responses, and the Role of International Co-operation and Trade Background Report.
12. WRAP (2010) Environmental Benefits of Recycling – 2010 Update 1. https://wrap.org.uk/sites/default/files/2021-02/WRAP-environmental-benefits-recycling-2010-update.pdf (accessed on 28 March 2018).
13. The UK Plastics Pact (2022) Eliminating Problem Plastics. https://wrap.org.uk/sites/default/files/2022-02/Eliminating-problem-plastics-v4.pdf.
14. Yang, D. et al. (2015) Microplastic pollution in table salts from China. *Environmental Science & Technology*, 49(22), 13622–13627. doi: 10.1021/acs.est.5b03163.

8

Waste Management Strategy of SME Sector towards Environmental Sustainability

Ipsita Saha
Jadavpur University

Amit Kundu
Techno India Group

Sadhan Kumar Ghosh
Jadavpur University

CONTENTS

8.1 Introduction

The wastes generated from industries have a dangerous impact on the environment. Green House Gases (GHGs) which consist of water vapor, methane, carbon dioxide, nitrous oxide, ozone and CFCs are emitted from the waste generated from the packaging industry, cement industries and electronic industries. Hollos, Blome and Foerstl (2012) had worked on these and examined that the sustainability of an organization is dependent upon various factors. Sustainable supplier co-operation has positive effects on a firm's performance across the entire sustainability index. Economic sustainability is extremely dependent on green practices throughout the supply chain. Gavronski, Klassen, Vachon and Nascimento had proposed that the factors of environmental perspective are accountable for the receptivity of supply chain strategies (Gavronski, Klassen, Vachon, and Nascimento, 2012).

The present study makes three significant contributions. First, it explains the role of environmental policy in enhancing the sustainability strategy of supply chain management. Second, it has developed and validated scales that can capture organizational processes which are used to enhance the adoption of sustainability practices among suppliers. Finally, this research signifies that waste prevention strategies have an important role in the growth of plant-level social climate. To establish a sustainable business model, organizations need to remodel the supply chain strategies with the advent of new technologies. This is the high time to strengthen the supply chain of SME sectors by enriching their supply chain as a resource to satisfy increased customer demand and to enhance firm's performance (Ketchen et al., 2008). In this chapter, we analyze the strategies taken by MEs for waste reduction and explore the impact of sustainability of their business models.

8.2 Literature Review

To achieve a competitive advantage, various companies have started strengthening their supply chain to enhance environmental sustainability. The aim of strategic advantage in manufacturing sector includes cost-effectiveness, productivity, deliverance, novelty and flexibility (Krause et al., 2001). Research suggests that though environmental sustainability is a prime parameter of industry norms (Ashby, 2018), it remains ignored throughout the supply chain. Decision-makers suggested that Green Supply Chain Management (GSCM) practices can be adopted in manufacturing industries to attain different layers of sustainability social responsibility, economic viability and environmental sustainability (Renukappa et al., 2012). The objective of GSCM is to eco-design products with reduced consumption of energy, recyclable components and reduced hazardous materials in the manufacturing process (Dey, La Guardia, and Srinivasan, 2011). In this context, Vickery et al. (1999) and Acar et al. (2012) had defined five supply chain flexibilities including new product, volume and distribution flexibility. Over the past decade, SCM and GSCM have been playing a pivotal role in the success of any organization (Aflaki, Kleindorfer, de Miera Polvorinos, and Saenz, 2012). In addition to that, Jayaraman, Singh and Anandnarayan (2012) suggested that incorporating environmental parameters within the operations of GSCM could result towards a sustainable business model.

On the contrary, Ageron et al. (2012) contributed inventory methods in quantitative models of operations management. Dinwoodie, Knowles, and Sansom (2012) intended a framework that can facilitate the business process in accordance with environmental management principles. The research further suggests that optimization phases of business process re-engineering can be considered for enhancing the adoption of sustainable practices (Acar et al., 2012). The balance between profitability and sustainability (Tang and Zhou, 2012) can be maintained if a firm can establish cohesion among financial, resource and development flows (Ageron et al., 2012; Rastogi, Siddique, Kumar, and Singh, 2013).

A study from Dey, LaGuardia, and Srinivasan (2011) suggests that the logistics functions play an important role as the enormity of costs involved and the opportunity to decrease the carbon footprint. Moreover, the concept of closed-loop supply chain (CLSC) management can determine the present environmental challenges and can provide recommendations to supply chain logistics operations (Huang, Yan, and Qiu, 2009). Prior to this work, Schoenherr (2012) enquired about some manufacturing sectors, and the result he found indicated that their ability to realize sustainable competitive advantage is being hampered by the lack of sustainable objectives.

8.3 Research Gap and Development of the Model

Though many of the researchers have talked about the environmental sustainability issues, a complete understanding towards environmental sustainability of the stakeholders of the manufacturing units of SME sector of India have not been studied earlier. In the present study, the parameters associated with technology intervention of improving the supply chain of manufacturing process, integrated GSCM and implementation of strategies in the waste reduction have been identified. The senior executives/managers of the manufacturing units of SME sector are willing to develop the sustainable business model. The identified parameters are stated in the form of statements (Table 8.1) that are used to construct the structured questionnaire to get the perception of the target group of the experimental study.

TABLE 8.1

Identified Parameters Associated with Environmental Sustainability Issues

Parameter Code	Description
BP01	The level of knowledge of the management regarding sharing of information among the employee for the reduction of waste
BP02	The perception of stakeholders regarding green initiative of the organization
BP03	Stakeholders' perception towards their organization in terms of efficiency of processing information regarding business environment related issues
BP04	Stakeholders' insight in terms of managing waste in the cloud storage in the organization
BP05	Role of management for the development of environmentally sustainable business model
BP06	Role of organization for the incorporation of environmental criteria to select suppliers in supply chain management
BP07	The intention of an organization for the reduction of hazardous waste to be environmentally sustainable

8.4 Research Discussion

An attempt has been made in the present study to understand the intention of the small and medium-sized manufacturing units for the adoption of process and technology to minimize the waste in the process of manufacturing through the successful implementation of ERP system. The study explores the emerging factors responsible for the development of environmentally sustainable business model. The development of eco-friendly production units has become one of the mandates as per the policy of industrial framework in India and also meets the global norms to grab the opportunity of export.

8.5 Research Methodology

8.5.1 Research Design

This chapter includes the respondents from small and medium industries of manufacturing sectors of India. The companies are selected on the basis of adaptation of sustainable practices to undertake the survey. Questionnaires were emailed to top managers of 300 manufacturing units, from which 190 responses were received. Among those, some were discarded due to missing responses in some queries. Ultimately, the sample size becomes 159.

8.5.2 Development of Questionnaire Based on Identified Parameters

From an environmental perspective, seven questions were framed that were validated by experts from industry and academia. Based on the above-mentioned parameters, the questionnaire was designed to ask for ratings on an ordinal Likert-type five-point scale to get the perception of environmental sustainability.

8.5.3 *Sample Size*

Considering the research context, it was decided to collect data from managers and business analysts who have at least 5 years of experience in ERP implementation process.

8.5.4 *Method*

An attempt has been made to evaluate the environmental sustainability strategy considering all possible parameters (Table 8.1) and estimate the most significant factors responsible for the development of sustainable production units through factor analysis.

8.5.5 *Reliability Measure*

The reliability of environmental sustainability parameters is measured with 7 items for 159 respondents, and the results (Cronbach's alpha=0.862) indicate that satisfactory internal consistency exists.

8.6 Analysis

8.6.1 *Analysis Stage I*

The construct's validity was tested by applying Bartlett's Test of Sphericity and The Kaiser–Mayer–Olkin (KMO) Measure of sampling adequacy analyzing the strength of association among variables. The results reveal that the value of KMO is 0.662 which is above 0.5. The results for Bartlett's Test of Sphericity (0.000) and KMO both were highly significant, and so it is concluded that factor analysis is suitable.

8.6.2 *Analysis Stage II*

Based on perceived data on all identified items, the factor analysis has been performed. The findings of this study indicated that three factors, namely Stakeholders' Intent for waste reduction strategy, Green Initiative towards environmental sustainability and Management Strategy for Waste Reduction, have emerged for establishing environmentally sustainable supply chain management and can explain the variation at the level of 59.464% (Table 8.2).

8.6.3 *Analysis - Stage III*

Three factors, namely, Stakeholders' Intent for waste reduction strategy, Green Initiative towards environmental sustainability and Management Strategy for Waste Reduction, have emerged as significant ones. The emerged factors with their corresponding parameters indicating their concerned significant loading have been presented in Table 8.3.

8.7 Conclusion

The present research has made an attempt to understand the most significant factors responsible for the development of a sustainable business model in the manufacturing units of SME sector. Industries have already introduced sustainable practices within their ERP system. Considering different manufacturing sectors, a novel approach has been taken to identify the major factors that are responsible for environmental sustainability. One of the deliverables of the manufacturing sector towards sustainability is *Stakeholders' Intent for Waste Reduction Strategy* (Factor 1), which clearly indicates that the intention

TABLE 8.2

Total Variance Explained by the Emerged Factors

| | Total Variance Explained | | | | | |
| | Initial Eigenvalues | | | Extraction Sums of Squared Loadings | | |
Component	Total	% of Variance	Cumulative %	Total	% of Variance	Cumulative %
1	2.053	29.323	29.323	2.053	29.323	29.323
2	1.127	16.102	45.425	1.127	16.102	45.425
3	.983	14.039	59.464	.983	14.039	59.464
4	.935	13.358	72.822			
5	.725	10.362	83.184			
6	.684	9.778	92.962			
7	.493	7.038	100.000			

Extraction Method: Principal Component Analysis.

TABLE 8.3

Output of Factor Analysis

Factor 1 (stakeholders' intent for waste reduction strategy)	BP03	Aggressiveness of management to follow business environmental policy	0.772
	BP01	Initiative of management towards employees regarding waste reduction techniques	0.660
	BP04	Use of technology for waste management	0.657
	BP06	Policy framework selection of green supply chain members	0.474
Factor 2 (green initiative towards environmental sustainability)	BP05	Management intention towards developing sustainable business model	0.631
	BP02	Perception of stakeholders regarding organizational green initiatives	0.586
Factor 3 (management strategy for waste reduction)	BP07	Intention of management for process improvement by reduction of hazardous waste	0.792

of the top-level management can only change the attitudes of their supply chain members and employees towards the ongoing practices for waste reduction initiatives within organizations, ultimately changing the orientation of the end-users, i.e. customers. Management of manufacturing units should understand that socio-cultural changes towards the environment and their sustenance have gradually changed the mindset of the customers. *Green Initiative towards Environmental Sustainability* (Factor 2) becomes the mandate from the government level as well as international manufacturing policy framework for the manufacturing units for running the business, and the policy framework has compelled them to undertake the necessary modification in the process as well as technology for better waste management. And, ultimately, *Management Strategy for Waste Reduction* (Factor 3) becomes another significant and challenging factor for the manufacturing units of the SME sector as a high level of capital investment is required to change the process of production and the investment is also becoming necessary for the sustenance of the business to comply the norms of environmental policy. It would be pertinent to take an effective operational strategy that would be both cost-effective and environmentally viable.

The findings show that these factors are significant to form an environmentally sustainable business model. In future, the findings can be used to evaluate the post-implementation of ERP on the basis of the outcomes of the strategy of waste reduction. In addition, these factors can be generalized for achieving environmental sustainability, especially in the SME sector.

BIBLIOGRAPHY

Acar, S, Eris, B, & Tekce, M. (2012), The Effect of Foreign Direct Investment on Domestic Investment: Evidence from MENA Countries. http://www.etsg.org/ETSG2012/Programme/Papers/143. pdf

Aflaki, S., Kleindorfer, P. R., & de Miera Polvorinos, V. S. (2012), "Finding and implementing energy efficiency projects in industrial facilities", *INSEAD Working Papers Collection*, Issue 60, 1–27.

Ageron, B., Gunasekaran, A., & Spalanzani, A. (2012), "Sustainable supply management: An empirical study", *International Journal of Production Economics*, 140, 168–182.

Ashby, A. (2018), "Developing closed loop supply chains for environmental sustainability: Insights from a UK clothing case study", *Journal of Manufacturing Technology Management*, 29(4), 699–722.

Bhardwaj, B.R., "Sustainable supply chain management through enterprise resource planning (ERP): A model of sustainable computing", *International Journal of Management Science and Business Administration* 1(2), 20–32.

Bouchery, Y., Ghaffari, A., Jemai, Z., & Dallery, Y. (2012), "Greening the supply chain: Operational adjustment versus technology investment", *Proceedings of the 4th International Conference on Information Systems, Logistics and Supply Chain Creative Logistics for an Uncertain World, ILS 2012*, Quebec (Canada), August 26–29.

Carter, C.R., & Ellram, L.M. (1998), "Reverse logistics: A review of the literature and frame work for future investigation". *Journal of Business Logistics*, 19(1), 85–102.

Dey, A., LaGuardia, P., & Srinivasan, M. (2011). "Building sustainability in logistics operations: A research agenda", *Management Research Review*, 34(11), 1237–1259.

Dinwoodie, T., Knowles, B., & Sansom, M. (2012), "Sustainable development of maritime operations in ports", *Business Strategy and the Environment*, 21(2), 111–126.

Gavronski, I., Klassen, R.D., Vachon, S., & Nascimento, L.F.M. (2012), "A learning and knowledge approach to sustainable operations", *Iuri Gavronski, International Journal of Production Economics*, 140(1), 183–192.

Goel, N., & Jindal, A. (2013), "Implementation of cost effective solution for e-Governance through APUS (Aadhar Card, PPP, Underprivileged, SMS)", *Proceedings of the 7th National Conference; INDIACom–2013*, New Delhi.

Gupta, M.C. (1995), "Environmental management and its impact on the operations function", *International Journal of Operations & Production Management* 15(8), 34–51. DOI: 10.1108/01443579510094071.

Hollos, D., Blome, C., &Foerstl, K. (2012), "Does sustainable supplier co-operation affect performance? Examining implications for the triple bottom line", *International Journal of Production Research*, 50, 2968–2986. DOI: 10.1080/00207543.2011.582184.

Huang, X.Y., Yan, N., & Qiu, R. (2009), "Dynamic models of closed-loop supply chain and robust H∞ control strategies", *International Journal of Production Research* 47(9), 2279–2300. DOI: 10.1080/00207540701636355.

Jayaraman, V., Singh, R. &Anandnarayan, A. (2012), "Impact of sustainable manufacturing practices on consumer perception and revenue growth: An emerging economy perspective", *International Journal of Production Research*, 50(5), 1395–1410. DOI: 10.1080/00207543.2011.571939.

Krause, D. R., Pagell, M., Curkovic, S. (2001), "Toward a measure of competitive priorities for purchasing", *Journal of Operations Management*, 19, 497–512.

Ketchen, D. K., Boyd, B., & Bergh, D. (2008), "Research methodology in strategic management: Past accomplishments and future challenges", *Organizational Research Methods*, 11(4), 643–658.

Kumar, S., Teichman, S., & Timpernagel, T. (2012), "A green supply chain is a requirement for profitability", *International Journal of Production Research*, 50(5), 1278–1296.

Mazeika Bilbao, A., Carrano, A. L., Hewitt, M., Thorn, B. K. (2011), "On the environmental impacts of pallet management operations", *Management Research Review*, 34(11), 1222–1236, DOI: 10.1108/01409171111178765.

Rastogi, R. Mohammed Ishaq Siddique, Ankit Kumar, & Gautam Kumar Singh. (2013), "Role of Information Technology in Disaster Management", *Proceedings of the 7th National Conference*; INDIACom-2013.

Renukappa, S., Egbu, C., Akintoye, A., & Goulding, J. (2012), "A critical reflection on sustainability within the UK industrial sectors", *Construction Innovation*, 12(3), 317–334, DOI: 10.1108/14714171211244578.

Sarkis, J. (2012) "A boundaries and flows perspective of green supply chain management", *Supply Chain Management: An International Journal*, 17 (2), 202–216. DOI: 10.1108/13598541211212924.

Schoenherr, T. (2012), "The role of environmental management in sustainable business development: A multi-country investigation", *International Journal of Production Economics*, 140(1), 116–128.

Tang, C.S., & Zhou, S. (2012) Research advances in environmentally and socially sustainable operations". *European Journal of Operational Research*, 223(3), 585–594.

Vickery, S., Calantone, R., & Dro¨ge, C. (1999), "Supply chain flexibility: an empirical study", *The Journal of Supply Chain Management*, 35(1), 16–24.

Walsh, H., & Dowding, T. J. (2012), "Sustainability and the Coca-Cola Company: The global water crisis and Coca-Cola's business case for water stewardship", *International Journal of Business Insights & Transformation, Special Issue*, 4, 106–118.

Warren, L. (2012) "Scenario analysis for S&OP", *Journal of Business Forecasting*, 31(1), 32–35.

Zhang, J. J., Joglekar, N. R., & Verma, R. (2012), "Exploring resource efficiency benchmarks for environmental sustainability in hotels", *Cornell Hospitality Quarterly*, 53(3), 229–241.

Zhu, Q., &Sarkis, J., (2004), "Relationships between operational practices and performance among early adopters of green supply chain management Practices in Chinese manufacturing enterprises", *Journal of Operations Management* 22, 265–289.

Zhu, D., Asnani, P. U., Zurbrügg, C., Anapolsky, S., & Mani, S. (2008), "Improving municipal solid waste management in India: A sourcebook for policy makers and practitioners", WBI Development Studies. Washington, DC, World Bank©. https://openknowledge.worldbank.org/handle/10986/6916.

9

A Robotic Trajectory Planning Technique for Smooth Maneuvering of Autonomous Waste-Management Equipment

Bhaskar Guin, Tarun Kanti Naskar, and Nipu Modak
Jadavpur University

CONTENTS

9.1 Introduction

Waste management and its disposal are of prime concern in the 21st century. With the rise in population and technological advancements, consumption of raw materials is supplemented by a substantial rise in waste production. Waste management involves the collection, segregation, transportation, and processing of wastes. These are mostly manual tasks engaging human labor, which poses a serious threat to the health, safety, and well-being of the workers [1]. Researchers across the globe are developing waste-management equipment capable of working autonomously or being operated remotely [2] to eliminate direct human exposure. It is a challenging task, as these equipment work in adverse environments and are required to perform very complicated maneuvering tasks [3]. Robotic systems are being used extensively in nuclear power plants for inspection, transportation, and handling of nuclear material [4]. The technologies used in these systems can be extended further for use in solid waste management [5]. Recent developments include automated waste segregation [6] employing advanced technologies to distinguish metallic, plastic, and biodegradable wastes [7]. Automated excavators [8] and transportation vehicles [9] ensure efficient and reliable waste handling and transportation. The future of waste management lies in harmonic interaction between man-machines and the environment [10].

In light of the present challenges in the automation of waste-management equipment, the chapter is aimed at presenting a novel trajectory planning technique for complex maneuvering tasks, while ensuring optimized motion smoothness that reduces dynamic stress and improves product lifespan. It is desired to guide the autonomous robotic equipment along the desired trajectory [11]. Parametric curves like Bezier [12], B-Splines, and NURBS [13] are often used to model robotic trajectories due to their scalable

DOI: 10.1201/9781003231608-10

nature. However, for autonomous robots maneuvering in an environment full of obstacles, the required trajectory is far more complex and needs higher-order curves [14,15] to accurately model them. The trajectories are further optimized to ensure motion smoothness by minimizing the peak or RMS (Root Mean Square) jerk [16]. A generalized trajectory planner can be developed using an nth-degree parametric curve passing through a set of precision points. Using a cycloidal displacement-time function, the peak jerk can be minimized to obtain the optimum smoothness. This allows a more realistic trajectory planner to be developed, which integrates flexibly with the state-space model of the robotic equipment.

9.2 Theoretical Formulation and Modeling

Autonomous equipment for waste management is required to operate in adverse conditions and to perform complex maneuvering tasks such as obstacle avoidance, maneuvering on rugged surface, etc. Hence, the trajectory planner for such an equipment must be robust and flexible so that it can accurately plan the trajectory and be easily integrated with the state-space model of that equipment. To develop the trajectory planner algorithm, a few necessary mathematical tools are presented.

9.2.1 Defining Bezier Curves

For maneuvering the autonomous equipment along complex trajectories, the trajectory must be expressible by a geometric curve and quantified by a mathematical equation. Bezier curves are particularly useful in this regard because they are parametric (independent of coordinate systems), scalable, derivable at all points along the curve and can be used to represent multiple-valued functions. Bezier curves are mathematically defined as:

$$P(u) = \sum_{i=0}^{n} B_i J_{n,i}(u), \quad 0 \le u \le 1 \tag{9.1}$$

where the Bernstein basis j or blending function is defined as:

$$J_{n,i}(u) = \binom{n}{i} u^i (1-u)^{n-i} \tag{9.2}$$

The degree of the polynomial defining the curve segment is $(n-1)$ where n is the number of control polygon vertices, B is the set of control polygon vertices, and u is the parameter. The 1st and 2nd derivatives of the curve can be written as:

$$P'(u) = \sum_{i=0}^{n} B_i J'_{n,i}(u), \quad P''(u) = \sum_{i=0}^{n} B_i J''_{n,i}(u) \tag{9.3}$$

where J' and J'' can be obtained by differentiating j with respect to u:

$$J'_{n,i} = \frac{i - nu}{u(1-u)} J_{n,i}(u), \quad J''_{n,i}(u) = \left(\frac{(i-nu)^2 - nu^2 - i(1-2u)}{u^3(1-u)^2} \right) J_{n,i}(u) \tag{9.4}$$

9.2.2 Joining and Subdivision of Bezier Curves

Two Bezier curve segments can be joined at a 'knot' point satisfying a geometric or parametric continuity relation. Let, $P(u_1)$ be the 1st Bezier curve of degree n and $Q(u_2)$ be the 2nd Bezier curve of degree m. For G^0 or C^0 continuity, $P(1) = Q(0)$. Using higher-order continuity allows smooth transitions from one segment to another.

Just as two curve segments can be joined, it is also possible to split an existing Bezier curve into two segments at any point along the curve. The task is to determine the control polygons for the two new curves which when combined will be identical to the original curve. Let the original curve be given by $P(u_0)$, the new curves after the split are $Q(u_1)$ and $R(u_2)$. Let B_i, C_i, and D_i be the vertices of the control polygon for curves P, Q, and R, respectively. For simplifying calculations, the point of split is chosen at $u_0 = 0.5$. C_i and D_i can then be expressed as:

$$C_i = \sum_{j=0}^{i} \binom{i}{j} \frac{B_j}{2^i}, \quad D_i = \sum_{j=i}^{n} \binom{n-i}{n-j} \frac{B_j}{2^{n-i}}, \quad i = 0,1,\ldots,n \tag{9.5}$$

Using De Casteljau's algorithm, it is possible to split the curve at any value of u in the range $0 \le u \le 1$. The principle equation:

$$B_i^k(t_0) = (1 - u_0) B_{i-1}^{k-1} + u_0 B_i^{k-1} \tag{9.6}$$

is used recursively to obtain the new control vertices C_i and D_i, respectively.

$$C_i = \left\{ B_0^0, B_1^1, \ldots, B_n^n \right\}, \quad D_i = \left\{ B_n^n, B_n^{n-1}, \ldots, B_n^0 \right\}. \tag{9.7}$$

where $k = 1,2,\ldots,n$; and $i = k, k+1, \ldots, n$; and u_0 is the parametric split point.

9.2.3 Computing Arc Length

Arc length is the distance between two points on the curve measured along the curve. The computation of arc length is especially important from the point of view that it can be used to calculate time derivative parameters like velocity and acceleration. The arc length from point $P(u_1)$ to $P(u_2)$ can be obtained by calculating the line-integral as:

$$s = \int_{u_1}^{u_2} \left| \frac{dP}{du} \right| du \tag{9.8}$$

where

$$\left| \frac{dP}{du} \right| = \sqrt{\left(\frac{dx(u)}{du} \right)^2 + \left(\frac{dy(u)}{du} \right)^2 + \left(\frac{dz(u)}{du} \right)^2}$$

and s is the arc length. Such an integral can be evaluated using numerical integration techniques such as the Gaussian Quadrature method. However, these methods can become quite difficult to implement and become computationally expensive in cases in which higher accuracy is desired. An alternative method is the 'adaptive subdivision' method. If a Bezier curve is subdivided into a finite number of segments, then the arc length of each segment will be greater than the chord length L_c, but less than the sum of the length of sides of the control polygon L_p.

Let P be a Bezier curve with control points B_i, $i = 0,1,\ldots,n$. Sum of lengths of sides of the control polygon L_p is given by:

$$L_P = \sum_{i=0}^{n-1} |P_{i+1} - P_i| \tag{9.9}$$

Chord length L_c between the end-points is given by:

$$L_C = |P_n - P_0| \tag{9.10}$$

'True-arc-length' L_B can be computed using a convex combination of L_c, L_p given by:

$$L_B = \frac{2L_c + (n-1)L_p}{(n+1)} \tag{9.11}$$

Using an adaptive approach [17] to compute arc-length reduces the computation time significantly. The numbers of subdivisions are not only dependent on the error tolerance ϵ but also on the nature of the curve. The choice of ϵ denotes how fast or slow the algorithm will converge to a solution. Taking a very small value of ϵ ensures higher accuracy at the expense of computation time. The following algorithm can be used to effectively compute arc length:

Step 1: Calculate L_P (9.9), and L_C (9.10).
Step 2: Calculate the error bound err $= L_P - L_C$.
Step 3: If (err $<\epsilon$).
   Then, calculate L_B, and STOP recursion.
Else, subdivide P into two equal halves and evaluate Steps 1–3 for each segment.

$$\text{Return} \sum L_{B\text{-segment}}$$

Step 4: End.

Using a function that computes the arc length along the space curve from points $P(u_1)$ to $P(u_2)$ according to the algorithm stated above, it is possible to solve two types of problems:

- Given the parametric values of u_1 and u_2, the arc length L_B can be found,
- Given the arc-length L_B^{req} and value of u_1 the value of u_2 can be found by solving the equation: $L_B^{req} - \text{ARCLENGTH}(u_1, u_2) = 0$, using any root search algorithm.

9.2.4 Curve Fitting Using *n*th-Order Bezier Curve

So far some of the key features and mathematical relations for Bezier curves have been discussed. Consider an autonomous excavator being used for waste handling activities, which is required to follow a complex path, as shown in Figure 9.1. Here we define a set of points in space through which the excavator maneuvering along the trajectory must pass through. These points can be called precision points d, defined as: $d_i = [x \ y \ z]^T; i = 0,1,\dots,n_d$; where n_d is the number of precision points. The trajectory is defined mathematically by fitting an *n*th order Bezier curve through the precision points. The fitted curve is of degree $(n-1)$, where n is the number of control polygon vertices. The choice of n depends on the complexity of the curve, usually a high value of n is suggested for curves of nonuniform curvature or having sharp bends and turns. The curve fitting is done using the method of 'total least squares', where the sum residuals in both x- and y- directions have been minimized. The method and algorithms presented in [18] are found to be very effective.

9.2.5 Time Frame in the Designed Trajectory

In the previous sections, 'Spatial Planning' of trajectory has been dealt with, i.e., a path in 3D space which has been constructed using a Bezier curve passing through a set of precision points d. However, space planning alone is insufficient to describe any kinematic parameters. So, the designed trajectory must be linked to a time frame. This can be achieved by defining a distance-time function as $s = f(t)$,

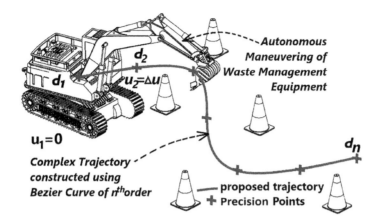

FIGURE 9.1 Mobile robotic equipment traversing along the complex trajectory.

where s is the distance = arc length and t is the time. The 'space curve' defines the path to be followed by an object, and the 'displacement-time' function relates time to the distance traveled. Kinematic parameters like velocity and acceleration can be computed by controlling the distance traveled in a given time interval. From a real-world perspective, a displacement-time function is constructed using a sinusoidal segment for acceleration, a linear segment for uniform velocity, and a sinusoidal segment for deceleration. Such a function resembles a system that starts at rest, then accelerates up to a particular velocity and continues to travel at that constant velocity, and finally decelerates to rest. Sinusoidal functions [19] are infinitely derivable leading to finite higher-order derivatives like jerk and ping. The displacement-time function after normalization can be expressed as:

$$s(t) = \begin{cases} \dfrac{\dfrac{2t_a}{\pi}\left(\sin\left(\dfrac{\pi t}{2t_a} - \dfrac{\pi}{2}\right) + 1\right)}{f} & 0 \le t \le t_a \\[3em] \dfrac{\left(\dfrac{2t_a}{\pi} + t - t_a\right)}{f} & t_a \le t \le t_d \\[3em] \dfrac{\left(\dfrac{2t_a}{\pi} + t_d - t_a + \dfrac{2(1-t_d)}{\pi}\sin\left(\dfrac{\pi(t-t_d)}{2(1-t_d)}\right)\right)}{f} & t_d \le t \le 1 \end{cases} \qquad (9.12)$$

where $\quad f = \dfrac{2t_a}{\pi} + t_d - t_a + \dfrac{2(1-t_d)}{\pi}$

t_a and t_d are the normalized blending times for acceleration and deceleration, respectively, and $s(t)$ is the normalized arc-length. Now it is possible to compute the kinematic parameters—velocity, acceleration, and jerk using the central difference method. If L_B^T represents the total arc length of the designed trajectory, then

$$\text{True displacement is given by} : \overline{s}(t) = L_B^T \cdot s(t) \qquad (9.13)$$

$$\text{Velocity is given by} : v = \dot{\overline{s}} = \dfrac{\overline{s}(t + \Delta t) - \overline{s}(t - \Delta t)}{2\Delta t} \qquad (9.14)$$

$$\text{Acceleration is given by: } a = \ddot{\overline{s}} = \frac{\overline{s}(t + \Delta t) - 2\overline{s}(t) + \overline{s}(t - \Delta t)}{\Delta t^2} \qquad (9.15)$$

$$\text{Jerk is given by: } j = \dddot{\overline{s}} = \frac{\overline{s}(t + 2\Delta t) - 2\overline{s}(t + \Delta t) + 2\overline{s}(t - \Delta t) - \overline{s}(t - 2\Delta t)}{2\Delta t^2} \qquad (9.16)$$

9.2.6 Optimizing the Designed Trajectory

The designed trajectory is now completely defined in terms of both space and time. Optimization is carried out to minimize the jerk for an nth-order Bezier curve following a predefined distance-time function. For the optimization problem, we define the following:

- **Objective Function**: Minimize the square of jerk j, since jerk value can be both positive and negative.
- **Design Variable**: Blending time t_a and t_d. For simplification, $t_d = (1 - t_a)$, i.e., equal blend times for acceleration and deceleration.
- **Constraints**: Limiting values of velocity and acceleration denoted by $v_{\text{limit}}, a_{\text{limit}}$, respectively.
- **Constants**: L_B^T (i.e., the designed trajectory is invariant of time) and t_{traj} (i.e., the total time taken to traverse the trajectory).

Thus, the optimization problem can be stated as:

$$\text{Minimize} \quad j^2(t)$$

$$\text{Subject to: } v \leq v_{\text{limit}} \quad \text{and} \quad a \leq a_{\text{limit}}$$

The optimization has been carried out using MATLAB® Optimization Toolbox. The solver selected for the purpose is fmincon() which can minimize constrained, nonlinear, multivariable problems. The problem was solved using MATLAB® version 2017b. The test problem and simulation results are discussed in the subsequent section.

9.2.7 Algorithm for Implementation

Step 1: Define a set of points d, set knots points k, degree and t_{traj} for each segment, values of $v_{\text{limit}}, a_{\text{limit}}$ based on the kinematic model of the robot, and step size stp.

Step 2: Use curve fit to construct the Bezier trajectory segments and join them at knot points using G^0 continuity.

Step 3: Define a displacement-time function for each segment based on the desired kinematic condition at the knot points.

Step 4: Calculate Arc Length L_B^i for each segment i.
    for $i = 1$ to (no of segments)
    for $t = 0$ to 1, (normalized time)

- calculate $\overline{s}$ using method prescribed in Section 9.2.3.
- calculate $\dot{\overline{s}}, \ddot{\overline{s}}$ and $\dddot{\overline{s}}$.
- optimize using fmincon and obtain $t_{ai}^{\text{opt}}, t_{bi}^{\text{opt}}$.

Step 5: Use $t_{ai}^{\text{opt}}, t_{bi}^{\text{opt}}$ in the respective displacement-time functions to obtain $\overline{s}, \dot{\overline{s}}, \ddot{\overline{s}}$ and $\dddot{\overline{s}}$.

9.3 Simulation Results and Discussion

The effectiveness of the algorithm described in Section 9.2.7 can be best described by implementing it in three different aspects, presented in the form of case studies.

Case Study 9.1

A simple planar trajectory has been constructed by joining five precision points using a 5th-degree Bezier curve segment. Boundary conditions:

- Travel time, $t_{traj} = 60$ s.
- Kinematic Limits, $v_{limt} = 300$ mm/s, $a_{limt} = 600$ mm/s^2.
- Normalized step size, stp = 0.0005.

RESULTS

- Opt. Blending Time, $t_a = 29.0017$ s, $t_d = 30.9983$ s.
- Path Length $L_B = 82.5109$ mm.
- Peak Velocity $v_{max} = 127.1918$ mm/s.
- Peak Acceleration $a_{max} = 413.3393$ mm/s^2.
- Peak Jerk $j_{max} = 0.67162$ mm/s^3.

The constructed trajectory using Bezier curve fit and the trace curve obtained is shown in Figure 9.2a. The trace curve depicts the actual path of the body obtained from displacement-time function, while the fitted curve is the assumed trajectory. The two curves coincide perfectly. By varying time t from 0 to t_{traj}, the kinematic parameters have been evaluated at each time step and the results are then plotted and shown in Figure 9.2b and c.

Case Study 9.2

For practical problems, it may be necessary to impose user-defined kinematic constraints at intermediate locations along the trajectory, such as an intermediate stop or slow-down. In such situations, the trajectory has to be split into multiple segments at that point, to be referred to as knot points. The planar trajectory constructed in this case has an intermediate stop point, hence the trajectory comprises two segments of varying degrees and arc length. Nine precision points are considered with the 6th point as the knot point. The degree of the 1st and 2nd segments is 6 and 4, respectively (Table 9.1). Boundary conditions:

- Travel time, $t_{traj}^{(1)} = 35$ s, $t_{traj}^{(2)} = 25$ s.
- Kinematic limits, $v_{limt} = 180$ mm/s, $a_{limt} = 800$ mm/s^2.
- Normalized step size, stp = 0.0005.
- $v = 0$, $a = 0$ at knot point.

RESULTS

By varying time t from 0 to t_{traj}, the kinematic parameters have been evaluated at each time step. The results are then plotted and shown in Figure 9.3. The knot point has been highlighted in each plot.

Case Study 9.3

In this case, the blend time t_a has been varied from 0 to $t_{traj}/2$. The variation of a jerk with blend time has been shown in Table 9.2. It is seen that smaller blend time leads to higher acceleration and hence a higher amount of jerk. The jerk decreases as the blend time approaches $t_{traj}/2$. It is interesting to note although the minimum jerk occurs at $t = t_{traj}/2$, it is unobtainable from the algorithm as it would imply eliminating the constant velocity section of the displacement-time function.

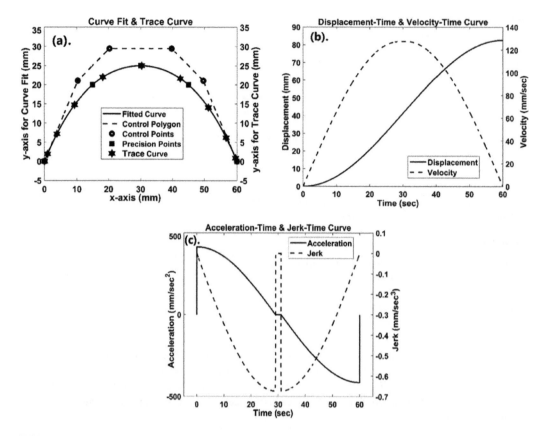

FIGURE 9.2 Plot of Case Study 9.1. (a) Curve fitting using 5th-degree Bezier curve with trace curve generated from displacement-time function, (b) displacement and velocity vs time, and (c) acceleration and jerk vs time.

The effect of changing v_{limit} and a_{limit} is also studied. Initially, the magnitude of v_{limit} is reduced keeping a_{limit} constant. Then, the magnitude of a_{limit} is reduced keeping v_{limit} constant. The result is tabulated and shown in Table 9.3. It is observed that the optimum blend time t_a approaches $t_{\text{traj}}/2$, and jerk reduces simultaneously.

TABLE 9.1

Optimization Results for Case Study 9.2

Parameter	Segment – 1	Segment – 2
Opt. blending time	$t_a = 17.4861\,\text{s}$	$t_a = 12.0038\,\text{s}$
	$t_d = 17.5139\,\text{s}$	$t_d = 12.9962\,\text{s}$
Path length L_B	107.6694 mm	57.5997 mm
Peak velocity v_{max}	169.0501 mm/s	88.4727 mm/s
Peak acceleration a_{max}	531.5076 mm/s^2	289.4347 mm/s^2
Peak jerk j_{max}	0.83555 mm/s^3	0.47343 mm/s^3

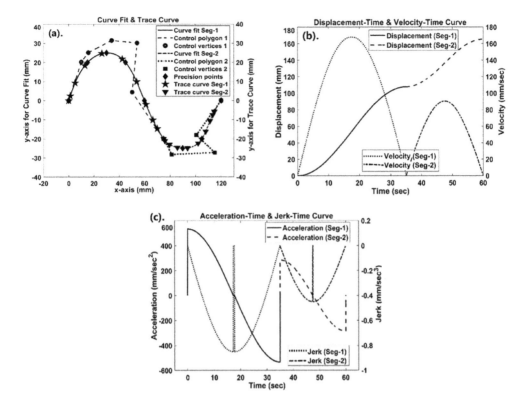

FIGURE 9.3 Plot of Case Study 9.2. (a) Curve fitting using 6th- and 4th-degree Bezier curve segments with trace curve generated from displacement-time function, (b) displacement and velocity vs time, and (c) acceleration and jerk vs time.

TABLE 9.2

Variation of Jerk with Blend Time

		Actuator Limits		Blend Time (s)		Calculated Values		Peak Jerk	
Sl. No.	Travel Time (s)	v_{limit} (mm/s)	a_{limit} (mm/s²)	t_a	t_d	v_{max} (mm/s)	a_{max} (mm/s²)	j_{max} (mm/s³)	Remarks
1	60	150	800	5	55	87.8302	1655.4743	15.598	$a_{max} > a_{limit}$
2				10	50	93.8826	884.8111	4.1694	Infeasible solution
3				15	45	100.8309	633.5354	1.9903	Feasible but not optimum
5				25	35	118.349	446.1641	0.84099	
6				29.5016	30.4984	128.3903	410.1637	0.65516	Obtained optimum
7				30	30	129.6078	407.1744	0.63959	Ideally optimum

9.4 Conclusion

The developed trajectory planner is extremely efficient and can be implemented easily by combining it with the state-space model of any autonomous equipment. The use of nth-degree Bezier curves makes modeling of complex trajectories easier. Minimizing the jerk yields a smooth trajectory improving maneuverability and reducing dynamic stresses. The use of knots allows the user to enforce local kinematic states such as using an intermediate stop. The proposed model and algorithm are fast and rapidly converge to a solution. The novelty of the work is thus the generation of nth-degree Bezier curves with knots, enforcing sinusoidal blending functions for smooth motion transitions and minimizing jerk which helps in designing trajectories suitable for maneuvering autonomous equipment.

TABLE 9.3

Variation of Jerk with Actuator Limits

Sl. No.	Travel Time (s)	Actuator Limits		Opt. Blend Time (s)		Calculated Values		Peak Jerk
		v_{limit} (mm/s)	a_{limit} (mm/s^2)	t_a	t_d	v_{max} (mm/s)	a_{max} (mm/s^2)	j_{max} (mm/sec^3)
1	60	120	800	25.6355	34.3645	119.6702	439.9618	0.80874
2		130		29.8426	30.1574	129.2208	408.0998	0.64442
3		140		29.028	30.972	127.2545	413.1673	0.67073
4		150		29.5016	30.4984	128.3903	410.1637	0.65516
5			790	29.5071	30.4929	128.4037	410.1294	0.65499
6			780	29.0084	30.9916	127.2078	413.2953	0.67139
7			770	29.0103	30.9897	127.2123	413.2831	0.67133

REFERENCES

1. A. K. Ziraba, T. N. Haregu, and B. Mberu, "A review and framework for understanding the potential impact of poor solid waste management on health in developing countries," *Arch. Public Heal.*, vol. 74, no. 1, pp. 1–11, 2016, doi: 10.1186/s13690-016-0166-4.

2. R. Sarc, A. Curtis, L. Kandlbauer, K. Khodier, K. E. Lorber, and R. Pomberger, "Digitalisation and intelligent robotics in value chain of circular economy oriented waste management: A review," *Waste Manag.*, vol. 95, pp. 476–492, 2019, doi: 10.1016/j.wasman.2019.06.035.

3. J. Petereit et al., "ROBDEKON: Robotic systems for decontamination in hazardous environments," in *2019 IEEE International Symposium on Safety, Security, and Rescue Robotics, SSRR 2019*, 2019, pp. 249–255, doi: 10.1109/SSRR.2019.8848969.

4. I. Tsitsimpelis, C. J. Taylor, B. Lennox, and M. J. Joyce, "A review of ground-based robotic systems for the characterization of nuclear environments," *Prog. Nucl. Energy*, vol. 111, pp. 109–124, 2019, doi: 10.1016/j.pnucene.2018.10.023.

5. M. A. Hannan, M. Abdulla Al Mamun, A. Hussain, H. Basri, and R. A. Begum, "A review on technologies and their usage in solid waste monitoring and management systems: Issues and challenges," *Waste Manag.*, vol. 43, pp. 509–523, 2015, doi: 10.1016/j.wasman.2015.05.033.

6. D. Bonello, M. A. Saliba, and K. P. Camilleri, "An exploratory study on the automated sorting of commingled recyclable domestic waste," *Procedia Manuf.*, vol. 11, pp. 686–694, 2017, doi: 10.1016/j.promfg.2017.07.168.

7. O. Adedeji and Z. Wang, "Intelligent waste classification system using deep learning convolutional neural network," *Procedia Manuf.*, vol. 35, pp. 607–612, 2019, doi: 10.1016/j.promfg.2019.05.086.

8. S. Dadhich, U. Bodin, and U. Andersson, "Key challenges in automation of earth-moving machines," *Autom. Constr.*, vol. 68, pp. 212–222, 2016, doi: 10.1016/j.autcon.2016.05.009.

9. A. Nieoczym, J. Caban, A. Dudziak, and M. Stoma, "Autonomous vans: The planning process of transport tasks," *Open Eng.*, vol. 10, no. 1, pp. 18–25, 2020, doi: 10.1515/eng-2020-0006.

10. I. Tiddi, E. Bastianelli, E. Daga, M. d'Aquin, and E. Motta, "Robot–city interaction: Mapping the research landscape: A survey of the interactions between robots and modern cities," *Int. J. Soc. Robot.*, vol. 12, no. 2, pp. 299–324, 2020, doi: 10.1007/s12369-019-00534-x.

11. Z. Wang, H. Li, and X. Zhang, "Construction waste recycling robot for nails and screws: Computer vision technology and neural network approach," *Autom. Constr.*, vol. 97, pp. 220–228, 2019, doi: 10.1016/j.autcon.2018.11.009.

12. C. S. Wu, Z. Y. Chiu, and J. S. Liu, "Time-optimal trajectory planning along parametric polynomial lane-change curves with bounded velocity and acceleration: Simulations for a unicycle based on numerical integration," *Model. Simul. Eng.*, vol. 2018, 2018, doi: 10.1155/2018/9348907.

13. G. Wu, W. Zhao, and X. Zhang, "Optimum time-energy-jerk trajectory planning for serial robotic manipulators by reparameterized quintic NURBS curves," *Proc. Inst. Mech. Eng. Part C J. Mech. Eng. Sci.*, 2020, doi: 10.1177/0954406220969734.

14. M. Boryga and A. Graboś, "Planning of manipulator motion trajectory with higher-degree polynomials use," *Mech. Mach. Theory*, vol. 44, no. 7, pp. 1400–1419, 2009, doi: 10.1016/j.mechmachtheory.2008.11.003.

15. G. J. Yang, R. Delgado, and B. W. Choi, "A practical joint-space trajectory generation method based on convolution in real-time control," *Int. J. Adv. Robot. Syst.*, vol. 13, no. 2, 2016, doi: 10.5772/62722.

16. Y. Li, T. Huang, and D. G. Chetwynd, "An approach for smooth trajectory planning of high-speed pick-and-place parallel robots using quintic B-splines," *Mech. Mach. Theory*, vol. 126, pp. 479–490, 2018, doi: 10.1016/j.mechmachtheory.2018.04.026.

17. B. Guenter and R. Parent, "Computing the arc length of parametric curves," *IEEE Comput. Graph. Appl.*, vol. 10, no. 3, pp. 72–78, 1990, doi: 10.1109/38.55155.

18. T. A. Pastva, "Bezier Curve Fitting (MSc thesis)," Naval Post Graduate School, Monterey, CA, 1998. [Online]. Available: https://apps.dtic.mil/dtic/tr/fulltext/u2/a350611.pdf.

19. R. Parent, *"Computer Animation: Algorithms and Thechniques*, 3rd edition. Morgan Kaufmann: Burlington, MA, pp. 61–108, 2012.

Section II

Circular Economy Adoption in Waste Management

10

Sustainable Waste Management towards Circular Economy in Nigerian Context: Challenges, Prospects and Way Forward

Oluwadare Joshua Oyebode

Afe Babalola University

CONTENTS

10.1 Introduction

Sustainable waste management is a method of preserving the environment and eliminating waste without a negative impact on the environment. Wastes can be reduced, reused and recycled by engineering technology for the benefit of the populace. A circular economy (CE) is an economic system geared towards the removal of wastes and the incessant utilization of resources. Circular systems employ reclaim, regenerative approach, allotment, revamp, restoration and re-manufacturing in the environment. Figure 10.1 indicates the difference between linear and circular waste disposal system.

Reuse refers to employing a product without reprocessing, either for its initial purpose or to satisfy a dissimilar performance. This conserves energy, time and different resources. It assists in job creation, entrepreneurship and contributes to the economy. This is totally different from usage and is often the breaking down of used things to make raw materials for the manufacture of the latest merchandise. Usage will forestall the waste of doubtless helpful materials and crop the consumption of up-to-date raw materials, thereby reducing energy usage, pollution (from incineration) and pollution (from land-filling). Usage aims at the environmental property by working material inputs into and redirecting waste outputs out of the economic system.

CE is environmental friendly and advantageous over linear economic system. CE helps reducing the resources used, conserve resources and reduce environmental pollution. However, it is argued by some that these assumptions are simplistic—that they disregard the complexness of existing systems and their potential trade-offs. As an example, the social dimension of a property appears to be solely marginally self-addressed in several publications on the CE. There are cases that will need totally different or further ways, like buying new, additional energy-efficient instrumentation.

The CE thought is gaining acceptance due to the imperative want for an additional approach to understand each economic development and environmental protection. The CE approach to resource-use

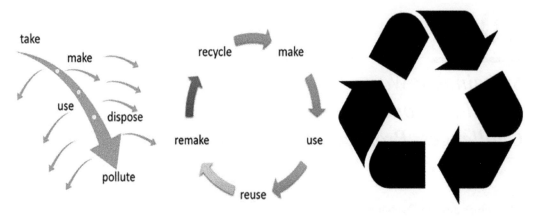

FIGURE 10.1 Difference between linear and circular waste disposal systems.

potency integrates cleaner production and industrial ecology during a very broader system encompassing industrial companies, networks or chains of companies, eco-industrial parks and regional infrastructure to support resource optimization. State-owned and private enterprises, government and private infrastructure and customers all have employment in achieving the CE.

10.2 Literature Review

The CE offers higher prospects for solid waste management and has been enforced successfully in its full theory, practice and policies in some developed locations of the world. The socio-economic disadvantages, short-skilled information and scarcity of data have hindered its appropriateness and implementation in low- and middle-income countries. This solid waste management principles are progressively being replaced with discussions on CE principles in so far deliberations on solid waste handling. This shift is supported by the worldwide adoption of the thought of property development. The CE offers higher prospects for solid waste management and has been enforced successfully in its full theory, practice and policies in some developed locations of the world (Ezeudu and Ezendu, 2019).

Waste management in Nigerian cities is actually monopolized by the agencies of state governments (sub-national governments) that have restricted capability to tackle the problems of solid waste management in their cities.

The CE concentrates on a regenerative approach to natural resource management. Many nations of the world have enraptured from the implementation of ancient waste management practices and policies and have absolutely adopted CE principles (Ellen et al., 2019; Mathews and Tan, 2011; Lehmann et al., 2014).

Three ways of aggregation of municipal solid waste are indicated by the use of six main technologies for accumulating MSW (Municipal Solid Waste) from the population and infrastructure. These technologies are employed in Russian cities and municipalities depending on the quantity of accumulated garbage and also the prospects of its removal (automobile accessibility to the village) (Figure 10.2).

Second, the role of informal waste usage activities has been thought to be a key element of the CE waste valorization system in developing countries. The safety problems can be resolved by establishing certain rules, such as informal staff having healthy operating conditions, safety kits, training and pension edges. Considering the extent of monetary condition and state within the country, there are nice prospects for attracting additional informal staff once the CE is formally declared through applicable policies and incentives (Figure 10.3). These measures are repeatedly stressed in previous analysis efforts (Scarlet et al., 2015; Ferronato et al., 2019; Nzeadibe, 2009).

A good example is the town of Ghorahi, Nepal, wherever a bunch of thoughtful, degreed and committed associates of municipal authorities and native stakeholders managed to develop an environmentally

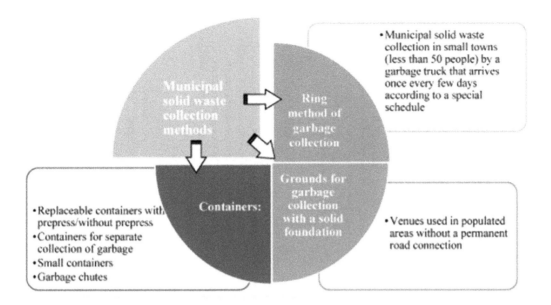

FIGURE 10.2 Main methods of municipal solid waste collection from the population and infrastructure.

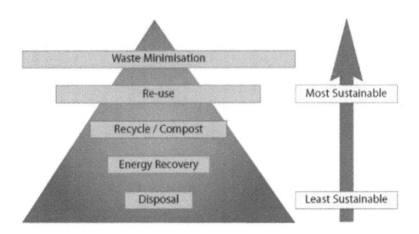

FIGURE 10.3 Waste minimization is the most sustainable compared with other methods.

sound waste process and disposal facility below the conditions of terribly restricted technical and monetary capacities, with no funding from abroad (Lamond et al., 2016; Wilson et al., 2013).

Environmentally sound treatment, disposal and resource recovery from waste need to be compelled in the systems to guarantee and capitalise on the synergism of actions toward strategic long-run goals of the system. In alternative words, it is necessary and helpful to verify the coherence of policy goals and cohesion among the instruments applied to achieve them (Howlett and Rayner, 2007; Marta Nilsson et al., 2012).

10.3 Methodology

The methodology adopted includes inspection of disposal sites, surveys and data from literatures. There is an urgent need for effective waste management in the majority of rural and urban centres in Nigeria.

TABLE 10.1

Relationship between Sustainable Development Goals and Solid Waste Management

Driver		Sustainable Development Goal (SDG)[a]	Specific Target	SWM-Related 'Virtual SDG'[b]
Protection of Public Health		SDG 11: Sustainable cities	11.1 Ensure access for all to adequate, safe and affordable basic services, upgrading slums	**Goal 1: Ensure access for all to adequate, safe and affordable solid waste collection services**
		SDG 3: Good health and well being	3.2 End preventable deaths of children under 5 years	Uncollected waste is: often dumped in waterways or burned in the open air, thus directly causing pollution and contamination. Waste also clogs the drains, which exacerbates floods, keeping stagnant water and contributing to water-borne diseases and malaria. Children are among the most vulnerable, so they are affected the most
			3.3 End malaria and combat water-borne diseases	
			3.9 Reduce illnesses from hazardous chemicals and air, water and soil pollution and contamination	
		SDG 11: Sustainable cities	11.6 Reduce the adverse environmental impact of cities; special attention to waste management	
Protection of the Environment	Local	SDG 12: Responsible consumption and production	12.4 Environmentally sound management of chemicals and all wastes in order to minimize their adverse impacts on human health and the environment	**Goal 2: Eliminate uncontrolled dumping and open burning** as the first stepping-stone to achieving environmentally sound SWM practices
		SDG 6: Clean water and sanitation	6.3 Improve water quality by reducing pollution, eliminating dumping and minimizing the release of hazardous materials	**Goal 3: Achieve environmentally sound management of all wastes, particularly hazardous wastes** (either chemical or biological hazardous wastes)
		SDG 15: Life on land	15.1 Ensure the conservation of terrestrial and inland freshwater ecosystem and their services	
	Global	SDG 7: Affordable and clean energy	7.2 Increase the share of renewable energy in the global energy mix	**Goal 3:** SWM technologies can derive renewable energy from (organic)waste
		SDG 13: Climate action	SDG 13: Take urgent action to combat climate change and its impacts	**Goal 3:** Adequate SWM practices can prevent emissions of large amounts of greenhouse gases[c]
		SDG 14: Life below water	14.1 Prevent marine pollution of all kinds, in particular from land-based activities, including marine debris	**Goals 1 and 2:** Extending waste collection to all and eliminating uncontrolled dumping will prevent waste (particularly plastics) ending up in the oceans

[a] Source of SDGs and targets. The wording of the targets is shortened.

[b] Source of SWM related 'Virtual SDGs'.

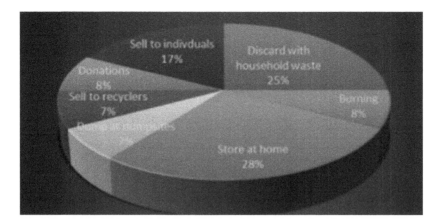

FIGURE 10.4 How waste is managed in some residential areas in Nigeria.

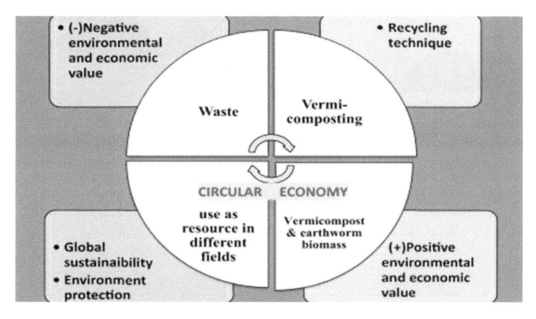

FIGURE 10.5 Circular economy negative and positive sides.

Challenges, prospects and ways forward were identified for sustainable waste management in the country. Figure 10.4 indicates how waste is managed in some residential areas in Nigeria. Figure 10.5 gives CE negative and positive sides. Figure 10.6 presents different ethnic groups and waste management cultures in Nigeria. Figure 10.7 gives the current waste management framework in Nigeria.

10.3.1 Challenges

There are a lot of challenges facing waste management in Nigeria. There are issues with the collection, inadequate treatment and disposal and recycling systems. There are problems of air and groundwater contamination. There are problems of flooding traceable to the dumping of refuse at canals and drainages in Nigeria. There are no formidable legal framework and implementation strategies for sustainable management and CE.

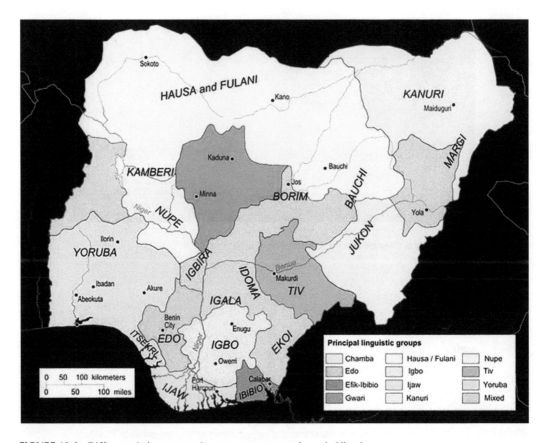

FIGURE 10.6   Different ethnic groups and waste management cultures in Nigeria.

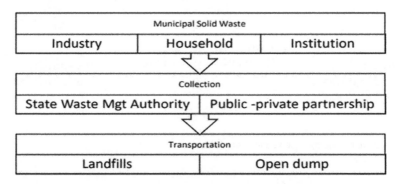

FIGURE 10.7   Current waste management framework in Nigeria (Ike et al., 2018).

Other challenges faced by waste management in Nigeria are as follows:

   i. insufficient funding
   ii. rapid population growth
   iii. poor investment in infrastructure development
   iv. low-skilled administrative and managerial personnel badly versed in technical issues
   v. lack of effective systems for data management
   vi. uncoordinated institutional functions
   vii. lack of adequate manpower in waste management engagements.

10.3.2 Prospects from Sustainable Waste Management towards Circular Economy

The major prospect is wealth and a hygienic environment for all and sundry. When waste is properly managed, it will be harmless to human and animal life—the ecology and environment generally. Public health will be preserved. Waste to wealth, health and energy will enhance environmental sustainability. Resource conservation will also be another prospect of sustainable waste management.

10.4 Way Forward

There is a need to attract private investment into waste management. Job creation and entrepreneurship programmes on waste management should be given priority.

Research and innovation in waste management should be supported by the government and all stakeholders. There is a need to create robust and workable sustainable waste management systems. The idea of CE should be given adequate awareness, planning and execution by institutions at all tiers of government. Manpower development, capacity building, stakeholder collaboration and consumer education campaign should be embraced and supported. Nigeria needs to attract recycling infrastructural investors for effective waste management. Policy and regulatory advocacy will yield tremendous benefits. Relevant attention should be given to all aspect of waste management and CE. Treatment, disposal, recycling systems and other facilities should be designed and put in place. Stakeholders' responsibility and legal requirements are very useful in CE.

10.5 Conclusion

Waste management requires adequate attention by all and sundry for the achievement of a cleaner environment, CE and effective waste management. Involvement of government agencies, laws and guidelines for wastes minimization, industries, institutions, building owners, appropriate capacity building and all stakeholders will yield tremendous rewards. Sustainable waste management and CE need adequate attention in Nigeria for health, wealth and energy. Waste management facilities need proper design and improvement. There issues are linked to planning, maintenance, recycling culture among Nigerians, inadequate storage and disposal facilities.

REFERENCES

Ellen MacArthur Foundation. (2019) Concept: What is circular economy? A framework for an economy that is restorative and regenerative by design. Available online: https://www.ellenmacarthurfoundation.org/circular-economy/concept (accessed on 9 September 2019).

Ezeudu, O.B., Ezeudu, T.S. (2019) Implementation of circular economy principles in industrial solid waste management: Case studies from a developing economy (Nigeria). *Recycling*, 4(4), 42.

Ferronato, N., Rada, E.C., Portillo, M.A.G., Cioca, L.I., Ragazzi, M., Torrenta, V. (2019) Introduction of the circular economy within developing regions: A comparative analysis of advantages and opportunities for waste valorisation. *Journal of Environmental Management*, 230, 366–378.

Howlett, M., Rayner, J. (2007) Design principles for policy mixes: Cohesion and coherence in 'New Governance Arrangements'. *Policy and Society*, 26, 1–18.

Ike, C.C., Ezeibe, C.C., Anijiofor, S.C., Daud, N.N. (2018) Solid waste management in Nigeria: Problems, prospects, and policies. *The Journal of Solid Waste Technology and Management*, 44(2), 163–172.

Lamond, J., Bhattacharya, N., Bloch, R. (2012) The role of solid waste management as a response to urban flood risk in developing countries, a case study analysis. In: *Flood Recovery, Innovation and Response III*, Proverbs, D., Mambretti, S., Brebbia, C.A., Eds. WIT Press: Southampton, UK, pp. 193–205. Available online: http://eprints.uwe.ac.uk/21458 (accessed on 1 December 2016).

Lehmann, M., Leeuw, B.D., Fehr, E., Wong, A. (2014) *Circular Economy: Improving the Management of Natural Resources*. World Resource Forum: Bern, Switzerland.

Mathews, J.A., Tan, H. (2011) Progress toward a circular economy in China. *Journal of Industrial Ecology*, 15, 435–475.

Nilsson, M., Zamparutti, T., Petersen, J.E., Nykvist, B., Rudberg, P., McGuinn, J. (2012) Understanding policy coherence: Analytical framework and examples of sector-environment policy interactions in the EU. *Environmental Policy and Governance*, 22, 395–423.

Nzeadibe, T.C. (2009) Solid waste reforms and informal recycling in Enugu urban area, Nigeria. *Habitat International*, 33, 93–99.

Scarlet, N., Motola, V., Dallemand, J.F., Monforti-Ferrario, F., Mofor, L. (2015) Evaluation of energy potential of municipal solid waste from African urban areas. *Renewable & Sustainable Energy Reviews*, 50, 1269–1286.

Wilson, D.C., Velis, C.A., Rodic, L. (2013) Integrated sustainable waste management in developing countries. *Waste and Resource Management*, 166, 52–68. Available online: http://www.icevirtuallibrary.com/doi/abs/10.1680/warm.12.00005 (accessed on 1 December 2016).

11

Indonesia Main (City-Level) Waste Bank for Waste Management toward Circular Economy: A Case Study of Bandung City

Ratnawati Kusuma Jaya and Sari Viciawati Machdum
University of Indonesia

CONTENTS

11.1 Introduction

The total population of Indonesia is experiencing growth from year to year, which is predicted to reach 305.6 million in 2035 (BPS, 2013). The population growth also has an impact on economic activities as well as the amount of waste generation. In 2025, a total of 4.3 billion urban residents are expected to produce around 2.2 billion tons of waste per year. Indonesia, which is densely populated, needs to highlight how important it is to be able to manage its waste by applying a sustainable development approach where economic, social, and environmental dimensions must go hand in hand and mutually succeed in development (Rogers, Jalal, & Boyd, 2008). The difficulties to have sustainable waste management are also faced by many cities in Indonesia. It is caused by many factors such as landfill site crisis, low capacity of local government, less awareness regarding solid waste management, no segregation habit, uncertain take back system policy, weak law enforcement, inadequate waste facilities and infrastructure, and low regional income and expenditure budget for waste management. In 2017–2018 Bandung City, one of the most populated cities in Indonesia, was noted as the second-highest city that has not been able to manage its waste, which is 264.09 tons/day (SIPSN, 2019).

The issue of waste is a global or international concern. This is stated in the 2030 Sustainable Development Goals (SDGs) target 12 point 5, namely "Substantially reduce waste generation through prevention, reduction, recycling, and reuse by 2030" (UN SGGs Agenda A/RES/70/1). This means that various ways to reduce the rate of growth in the amount of waste can be done through prevention, reduction, recycling, and reuse of waste so that it can be expected that by 2030 the world will be able to prevent unmanaged waste and achieve the target number 12, namely "Ensure sustainable consumption and production".

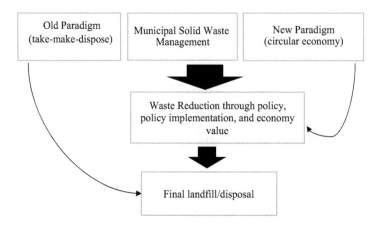

FIGURE 11.1 Waste paradigm shift framework. ("Shifting Waste Management Concept", Sitanggang et al. (2017) in Indartik et al. (2018), modified by the author.)

In response to this, the Government of Indonesia issued Presidential Regulation Number 97 of 2017 containing a target of reducing household waste and similar waste by 30% and waste handling by 70% (Indonesia Ministry of Environment and Forestry, 2018). This target also changes the old paradigm of collect-haul-dispose or take-make-dispose and the treatment that relies on waste reduction and handling with 3R Model (Reduce, Reuse and Recycle) into a new paradigm which is circular economy (CE) as an alternative to solve the waste problem. The process of changing the paradigm in waste management policy is illustrated in the framework below (Figure 11.1).

To achieve the national targets for waste reduction, the Government of Indonesia focuses on waste generation limitation (reduce), recycling waste, and reusing the waste from the sources through community participation in order to shape behavioral and cultural change in waste management, which has become the community movement. Although currently Indonesia still focuses a lot on three waste management strategies in the CE approach in the form of reduce, reuse, recycling (3R), according to Morseletto's research (2020), it will be more effective if the CE strategy can also focus to remanufacture, refurbish, repair, reuse, refuse, rethink and reduce. So, in accordance with the initial concept of waste reduction and handling which must involve all stakeholders, the CE approach needs to be applied starting from the production process to the waste management process.

In the new paradigm of waste management system, waste banks can be an instrument in the process of community education in sorting waste that is expected to also be able to play a role in the community-based waste-reduction process. Mirroring the conventional bank, waste bank means a place for sorting and collecting waste that can be recycled and/or used which has economic value. Waste bank is a campaign for reducing and handling waste by buying back waste in terms of a deposit-like banking system (Periathamby & Tanaka, 2014). Economic incentives in the waste bank system are expected to be the initial trigger for the community to be able to select and utilize the waste they have.

As said by the Minister of Environment of the Republic of Indonesia in 2018 "Circular economy through the waste bank mechanism is increasingly becoming a hope. Currently the number of waste banks has reached 5,244, spread across 34 provinces and 219 cities throughout Indonesia, involving more than 179 thousand customers" (Hadi, 2018). Waste banks are said to be part of the CE ecosystem in Indonesia where the waste bank also plays a role in the entire waste management process (Figure 11.2).

Waste bank is expected to be able to trigger more circular waste management based on the built-in economic incentives. The CE framework is based on 10 CE strategies – recover, recycling, repurpose, remanufacture, refurbish, repair, reuse, reduce, rethink and refuse (Morseletto, 2020), which are expected to be able to be adopted by waste banks as a technical unit for community-based waste management. However, waste bank also faced some challenges in the field. As waste bank sells their goods to the informal sectors, sometimes they get erratic price and low-profit margins. To finish this problem, the Indonesian Government then lowered to the provincial and district/city levels obliging them to have at least one main (city-level) waste bank (Bank Sampah Induk/BSI).

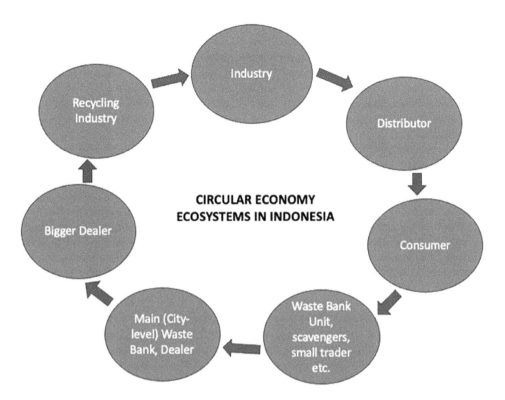

FIGURE 11.2   Waste bank and circular economy ecosystems in Indonesia. (Satori, Mohamad (2020) "Peran Bank Sampah dan Usaha Sektor Informal dalam Mewujudkan Circular Economy di Indonesia", modified by the author.)

Waste banks generally play a role in creating a clean and green environment as well as realizing government programs and empowering communities (Wulandari, Utomo, & Narmaditya, 2017). The waste bank is also considered to have triggered the development of independence in a community, both economic independence and intellectual independence which is formed by sorting and managing waste in the household environment (Wijayantia & Suryania, 2015). Some researchers also said that waste banks are a solution to managing waste through community participation (Purba, Meidiana & Adrianto, 2014).

11.2 Theoretical Framework

11.2.1 Waste Bank: An Overview

Similar to the conventional bank concept, a waste bank is an intermediary institution that has a function as a place to store and lend money and conduct financial transactions. What distinguishes a waste bank from a bank, in general, is not the money that is deposited by the community to the waste bank, but the waste itself. The waste bank then becomes an intermediary in the process of converting the waste into cash. Apart from cash, the Garbage Bank also has various other programs such as garbage alms, smart cards or money from waste savings that can be stored on cards for shopping and toll payments, as well as other programs such as saving gold from waste products and so on. The waste bank management model is almost the same as a commercial bank where there are customers, bookkeeping and management (Wulandari, Utomo, & Narmaditya, 2017).

The waste bank itself can be present at the top-down (initiated by the government) or bottom-up (initiated by the community). The waste bank that is managed directly by the government is called the Main Waste Bank (BSI) (Yustiani & Abror, 2019). However, there are also main waste banks that are bottom-up that are managed or initiated by the private sector or community groups (non-governmental organizations) (Auliani, 2018).

The background of the regulations for the establishment of the main waste bank is Law Number 18 of 2008 concerning Waste Management, Law Number 32 of 2009 concerning Environmental Protection and Management and Government Regulation Number 81 of 2012 concerning Management of Household Waste and Waste Similar to Household Waste. Indonesia Minister of Environment Regulation Number 13 of 2012 concerning Guidelines for the Implementation of Reduce, Reuse, Recycle through Waste Banks, Law Number 7 of 1992 concerning banking which was later amended to Act Number 10 of 1998, and Law Number 1 of Year 2013 concerning Micro Finance Institutions (KLHK, 2018).

Waste bank teaches people to sort the waste, raising public awareness to process waste wisely in order to reduce waste going into the landfill (Asteria and Heruman, 2016). Innovation of waste with waste banks has become innovation program at the grassroots level to increase the income of the urban poor (Winarso & Larasati, 2011; Asteria and Heruman, 2016). Purba et al. (2014) explained that the development of this waste bank will also assist local governments in empowering communities to manage waste wisely and reduce waste transported to the final disposal (landfill). Waste Bank Unit (BSU) requires intensive and continuous monitoring and guidance. This is because waste bank units face various problems in the form of management, finance and facilities, price treatment and business protection. With the formation of the Main Waste Bank (BSI) which can serve as the protector and supervisor of the BSU, BSI has the responsibility to align the socio-economic and environmental interests in its territory.

Meanwhile, the main waste bank (BSI) functions are to (i) set a target for reducing waste generation at the district scale based on waste banks, (ii) compile a waste management program, (iii) conduct data collection and guidance for unit waste banks through mentoring, technical capacity building, and unit waste bank management (BSU) and growing the number of BSU, (iv) record the amount of managed waste and its residue into the logbook, (v) report to the Ministry of Environment and Forestry regarding the status of the waste bank unit in the city, (vi) maintain price stability and (vii) facilitate after-sales of recycled products (KLHK, 2018).

Main Waste Bank (BSI) also has a role to (i) foster unit waste banks (BSU) to become agents of changes in consumption patterns and household waste management through the application of the 3R principle, (ii) foster unit waste banks that enable administrators, managers and employees to develop and innovate through improvement programs—Human resource capacity, (iii) foster unit waste banks to carry out BSU management properly in order to improve their performance to become a small and medium business unit, (iv) protect BSU from the practices of waste stalls outside the BSU, (v) provide recommendations and reference market prices for waste sales plastic/paper/can/iron and (vi) distribute recycled products and handicrafts made from recycled waste.

11.2.2 Circular Economy

Human activities in fulfilling their needs often cause environmental degradation, habitat destruction and endanger ecosystems, which ultimately endanger humans themselves. This makes the world try to find new strategies or approaches that are more sustainable. Then in 2013, the Ellen MacArthur Foundation mentioned a concept called the CE as a system of consumption and production which is expected to be able to minimize the loss of material and energy through reuse, recycling and recovery. CE itself is a transformational change throughout the value chain to maintain both types of material and maintain its value as long as possible (Bicket et al., 2014 in Rizos, et al., 2015). CE is the opposite of the linear economic development or paradigm (Kirchherr, Reiker & Hekkert, 2017), namely the 'take-make-dispose' paradigm which consistently increases the amount of waste (Day, 2015 in Pribadi, 2017).

This is because a lot of waste is wasted in vain even though there is still a lot of material in it which is still useful and can be reused (Ellen MacArthur Foundation, 2015 in Pribadi, 2017). The new paradigm originating from the new economic paradigm then emphasizes the concept of the creation of a shared value or the co-creation of a new value (Porter & Kramer, 2006 in Mura et al., 2020). Although in its development, CE is defined in many theories. CE which is a concept or strategy from the direction of sustainable development seeks to minimize negative consequences of economic growth such as depletion of resources and environmental degradation by means of resource efficiency by reducing waste and creating long-term value of a material or product, a reduction in the use of natural resources, within the limits of environmental protection that prioritizes socio-economic benefits (Murray et al., 2017; Babbitt et al., 2018; Hofmann, 2019 in Morseletto, 2020).

CE is a process of production and consumption that requires minimal use of natural resources and environmental damage by trying to make a material last longer, minimize consumption and waste material and energy. The long life of a material is done by transforming material from one product to another, designing a product to be more sustainable in its lifetime, minimizing waste production, reusing an item, reducing individual ownership and popularizing joint ownership of goods and services (Gillian Foster, 2020). In CE, it is believed that the stimulation of a sustainable culture must start at the level of politics, business entities and society. The focus on public awareness in CE is very important. According to Foster (2020), CE as a sustainable approach that must be included in the social structure and promote the common good in creating a more sustainable earth.

The concept of CE is still developing today (Ghisellini et al., 2016; Kalmykova et al., 2018; Laurenti et al., 2018; Reike et al., 2018 in Morseletto, 2020). Experts classify theoretical and empirical studies of CE by origin, principle and model or implementation. Some reviews are based on the supporters and barriers of CE implementation and its implementation methods (Foster, 2020). Various studies look at strategies in CE adopted by governments and organizations (e.g. Reduce, Reuse and Recycle/3R), see specific solutions and have a specific geographic focus (e.g. in countries or regions) and look at the industrial sector in implementing CE (e.g. management energy or waste).

Regarding the waste management system, the waste management paradigm is indeed developing in line with global targets and developments. Starting from end-pipe-solution or 'take-transport-waste', to Reduce, Reuse, Recycle (reduce, reuse, recycle), and Extended Producers Responsibility (EPR), and now the CE or CE. In line with the projection that Indonesia's waste generation in 2025 is estimated to reach 71.3 million tons per year, the waste problem is certainly a significant problem. Therefore, it can be understood that currently, the problem of waste and the environment is the concern of the government and the people of Indonesia so that a national strategy for Indonesian waste management will emerge in 2025. Therefore, Indonesia has also begun to see and implement the CE concept in its waste management.

In the picture below, it can also be seen that the main waste bank plays a role in the waste processing stage as the key to implementing a CE in the waste management business process (Figures 11.3 and 11.4).

The CE concept applied in a policy is expected to prevent environmental damage and conserve scarce resources through effective waste management. The CE concept seeks to shift the narrow waste recycling thinking to a larger thing, namely broad efficiency at all stages of production, distribution and consumption. Thus, the CE is expected to be able to increase the reduction of environmental degradation and create value in terms of politics, the business world and society at large (Bastein et al., 2013).

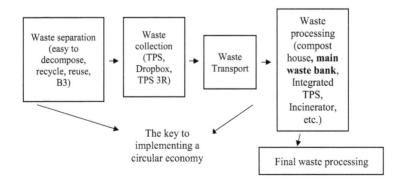

FIGURE 11.3 Waste management business process. (KLHK (2018), modified by the author.)

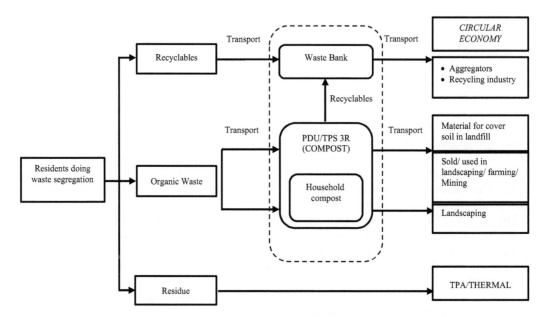

FIGURE 11.4 Waste management at source, waste bank for circular economy. (Modified from Plan Miljoee & ESP3 DANIDA (2018), "Best Indonesian Practices in Circular Economy and SWM".)

11.2.3 Community Empowerment

Society or community is "a combination of social unity and a system that performs major functions that have locality relevance" (Warren, 1978 in Lee, 2001). Population size, similarity in identity, relationships and local culture are also aspects of society or community (Ife, 2013; Popple & Stepney, 2008).

According to Adams (2003) in Qodir (2011) empowerment means active community activities that are transformational. According to Payne in Adi (2008), empowerment is helping clients to be empowered or to have the strength to be able to act and make decisions for themselves through increasing self-confidence and capacity to use their power or strength. Empowerment is often focused on society or people who are weak and disadvantaged (Ife, 1995). Another goal is for such weak or powerless people to have freedom, such as financial freedom, freedom from ignorance and hunger. This empowerment can take the form of increasing income, primary needs, and contributing or playing a role in the development process (Suharto, 2005 in Qodir, 2011).

According to Chen (1997), there are four reference conditions for changes in empowerment. First, material conditions, namely conditions that can empower access and control material resources, income, and satisfaction with the fulfillment of basic needs (Chen, 1997: 4). Second, cognitive conditions, namely increased knowledge, skills, and awareness of the environment. Third, the three perceptual conditions or self-esteem, namely the increase in individual perceptions of self-understanding of their interests, values, confidence in their capacities and determination of future visions. Finally, the relational condition, namely the ability to make decisions both within the family and in the community, the ability to lead, have bargaining power, be able to participate in organizations or institutions, be independent or not depend on others and be able to mobilize themselves independently.

Community empowerment is often interpreted as the concept of independence, participation, networking and justice (Qodir, 2011). According to Ife (2013), community empowerment is a process, in which every program of community empowerment is an important part of achieving the empowerment process itself. Focus on achieving a good process (means) is what is to be achieved, not the results or output (ends) to be achieved. Focusing on a good process can produce good results or output, but if we focus on good or maximum results, this does not necessarily result in a good process.

According to Ife (2013), to be able to produce a good process, we need to pay attention to the integrity of the process. That is, the process must pay attention to the principles of social and environmental justice because according to Ife (2013) community empowerment principles reflect sustainability and social justice. Of course, participation is very important in the community empowerment process, Ife (2013) states that the way to increase community participation is to make issues and activities important to be felt by the community, making people feel that their participation or presence of collective participation will make changes rather than they do it individually, every type of participation that is carried out must get attention and appreciation, the community must be ensured to be able to participate and get support for the participation they do, lastly, the structures and processes must be made as inclusive as possible or no one is alienated.

11.3 Methods

This research is descriptive qualitative research. The approach in this study is a qualitative approach and describes the role of the main waste bank in reducing waste and improving community economics, knowledge, perception and relation. The survey consists of in-depth interview and observational method on the object of research. According to Alkin (2011), the observation method will produce detailed information about trends, related stakeholders, social situations and conditions and existing activities. This research also uses observation methods to the activities and systems that run at the case study's location to understand and get deeper information about the activities carried out (Figure 11.5).

This research is a case study research located in Resik Waste Bank, a Main (City-Level) Waste Bank in Bandung City, Indonesia (Figure 11.6).

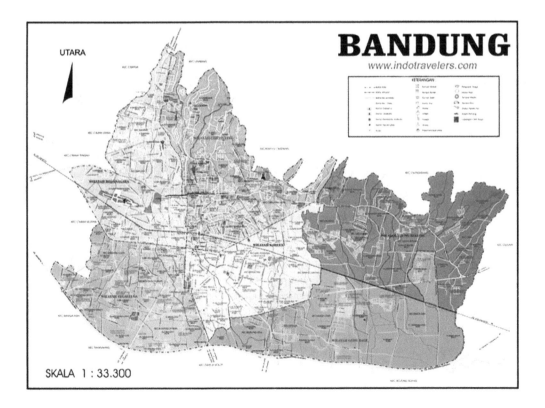

FIGURE 11.5   Map of Bandung City area. (Google.com.)

FIGURE 11.6   Resik Waste Bank.

11.4 Results and Discussion

Resik Waste Bank is a Main Waste Bank (BSI) in Bandung City. The Resik Waste Bank was established on December 11, 2014 based on the decision of the director of the Bandung City Cleanliness Regional Company (PD) director letter no.658.1/kep.240- PDKBR/2014 and officially became BSI of Bandung City in 2019. The objective of the Resik Waste Bank is to raise awareness of the local community about the importance of managing their waste properly by sorting, reducing and processing their own waste and making them aware of a healthy, tidy and clean environment (Puspita, 2016).

The number of Resik Waste Bank customers continues to grow from year to year. It was recorded that at the end of 2016, Resik Waste Bank customers totaled <300 customers, and at the end of 2017, there were 657 customers. Currently, the number of Resik Waste Bank customers is 3,497. The members of the Resik Waste Bank consist of the general public, government offices and the commercial sector such as malls, hotels, restaurants, schools, and universities. Usually in a year, the Resik Waste Bank is able to process up to 2,130 tons of waste, but during the pandemic period in 2020, the Resik Waste Bank was only able to process 135 tons of waste.

The types of products that can be purchased by the Resik Waste Bank from customers are: (i) Mixed Buckets, consisting of plastic bottles such as clear mineral water bottles, glass plastic such as clear mineral water glasses, used shampoo bottles or toys and plastic bottles, bottles and/or colored drink glasses; (ii) Plastics, consisting of clear plastic/PP/polypropylene, Kresek/HD/High Density, and PE/Polyethylene; (iii) Paper, consisting of used newspapers and boxes; (iv) Metal, consisting of frozen iron, thin iron, aluminum such as beverage cans and aluminum PC or aluminum used for kitchen utensils, water taps and the like; and (v) Bottles made from glass.

The waste collection system by the Resik Waste Bank is carried out in two ways—(i) the customer can come directly to the Resik Waste Bank and deposit the waste or vice versa, the Resik Waste Bank provides a waste pick-up system to the Unit Waste Bank with a minimum of 50 kg of waste to be transported. The waste that has been collected from customers is then sold by the Resik Waste Bank to a larger Bandars or traders (informal sector), factories, and also taken to the recycling center for further processing. Unfortunately, the amount of waste that enters the recycling center is still very small at 12 tons per month. Bandung City produced around 1,500 tons/day and Resik Waste Bank as the main waste bank in Bandung City collected around 6 tons of waste per day and it is only as 0.4% from overall waste generation per day in Bandung City. Based on this, the main waste bank is considered to have not maximally reduced waste from its source in the city of Bandung.

Apart from the low number of recycling and waste collection, based on the results of interviews and field observations, the Resik Waste Bank has not been able to fully play its role as the main waste bank in Bandung. The findings are shown in Table 11.1.

However, Resik Waste Bank found contributed to empowering their community or members. They have impacted the community in terms of building knowledge, changing perception, economy and relation. Even though the concept of CE and 3R practices are socialized by Main Waste Bank to the members, the members are still not aware of other CE principles or strategies such as recovery, repurpose, remanufacture, re-furbish, repair, rethink, and refuse, and only a few people understand and practice it. Unfortunately, the idea and principles of CE are not a concern of unit waste bank officials as well as the main waste bank itself, but the Resik Waste Bank has modalities to scale up their ability to be more impactful to the waste bank units in Bandung City if their role as the main waste bank is effectively being implemented.

Knowledge: Waste bank members are now aware of the importance of a clean environment and the negative effects of garbage if it is littered. They also know the concept of reduce, reuse, recycle (3R), but it is deeper in the advanced stages of the CE such as rethinking, redesign and so on.

Perception: There is a change in perception among members of the waste bank. Currently, they consider garbage as not something disgusting but has economic value.

Economy: Members of the waste bank get economic benefits by storing waste in the waste bank. Generally, members of the Waste Bank take the money they have saved on big holidays. However, the sorters from the informal sector who are already members of the waste bank deposit and take their money every day because they make waste their livelihood.

TABLE 11.1

Findings of Interviews and Field Observations at the Resik Waste Bank

Roles	Existing	Change
1. Empowering unit waste banks (BSU) to become agents of changes in consumption patterns and household waste management through the application of the 3R principle	Conduct one socialization program to the prospective waste bank unit, if Resik Waste Bank is invited	Need to have sustainable program to turn the Waste Bank Unit become an agent of change in changing consumption patterns and implementing the 3R principles in waste management
2. Enabling administrators, managers and employees of the waste banks by develop and innovate through improvement programs Human resource capacity	The Resik Waste Bank does not yet have a training or capacity building program for either the management or the manager	Training program and capacity building to all managers or management team of the waste bank are needed
3. Empowering unit waste banks to carry out waste bank management properly in order to improve their performance to become a small and medium business unit	Resik Waste Bank currently only carries out waste transportation from its unit and provides information on types of waste and how to sorting the waste	Need a regular meeting between the main (BSI) and the waste bank units (BSU) for training to become a small and medium enterprise unit
4. Protecting the Waste Bank Unit from trash stall practices and providing recommendation and reference to market prices for the sale of plastic/paper/can/metal waste	Resik Waste Bank mostly selling its goods to larger dealers (informal sectors) and factories, and has not been able to negotiate the price with them. So there is a difficulty in controlling the selling and buying prices	Need to increase the waste transported to their recycling center to have more valuable value of waste and more directly connected to factories
5. Distributing recycled products and handicrafts made from recycled waste	The Resik Waste Bank has difficulty marketing recycled products from unit waste banks due to its low quality and designs that are less attractive to the market	Need to focus in building relation and cooperation with upcycled businesses

Relationship: Members of the waste bank revealed that the relationship between neighbors and members of the waste bank is getting closer. They gather together when the Waste Bank picks up trash to be weighed together, making their relationship closer.

11.5 Conclusion

Resik Waste Bank faces various obstacles in carrying out its role as the main waste bank in Bandung City. The profit owned by the main waste bank is still limited to covering its operational costs. Currently, the government is still providing funding support. Low-value waste is also bought by the Main Waste Bank in order to reduce waste going to the final landfill, so that the profit margin they have is very small. It is because the main waste bank has an empowerment and educational role in the society.

The Main Waste Bank still distributes very little of its waste to the recycling center owned by the Bandung City Government; this is due to the lack of infrastructure and Human Resources of the Main Waste Bank and Recycling Center. The Main Waste Bank needs a faster turnover of money so that most of their goods are sold to the bigger Dealer or informal sectors. The main objective of the Waste Bank Unit is to educate and empower the community, especially its units so that there can be changes in behavior and social movements in the community in handling waste.

Based on that issue, a policy to support the role main waste bank in assisting and empowering the waste bank unit is needed in order to solve the problem of waste bank unit' sustainability and main waste bank role effectively. The Indonesian Government should support the main waste bank for the upstream to downstream infrastructure and connection regarding the waste flow. It is also necessary to strengthen the institutional capacity and knowledge of officials working in the main waste bank with the principles of CE and to increase the capacity of the recycling center or recycling industry. Indonesia also needs to multiply and provide incentives to SMEs in the field of upcycling waste and connect it with the main waste bank as a distributor of recycled materials. The main waste bank needs to work a lot with SMEs in upcycling waste, an online-based application in picking up and distributing waste and the government must aggressively promote the use and trends of recycled/upcycle products to societies.

BIBLIOGRAPHY

Adams, R. (2003). *Social Work and Empowerment*, 3rd edition. New York: Palgrave Macmillan.
Adams, R., Dominelli, L., Payne, M., Campling, J. (2003). *Social Work: Themes, Issues and Critical Debates*. London: MacMillan Press Ltd.
Adi, I.R. (2008). *Intervensi Komunitas Pengembangan Masyarakat sebagai Upaya Pemberdayaan Masyarakat (Ed.1)*. Jakarta: Rajawali.
Adi, I.R. (2012). *Intervensi Komunitas & Pemberdayaan Masyarakat Sebagai Upaya Pemberdayaan Masyarakat (Revisi Ed.)*. Jakarta: PT RajaGrafindo.
Adi, I.R. (2013). *Kesejahteraan Sosial (Pekerjaan Sosial, Pembangunan Sosial, Dan Kajian Pembangunan*. Jakarta: PT RajaGrafindo.
Alipour, S., Rahmati, F.S., & Akbarbeyki, M. (2015). "The role of citizenship education and participation in waste management". *Science Journal (CSJ)*, 36(3), 1–18.
Alkin, M.C. (2011). *Evaluation Essentials from A to Z*. New York: The Guilford Press.
Anbumozhi, V. & Kimura, F. (2018). Industry 4.0: What does it mean for the circular economy in ASEAN?". In: *Industry 4.0: Empowering ASEAN for the Circular Economy*. Jakarta: ERIA, pp. 1–35.
Asteria, D. & Heruman, H. (2016). "Bank Sampah Sebagai Alternatif Strategi Pengelolaan Sampah Berbasis Masyarakat di Tasikmalaya". *Jurnal Manusia dan Lingkungan*, 23(1), 136–141.
Auliani, R. (2018). Penentuan Kapasitas Optimal Bank Sampah dalam Rangka Peningkatan Keuntungan Pengelolaan Sampah di Kota Medan. Tesis Magister. Pengelolaan Sumber Daya Alam dan Lingkungan. Universitas Sumatera Utara, 183 p.
Babbitt, C.W., Gaustad, G., Fisher, A., Chen, W.Q., & Liu, G. (2018). "Closing the loop on circular economy research: from theory to practice and back again". *Resources, Conservation & Recycling*, 135, 1–2.
Badan Pusat Statistik (BPS). (2013). *Proyeksi Penduduk Indonesia 2010–2035*. Badan Pusat Statistik: Jakarta.

Bastein, T., Roelofs, E., Rietveld, E., & Hoogendoorn, A. (2013). *Opportunities for a Circular Economy in the Netherlands*. Netherlands: TNO.

Bandung City Environmental Management Board. (2014). Environmental Management (Solid Waste Management) in Badung City, Indonesia.

Bicket, M., S. Guilcher, M. Hestin, C. Hudson, P. Razzini, A. Tan, P. ten Brink, E. van Dijl, R. Vanner, E. Watkins & S. Withana. (2014), "Scoping study to identify potential circular economy actions, priority sectors, material flows & value chains". *Study Prepared for the EU Commission, DG Environment*.

Bryman, A. (2012). *Social Research Method*, 4th Edition. New York: Oxford University Press.

Chen, M. (1997). *A Guide for Assessing the Impact of Microenterprise Services at the Individual Level*. Assessing the Impact of Microenterprise Services (AIMS) Management System International, Washington DC.

Damanhuri, E. (2010). Recycling in Indonesia: Integration of formal-informal sectors issues. *Workshop of ERIA Working Group (3R Policy in Southest and East Asia)*, Kuala Lumpur.

Deutsche Gesellschaft für Internationale Zusammenarbeit (GIZ) GmbH. (2019). Rethinking Plastics – Circular Economy Solutions to Marine Litter.

Esa, M.R. (2016). "Developing strategies for managing construction and demolition wastes in Malaysia based on the concept of circular economy". *Journal of Material Cycles and Waste Management Springer Japan*, 18, 1–11.

European Comission. (2017). Circular Economy Strategy. https://ec.europa.eu/environment/circular-economy/index_en.htm. Accessed on 5 January 2020.

Farizal, R.A., et al. (2017). "Indonesia's municipal solid waste 3R and waste to energy program". *Makara Journal of Technology*, 21/3, 153–159.

Ferronato, N., et al. (2019). "Introduction of the circular economy within developing regions: A comparative analysis of advantages and opportunities for waste valorization". *Journal of Environmental Management*, 230, 366–378.

Firmasyah, A., et al.. (2016). "Inovasi Pengelolaan Sampah Berbasis Masyarakat". *Prosiding Seminar Nasional Hasil-Hasil PPM IPB 2016*, 184–197. ISBN: 978-602-8853-29-3.

Foster, G. (2020). "Circular economy strategies for adaptive reuse of cultural heritage buildings to reduce environmental impacts". *Resources, Conservation & Recycling*, 152, 104507.

Ghisellini, P., Cialani, C., & Ulgiati, S., (2016). "A review on circular economy: the expected transition to a balanced interplay of environmental and economic systems". *Journal of Cleaner Production*, 114, 11–32.

Hadi, D.W. (2018). "UE: Circular Economy Indonesia Jadi Harapan Global". http://ppid.menlhk.go.id/siaran_pers/browse/1585. Accessed on 5 January 2020.

Halimatussadiah, A. (2012). Social capital to strengthen collective action in environmental protection. *Conference as Poster Presentation: Presented in HDCA (Human Development and Capability Association)*, Jakarta, September, 5th–7th 2012.

Harrison, K.L. (2003). *Does Social Capital Count? The Case of Solid Waste Management and Public Safety in Dobsonville and Yeoville, South Africa*. Ann Arbor, MI: ProQuest LLC 2014.

Hofmann, F. (2019). "Circular business models: business approach as driver or obstructer of sustainability transitions?". *Journal of Cleaner Production*, 224, 361–374.

Hoornweg, D., et al. (2012). "What a waste: A global review of solid waste management". Urban Development Series: Knowledge Papers No.15. Washington, DC: World Bank.

Ife, J. (1995). *Community Development: Creating a Community Vision, Analysis, and Practice*. Melbourne: Longman Australia Pty. Limited.

Ife, J. (2009). *Human Rights from Below: Achieving Rights through Community Development*. New York: Cambridge University Press.

Ife, J. (2013). *Community Development in an Uncertain World (Vision, Analysis and Practice)*. Melbourne: Cambridge University Press.

Indartik, I., et al. (2018). "Penanganan Sampah Rumah Tangga di Kota Bandung: Nilai Tambah dan Potensi Ekonomi". *Jurnal Penelitian Sosial dan Ekonomi Kehutanan*, 15(3), 195–211.

Indonesia Ministry of Environment and Forestry. (2018). *Jurnal Penelitian Sosial Dan Ekonomi Kehutanan*, 15, 3.

Indonesian Presidential Regulation. (2017). National Policy and Strategy Household Waste Management and Waste Household Kind, No. 97.

Kalmykova, Y., Sadagopan, M., Rosado, L., (2018). "Circular economy – from review of theories and practices to development of implementation tools". *Resources, Conservation & Recycling*, 135, 190–201.

Kaza, S., et al. (2018). *What a Waste 2.0: A Global Snapshot of Solid Waste Management to 2050*. International Bank for Reconstruction and Development/The World Bank: Washington, DC.

Kementerian Lingkungan Hidup dan Kehutanan (KLHK) Republik Indonesia. (2018). "Household waste management in Bandung City: Added value and economic potential." *Jurnal Penelitian Sosial Dan Ekonomi Kehutanan*, 15(3), 195–211.

Khatimah, H. (2018). *Evaluasi Program Pengelolaan Bank Sampah Unit Mawar Merah di Kelurahan Tamalanrea Indah, Kecamatan Tamalanrea Kota Makassar*. Makassar: Universitas Hasanuddin.

Kholil, A., et al. "Waste management based on 3R in Mutiara Waste Banks Bekasi City Indonesia". *World Environment 2018*, 8(3), 71–76. DOI: 10.5923/j.env.20180803.02.

Kirchherr, J., Reiker, D., & Hekkert, M. (2017). "Conceptualizing the circular economy: An analysis of 114 definitions". *Resources, Conservation & Recycling*, 127, 221–232.

Kotari, C.R. (2004). *Research Methodology Methods and Techniques*. New Delhi: New Age International (P) Limited Publisher.

Kumar, R. (2011). *Research Methodology*. Thousand Oaks, CA: Sage Publication.

Kusuma, L.T.W.N. et al. (2015). "The Quality Function Deployment (QFD): Based Evaluation of Malang Waste Bank Management". *Proceedings of the 2015 IEEE IEEM*, Singapore.

Kusumowibowo, T.S. & Tri, J.W.A. (2019). "Circular economy model of Indonesian construction industry waste based on systems dynamics". *Advances in Engineering Research, 187. Third International Conference on Sustainable Innovation 2019 – Technology and Engineering (IcoSITE 2019)*. Atlantic Press.

Laurenti, R., Singh, J., Frostell, B., Sinha, R., & Binder, C. (2018). "The socio-economic em-beddedness of the circular economy: an integrative framework". *Sustainability*, 10(7), 2129.

Lee, J.A.B. (2001). *The Empowerment Approach to Social Work Practice: Building the Beloved Community Second Edition*. New York: Columbia University Press.

Lyons, Y., et al. (2019). *A Review of Research on Marine Plastics in Southeast Asia: Who Does What?*. Singapore: CIL NUS.

Material Presentation by the Director of Waste Management, Directorate General of Waste, Waste and B3 Management. (2017). "Waste Bank and Main Waste Bank Policy in Waste Management". Ministry of Environment and Forestry (KLHK) Republic of Indonesia: Banjarmasin.

Material Presentation by the Director of Waste Management, Directorate General of Waste, Waste and B3 Management. (2017). "Main Waste Bank Policy". Ministry of Environment and Forestry (KLHK) Republic of Indonesia: Jakarta.

Material Presentation by the Director of Waste Management, Directorate General of Waste, Waste and B3 Management. (2018). "National Waste Management Policies and Strategies". Ministry of Environment and Forestry (KLHK) Republic of Indonesia: Semarang.

Morseletto, P. (2020). "Targets for a Circular Economy". *Resources, Conservation & Recycling*, 153, 104553.

Mura, M. et al., (2020). "Circular economy in Italian SMEs: A multi-method study". *Journal of Cleaner Production*, 245, 118821.

Murray, A., Skene, K., & Haynes, K. (2017). "The circular economy: An interdisciplinary ex-ploration of the concept and application in a global context". *Journal of Business Ethics*, 140(3), 369–380.

Mustofa, M.U. (2016). "Deradikalisasi Semu: Strategi Derutinisasi Penanganan Sampah Analisis Strukturasi dalam Isu Penanganan Sampah di Kota Bandung Oleh Walikota Periode 2013–2018". *Jurnal Wacana Politik*, 1(2), 152–165.

Nieuwenhoven, M.V. (2018). "From waste management to circular economy in Indonesia context". Holland Circular Hotspot, Kedutaan Besar Belanda: Jakarta.

Nurdiana, J., Michiel, H., & Maria, L.F. (2019). *Mapping Indonesia's Transition Towards Circular Economy*. Mexico: Memorias De Congreso Nacional AMICA.

Pargal, S., et al. (1999). *Social Capital in Solid Waste Management: Evidence From Dhaka, Bangladesh*. Washington, DC: The World Bank.

Periathamby, A. & Tanaka, M. (2014). *Municipal Solid Waste Management in Asia and the Pacific Islands: Challenges and Strategic Solutions*. Singapore: Springer.

Plan Miljoee & ESP3 DANIDA (2018) "Best Indonesian Practices in Circular Economy and Solid Waste Management".

Popple, K. & Stepney, P. (2008). *Social Work and the Community: A Critical Context for Practice*. New York: Macmillan International Higher Education.

Porter, M. & Kramer, M. (2006). "Strategy & Society: The Link between Competitive Advantage and Corporate Social Responsibility". *Harvard Business Review*, 84, 78e92.

Posavac, E.J. (2015). *Program Evaluation: Methods and Case Studies*. Boston, MA: Pearson Education Inc.

Pribadi, F.S. (2017). *The Integration of Circular Economy and into Municipal Waste Management in Metro City*. Indonesia: University of Twente.

Purba, H.D. Meidiana, C., & Adrianto, D.W. (2014). "Waste Management Scenario through Community Based Waste Bank: A Case Study of Kepanjen District, Malang Regency, Indonesia". *International Journal of Environmental Science and Development*, 5(2), 1–5.

Puspita, M. (2016). *Analisis Strategi Pengelolaan Sampah Kota Bandung Melalui Bank Sampah (Studi Kasus: Bank Sampah Resik)*. Depok: Universitas Indonesia.

Puspitawati, Y. & Rahdriawan, M. (2012). "Kajian Pengelolaan Sampah Berbasis Masyarakat dengan Konsep 3R di Kelurahan Larangan Kota Cirebon". *Jurnal Pembangunan Wilayah dan Kota*, 8(4), 349–359.

Qodir, A. (2011). *Analisis Kelembagaan dalam Upaya Pembangunan Kesejahteraan Masyarakat (Studi Kasus Peranan Koperasi Jasa Keuangan dalam Pelaksanaan Program Pemberdayaan Ekonomi Msyarakat Kelurahan di Kelurahan Kebon Kosong Kecamatan Kemayoran)*. Depok: Universitas Indonesia.

PD Kebersihan Kota Bandung. (2015). *Performance Report of Regional Cleanliness Company of Bandung City (PD Kebersihan Kota Bandung) Year 2015*. https://ppid.bandung.go.id/knowledgebase/laporan-kinerja-pd-kebersihan-tahun-2015/. Accessed on 10 August 2016.

Reike, D., Vermeulen, W.J., & Witjes, S. (2018). "The circular economy: new or refurbished a CE 3.0?—exploring controversies in the conceptualization of the circular economy through a focus on history and resource value retention options". *Resources, Conservation & Recycling*, 135, 246–264.

Rinaldoa, E. & Ami Yayuk S.R. (2019). "Penanganan Sampah Secara Kolaboratif antara Masyarakat dan Petugas Penanganan Prasarana dan Sarana Umum (PPSU) Tingkat Kelurahan (Kasus Penanganan Sampah di Kelurahan Jembatan Lima Kecamatan Tambora Kota Administrasi Jakarta Barat". *Jurnal Inspirasi BPSDM Provinsi Jawa Barat*, 10(1).

Rizos, V., et al. (2015). "The circular economy: Barriers and opportunities for SMEs". *GreenEcoNet, a Project Financed by the 7th EU Framework Programme for Research*, CEPS Working Document, ISBN 978-94-6138-479-9

Rogers, J., Jalal, J., & Boyd, J. (2008). *An Introduction to Sustainable Development*. London: Earthscan.

Rubin, A. & Babbie, R. (2011). *Research Method for Social Work*. USA: Linda Schreiber.

Salim, R. (2013). "Waste Not, Want Not: 'Waste Banks' in Indonesia," The World Bank, [Online]. Available: http://blogs.worldbank.org/eastasiapacific.

Sarani, M.R., et al. (2017). "The concept of right and its three generation". *International Journal of Scientific Study*, 5, 1–9.

Sarani, M.R., et al. (2017). "The concept of right and its three generation". *International Journal of Scientific Study*, 5, 37–41.

Satori, M. (2020). "Peran Bank Sampah dan Usaha Sektor Informal dalam Mewujudkan Circular Economy di Indonesia".

Singh, J. & Ordonez, I. (2015). "Resource recovery from post-consumer waste: Important lessons for the upcoming circular economy". *Journal of Cleaner Production*, 134, 342e353.

Sistem Informasi Pengelolaan Sampah Nasional (SIPSN) Kementerian Lingkungan Hidup dan Kehutanan. (2019). 20 November 2019. https://sipsn.menlhk.go.id/sipsn/public/data/timbulan.

Small, P. (2018). Four Differences between Research and Program Evaluation. https://managementhelp.org/blogs/nonprofit-capacity-building/2012/01/08/four-differences-between-research-and-program-evaluation, 20 September 2018. Accessed on 20 September 2018.

Suharto, E. (2005). Kebijakan Sosial, Sebagai Kebijakan Publik. Bandung: Alfabeta.

Suparmini, Purnawan Junaidi. (2018) "The role of waste bank partnership in efforts to decrease waste volume in urban: A case study at a Waste Bank in Kalibaru, Cilodong, Depok City". *IOP Conference Series: Earth and Environmental Science*, 126, 012125.

Suryani, A.S. (2014). "Peran Bank Sampah dalam Efektifitas Pengelolaan Sampah (Studi Kasus Bank Sampah Malang)". *Aspirasi*, 5(1), 71–84.

Sutomo. (2013). *Community-Driven Waste Management: How Sustainable are Waste Banks in Yogyakarta?*. Yogyakarta: Universitas Gadjah Mada.

Tisseran, A. et al. (2017). "Solid waste and the circular economy: A global analysis of waste treatment and waste footprints". *Journal of Industrial Ecology*, 21(3), 628–640.

United Nations. (2015). A/Res/70/1. *Transforming Our World: The 2030 Agenda for Sustainable Development*.

Warren, R.L. (1979). *The Community in America*. Chicago: Rand McNally.

Winarso, H., dan Larasati, A. (2011). "Dari Sampah Menjadi Upah: Inovasi Pengolahan Sampah di Tingkat Akar Rumput Kasus Program Bank Sampah "Sendu" di Kelurahan Pasar Minggu Jakarta Selatan". *Jurnal Manusia dan Lingkungan*, 18(1), 43–59.

Widiyanto, A.F. & Rahab, R. (2017). "Community participation in bank of garbage: Explorative case study in Banyumas regency". *Masyarakat, Kebudayaan dan Politik*, 30(4), 367–376.

Wijayantia, D.R. & Suryania, S. (2015). Waste bank as community-based environmental governance: A lesson learned from Surabaya. *Procedia - Social and Behavioral Sciences* 184, 171–179.

Wulandari, D. et al. (2017). "Waste bank: Waste management model in improving local economy". *International Journal of Energy Economics and Policy,* 7(3), 36–41.

Yuliarso, M.Z. & Purwani, D.A. (2018). "Perubahan Sosial Masyarakat melalui Gerakan Bank Sampah Gemah Ripah: Kajian dI Desa Badegan Kabupaten Bantul Yogyakarta". *Jurnal Agrisep*, 17(2), 207–218.

Yustiani, Y.M. & Abror, D.F. (2019). "Operasional Bank Sampah Unit dalam Pengelolaan Sampah Perkotaan". *Jurnalis*, 2(2), 82–29.

Zahra, M. (2017). *Presiding over a Flood of Waste: A Case Study on a Local Movement for Domestic Waste Management at the Household Level, Bandung City, Indonesia*. Sweden: Lund University.

Zakianis, Z. & Pratiwi P. (2018). "Citizens' participation in household solid waste management and monitoring of household solid waste separation in Kelurahan Abadijaya, Kecamatan Sukmajaya". *ASEAN Journal of Community Engagement*, 2(2), 221–236.

12

Sustainable Methodology for Mechanised Segregation of Legacy Waste and Reclamation of Landfill Area

Harish Hirani
CSIR-Central Mechanical Engineering Research Institute

CONTENTS

12.1 Introduction

The solid waste (SW) collected from households, markets, institutions, commercial establishments and roadside sweeping must be transferred to the treatment plants and/or landfills. With the ever-increasing population and urbanisation, the volume of generated SW is continuously increasing, requiring additional huge chunks of precious land per year, which is a huge problem for developing countries like India. With a continuous increase in waste generation, many municipalities simply dump the waste, as illustrated in Figure 12.1 and label those dumpsites as landfills, which is becoming a major societal challenge.

Unscientific landfilling is a threat to water supplies when water percolates through waste, picking up a variety of substances such as metals, minerals, organic chemicals, bacteria, viruses, explosives, flammables and other toxic materials. This contaminated water is called leachate and produced when the waste becomes saturated with water (USEPA Office of Water, 1994). Leachate generated from landfill sites affects surface water quality through surface runoff and groundwater quality in the adjacent areas through percolation into the subsoil (Hossain et al., 2014). Similar observations were made by a study where the surface water quality from surrounding ponds of a closed dumping site (Maiti et al., 2016). It was found that the Total Dissolved Solids (TDS) concentration in the surrounding water bodies was high because of the leaching of pollutants into the water. Concentrations of sulphate and chloride were also higher, which could be attributed to surface water contamination. The higher Biological Oxygen Demand (BOD) and Chemical Oxygen Demand (COD) values indicated the presence of organic matter in water. Naveen et al. (2018) conducted a survey on contamination of surface and groundwater bodies by leachate leakage from a landfill. In this study, the researchers found that municipal solid wastes (MSWs) were deposited in unscientific landfills that had resulted in steep and unstable slopes, and leachate runoff into nearby water bodies such as ponds and open wells. High concentrations of organic, inorganic pollutants and heavy metals were found in these water samples making them unsuitable for drinking.

DOI: 10.1201/9781003231608-14

FIGURE 12.1   Unscientific dumping of solid waste.

This landfill area was constructed with liners to restrict leachate, but unsegregated and mismanagement of landfill led to punctures in the liner system in the landfill.

Considering these aspects, it can be said that the traditional economic management process of dumping waste in landfills is completely unsustainable.

Most of ULBs (local bodies that administer a city/town of specified population) spend about Rs. 500 to Rs. 1,500 per tonne on SW collection and transportation, but hardly any amount is spent on scientific treatment and disposal of waste. In absence of scientific methods to process and dispose-off SW (as shown in Figure 12.1), environmental, ecological and health implications for human society emerge.

Inadequate number of waste-processing plants and insufficient landfills pose problems of partial collection (<70%) of SW and a rise in the number of overflowing containers, thereby causing foul-smelling waste litter on streets, increasing ground-level pollution and frequently choking of water drainage system. There are numerous alternatives available for designing landfills, which allow complete separation of biodegradable materials from the environment by using appropriate technology or extracting landfill gas (LFG) so that air pollution reduces and lessens the spread of doors from the landfill.

LFG typically contains majorly methane and carbon dioxide, and small amounts of ammonia, sulphides, hydrogen, carbon monoxide, etc. (www.atsdr.cdc.gov). LFG can be extracted from engineered landfills using a series of wells and a blower vacuum system and collected to a central point where it can be treated depending on the ultimate use for the gas (www.epa.gov). However, the collection of LFG from an unscientifically managed active landfill site is a challenging task. Kashyap et al. (2016) has demonstrated the recovery of LFG from an unscientifically managed MSW landfill site at Ghazipur, Delhi. The feeder pipe collected LFG from the well and transferred the same to the header pipe.

Such plants enable transforming an open dump into a controlled landfill where gas energy can be collected and the risks to the environment are minimised. Similarly, appropriate treatment of leachate can be used to reduce the negative impacts on underground water. However, due to financial limitation and huge associated capital expenditure, most of the ULBs refuse to improvise existing landfills into sanitary landfills, which contains methane collection and leachate-treatment plants (LTPs). Few ULBs, who have implemented such scientifically designed plants experienced failure due to issues related to the unsustainability of collection and then usage of methane, relatively high cost of leachate-treatment plant and lack of sufficient knowledge/manpower to operate/maintain the sanitary landfills. It is worthy to note that cities having the tag of "Smart City" do not have an appropriate SW management system.

It is worth mentioning that three Indian cities, viz. Indore, Surat and Navi Mumbai were ranked 1st, 2nd and 3rd, respectively, in Swachh Survekshan 2020 by MoHUA, Govt. of India. Indore generates about 1,115 MT of SW daily of which close to 90% is collected. Source segregation of different types of SW has been made mandatory to ensure the waste-to-wealth concept. The collected waste is taken to ten stations across the city, where the waste workers ensure that the collected waste is properly segregated. The waste is then taken to the waste-processing facility. About 550 MT of organic waste which is generated daily is converted to compost and sold to farmers as manure. Recyclable waste is given to companies that use recycled material. The non-recyclable waste is processed in the cement industry as fuel or is used in the construction of roads. The construction waste is crushed and used as raw material for making bricks, paver tiles and material to edge footpaths. Further details of MSW management in the city of Indore can be found in Singh (2021) (www.indiaspend.com; www.smartcityindore.org).

Surat generates about 2,200 MT of SW daily. The collected waste is segregated in the centralised waste-processing treatment plant at Khajod. The segregated wet organic material consisting of 30%–35% of the total waste is processed through composting, the dry organic material of about 35%–40% is used in RDF making and the rest material of about 3%–5% consists of plastic, MS and rubber items are sent for recycling. The construction waste is recycled to various construction materials such as sand, block and paver block. Further details of MSW management in the city of Surat can be found in www.suratmunicipal.gov.in.

The city of Navi Mumbai collects about 700 MT of SW daily on door-to-door basis (www.nmmc.gov.in). The segregated waste is collected by local civic authorities and the wet waste is processed through composting (www.hindustantimes.com). Bulk waste generators with waste generation >50 kg/day are required to process their waste within their own premises (www.thehindu.com).

It is worth noticing that problems related to waste management are many. The capacity limitations of municipal corporations, financial constraints, insufficient formal manpower and lack of awareness among the people are the main hurdles in effective management of the SW. Therefore, the purpose of the present chapter is to provide a comprehensive methodology involving the public and encouraging a more decentralised approach to collect, segregate, transport, process and safely dispose of the generated SW. An integrated concept of (technical, economical and social) SW management is proposed.

12.2 Sustainable Methodology

One of the major problems of landfill form of waste management is dumping of organic wastes, which is a large fraction (40%–60%) of SW. Unscientifically disposal of organic waste in landfill sites causes:

- **Greenhouse Gas (GHG) Emissions and Smoke**: Degradation of organic waste releases highly combustible methane gas into the atmosphere. Methane contributes to global warming. Sometimes fire breaks out at landfill and becomes difficult to extinguish because of the availability of plastic and other highly flammable substances in the landfill and pollute air quality by releasing carbon monoxide, hydrocarbons, particulate matter, nitrogen oxides and dioxins/furans. Health problems related to various emissions from landfills include high PM10 (particulate matter consisting of inhalable particles, with diameters that are generally $10\,\mu m$ and smaller.) exposure, breathing problems, bacterial infections, asthma, elevated cardiovascular risk, eye irritation, stomach problem and other infections (Swati et al., 2018).
- Breeding hubs for pathogens, bacteria, flies, rodent and pests and becoming the root of many diseases. Liquid seeping through the degrading organic material, called leachate, is toxic and causes land and groundwater pollution. There are concerns regarding the flow of toxins in the food chain of birds and animals (Swati et al., 2018).

This means ordinary SW collected without separation and brought directly to the landfill site requires: technology to collect biogas released from SW, efficient usage of the collected biogas and in-situ LTP. These plants at landfill sites will increase the initial as well as running including maintenance costs. Further recovery cost from collected biogas and recyclable metal and plastic is negligible. The cost of waste collection, transportation, landfilling, gas collection and leachate treatment makes landfilling option of SWM unsuitable as it poses a serious threat to health and the environment.

One of the major findings from the literature is that if organic wastes (40%–60% of SW) are separated at the source and processed the same by "Composting", fertilisers can be produced. Unfortunately, scaling up the "composting" process does not yield socio-economic returns as it requires more land space, more labour and pasteurisation for effective disinfection (pib.gov.in, 2020). During the rainy season, managing composting becomes difficult.

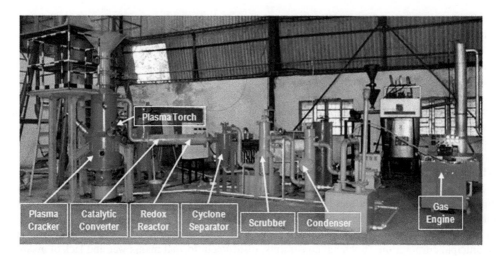

FIGURE 12.2 Plasma gasification plant.

One of the sustainable methods to treat waste is to convert it into cost-effective renewable energy sources. Safe disposal of SW is possible by utilising a plasma arc gasification process (shown in Figure 12.2) in which a large volume reduction of waste up to 95% is possible (Hazra et al., 2019; Byun et al., 2012; Seo et al., 2018). This process can be labelled as an integrated Municipal Solid Waste disposal system (iMSWDS) since complete waste is processed to high-temperature plasma arc, produced inside the furnace by high-intensity current (~30 V and 250 A) using plasma torches that breaks down the SW into elemental form. The produced syngas is cleaned using a series of gas purification system comprising of catalytic converter, redox reactor, cyclone separator, and scrubber and injected in gas engines for the generation of electricity (Byun et al., 2011).

This method for the treatment of solid waste offers the following advantages:

a. High energy density and temperature inside the furnace resulting in fast reaction times, thus allowing a large volume of waste to be treated within a relatively small furnace.

b. High heat flux density at the reactor boundary leading to rapid achievement of steady-state condition resulting in a reduction in start-up and shutdown times as compared to thermal incineration.

c. Energy generation from the produced syngas comprising mainly of CO and H_2 in steam power plants leading to economic benefit.

d. Lesser emission of pollutants in the atmosphere like oxides of nitrogen and sulphur because the gasification occurs in the absence of oxygen.

e. Large volume reduction of SW is possible thereby reducing the demand for land required for the dumping of SW in landfill sites.

The residual vitrified slag after cooling can be utilised as construction materials (Byun et al., 2010). However, despite several advantages as listed above in (a) – (e) this technology has not gained much popularity in India due to the high capital and recurring costs. This technique requires a very high electrical energy input (~1.5 kWh/kg of waste processed for small plants <100 MT capacity) and ~1.2 kWh/kg of waste processed for plants with >100 MT capacity). In addition, the high rate of electrode consumption (~500 mg/kg of waste processed) further leads to an increase in recurring expenses making the process expensive (pib.gov.in, 2020).

To make a cost-effective CE-iMSWDS, first it is necessary to segregate the MSW and then create value-added end-products by processing each type of waste. In addition, the operational cost of a plant can be reduced by the installation of roof-mounted solar panels, for providing the energy requirements of the system and which can also feed the surplus energy supply onto a mini-grid (www.drishtiias.com).

FIGURE 12.3 Decentralised waste management plant of 1.0 tonne/day capacity.

To save associated transportation costs (negligible transport to reduce the burden of imported diesel) and environmental (nitrous oxide, particulates, CO_2) pollution, the decentralised plant (shown in Figure 12.3) of 0.5–5.0 ton/day capacity can be installed in various localities. Such plants are equipped with disinfection capabilities to help break the COVID (or similar disease) chain through UV-C lights (wavelength of 254 nm) and hot-air convection methods. Such plants open up the opportunities to fulfil the dream of city with "Zero Waste and zero landfill", become "Source of Job Creation", provide support to micro-enterprises and bring sustainability (pib.gov.in, 2020).

The CE-iMSWDS must follow a scientific way for disposal of SW. The primary focus of CE-iMSWDS is to segregate the common municipal garbage using segregation techniques. The mechanised segregation system separates SW into categories such as metallic wastes, biodegradable wastes, non-biodegradable wastes and inert wastes (pib.gov.in, 2020). Few cities, where segregation at source has been imposed seriously, may not require segregation facilities for daily collected waste, but to make "Zero landfill", those cities also require "Segregation of legacy waste collected from landfills". It is worthy to note that there are a number of rag pickers, who scrounge the landfills for segregating and selling off the waste materials as a profession. An approach to engage those rag pickers in training and subsequently operating "Mechanised segregation of legacy waste collected from landfills" and allowing them to sell the recyclables they collected will be beneficial to society and provide a Robust Ecological Model. Such an approach will substantially increase recycling of wastes.

12.2.1 Mechanised Segregation of Legacy Waste

The legacy waste may contain metals, fines (degraded organic waste), polymers and inert materials. To segregate the legacy waste, a vibrating feeder chute having an opening from the top can be utilised (pib. gov.in, 2020).

The material fed to the chute passes through a rotary trommel, having 8-mm pores all over its screen, to separate the dust or biodegraded residue from the waste. The material, having a size more than 8 mm, is then collected in the drum of the inclined bucket conveyor and then send to the Eddy Current Separator. Then the material will be passed through a rotary magnetic drum separator. The iron components (Fe) will be separated out and collected in a hopper for re-use. The rest material will be passed through an Eddy Current Separator, and non-ferrous metallic parts (Al, Cu) will be separated out. After this separation, the rest of the material (Plastics, polymer, papers, etc.) is fed inside the air separation unit via an inclined bucket conveyor. Air separation will be used to separate the light fractions (plastic, polymer waste) from the heavy fractions (inert materials). The materials will be pushed into the chamber for gravity falling from a height using conveyor and conical hopper. The chamber will be connected to two

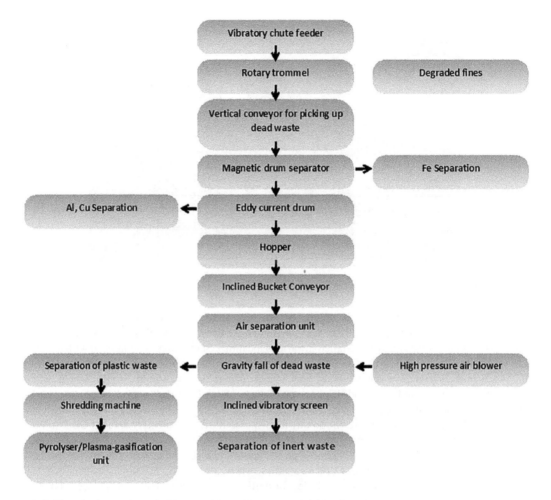

FIGURE 12.4 The process flow diagram of the mechanised segregation unit (legacy waste).

nos. of air blower to collect the lighter particle (plastic, paper, etc.), whereas the heavier particle (inert, etc.) will fall down due to gravity. Then the material will be passed through an inclined vibratory screen of perforation size 80 mm×80 mm where the heavier particle (inert etc.) will fall through the perforated screen and the flat product (plastic, paper, diaper, sanitary pads, etc.) are collected over the screen. The inclination and vibration of the screen help in travelling the collected material to the attached hopper in front of the screen. The polymer waste can be directly fed into the shredder. The shredded material will be sent to the pyrolysis process. The process flow diagram of the mechanised segregation unit is shown in Figure 12.4. The fabricated mechanised segregation unit of legacy waste of capacity 100 kg/h has been shown in Figure 12.5.

12.2.2 Disposal of Biodegradable/Biodegraded Waste

The biodegradable waste decomposes with time, by releasing carbon dioxide, methane, nitrous oxide, etc. In a few days' time, the weight of biodegradable material becomes half as it releases GHG gases in the atmosphere if not treated scientifically. But with the usage of appropriate technology, biogas liberated from biodegradable can be collected, stored and used as fuel for cooking purposes. The gas can also be utilised in gas engine for the generation of electricity. The residual slurry from biogas plant is converted to compost in a natural process known as vermi-composting by introducing earthworms (pib. gov.in, 2020). In other words, biodegradable waste is harmful if not treated appropriately, but it becomes

FIGURE 12.5   Mechanised segregation unit for legacy waste at CSIR-CMERI (capacity: 100 kg/h).

FIGURE 12.6   Small-scale bio-methanation plant.

a wealth with the usage of appropriate technology. For the best results, the movement of biodegradable material should be minimised. This is the main reason for requesting vegetable markets, restaurants, food courts and other sources having separate organic waste to employ small-scale bio-methanation plant (shown in Figure 12.6).

In addition, biomass waste such as dry leaves, dead branches, dry grass, etc. can be disposed of energy efficiently by first shredding it to a suitable size followed by mixing it with the slurry of the biogas digester to make a briquette that can be utilised as fuel for cooking. The ash produced from burning of briquette is mixed with cement and water in an appropriate proportion for the production of bricks (pib. gov.in, 2020).

Another usage of briquettes is utilisation in gasifiers for the production of syngas (mainly a mixture of CO and H_2). The syngas has a calorific value equivalent to that of biogas and hence can be utilised for the generation of electricity (pib.gov.in, 2020).

12.2.3 Disposal of Polymeric Waste

The polymer waste, consisting of polybags, plastics bottles/containers, HDPE, PVC, sanitary waste, etc. can safely be disposed-off through two main processes, i.e. pyrolysis and plasma gasification. As it was mentioned earlier in the present manuscript that though plasma gasification safely disposes off the SWs, this technology is not economically viable. Therefore, the pyrolysis process is preferable as far as possible, and only in extreme situations when other technologies cannot provide a safe technical solution, plasma gasification should be used.

In the pyrolysis process, the polymer waste is converted to syngas, oil and char by heating it in temperature range from 400°C to 800°C in an anaerobic environment in presence of suitable catalyst. As a result of heating, the volatile matters from the polymer evaporate which on condensation provide pyrolysis oil. The non-condensed syngas and pyrolysis oil can be reused for heating purposes to achieve energy-efficient pyrolysis process. The solid residue, known as char, can be mixed with the biogas slurry for the production of briquette. These briquettes are being utilised for heating the reactor of the pyrolysis process thereby making the process self-sustainable in terms of fuel. Different heating mechanisms such as solid fuel based, liquid fuel based and gaseous fuel based in standalone mode or in combined mode can be developed for carrying out the process (pib.gov.in, 2020).

The sanitary items including masks, sanitary napkins, diapers etc. (which constitute <5% of MSW) can be disposed of utilising high-temperature plasma. The advantages of using combination of pyrolysis and plasma treatment are:

- Safe disposal of polymeric waste
- Energy recovery system—contributing towards saving of resources
- Significant reduction in the volume of waste streams, which reduces the need for landfills to almost nil.

12.2.4 Management of C&D Waste and Reclamation of Landfill Area

With an ever-increasing rise in construction activities throughout the developing nations, the quantity of construction and demolition (C&D) of waste is continuously increasing. In absence of appropriate processing technology, C&D wastes are dumped into water bodies and wetlands around urban centres. The dumping of these wastes also blocks the natural permeation of rainwater from the surface to recharge the groundwater table.

To manage the disposal of C&D debris, waste can be crushed in a jaw crusher and then segregated in different pore sizes (1–5 mm) in the trommel. The finer particles can be mixed with cement and water in appropriate proportion (pib.gov.in, 2020). The prepared mixture is then filled in the mould cavity of the brick press. Then it may be pressed using a hydraulic press to give it a proper shape and compaction. The bricks are cured to achieve the desired strength. Finally, the prepared bricks are tested for engineering properties. The common bricks have a minimum crushing strength of 3.5 N/mm². The crushing strength (compressive strength) of the bricks prepared by CMERI i-MSWDS is more than 3.5 N/mm².

It is worth noting that most (60%–70%) of the waste accumulated in landfill contain fines (particle size <8mm), as shown in Figure 12.7. On analysis, it has been found that the residual slurry of landfill cannot be used as organic manure as it contains heavy metals (indicated in Table 12.1) like phosphate, chromium, etc. which are harmful to agriculture. However, such residual slurry can be used for the preparation of concrete, which allows saving of sands and aggregates, required for concrete.

Natural sand is an important constituent in any civil construction work. Tonnes of sands are excavated from river beds and transported daily to the urban cities to meet the demand of booming construction industries and real estate demand. This continuous extraction of natural resources leads to natural imbalance like soil erosion and flood. In some regions, river sand is so over-exploited that it endangers the stability of riverbanks. To overcome this exploitation of natural resources, there is a need to explore and utilise the remains of legacy waste for use as a substitute for natural sand in construction industries. The physical properties of legacy wastes like bulk density, specific gravity, particle size and texture are comparable with natural sand that make it more lucrative for use as an alternative to sand. This not only helps to clear the heap of legacy wastes, free the environment from landfill but also saves exploitation of natural river sand. Further, to reclaim the landfill area, first it is necessary to remove the waste accumulated over the years and segregate the same into different components, and to devise appropriate technology for disposal of different components of waste with maximum resource recovery. The mechanised segregation system for legacy waste must have the following features:

- Waste segregation efficiency higher than 95%.
- UV-C lights installed in the segregation system ensure the first line of disinfection of waste.
- Moisture removal from waste with hot-air convection at temperatures of 65°C–70°C utilising waste heat from auxiliary pyrolysis process ensures second line of disinfection of waste and helps in achieving higher segregation efficiency.

One method of reclaiming landfills is to start making boundaries (as shown in Figure 12.8) for landfill using concrete by recycling landfill waste. A large portion of legacy waste fines can be used for roads

FIGURE 12.7   Waste collected from landfill.

TABLE 12.1

Test Results of Waste Collected from Landfills and Segregated Using Trammel

Sl. No.	Test Name/Parameters	Test Results	Limits[a]
1	Colour	Dark brown	
2	Odour	Absence of foul smell	
3	Particle size	93% passes through 4 mm IS sieve	Minimum 90% material should pass through 4.0-mm IS sieve
4	pH	7.92	6.5–7.5
5	Bulk density (g/cm²)	1.12	<1.0
6	Moisture, % by mass	13.40	25.0
7	Total organic carbon, % by mass	5.36	14.0
8	Total nitrogen (as N), % by mass	0.40	0.5
9	Total phosphate (as P_2O_5), % by mass	0.89	0.5
10	Total potassium (as K_2O_5), % by mass	0.45	0.5
11	C/N ratio	13.4	<20
12	Arsenic (as As), mg/kg	0.0017	10.0
13	Cadmium (as Cd), mg/kg	1.17	5.0
14	Chromium (as Cr), mg/kg	62.61	50.0
15	Copper (as Cu), mg/kg	50.59	300.0
16	Lead (as Pb), mg/kg	38.92	100
17	Mercury (as Hg), mg/kg	0.10	0.15
18	Nickel (as Ni), mg/kg	22.72	50.0
19	Zinc (as Zn), mg/kg	39.60	1000.00

[a]   The limit is published in The Gazette of India: Extraordinary [PART II-SEC. 3(ii)] and FCO Standard.

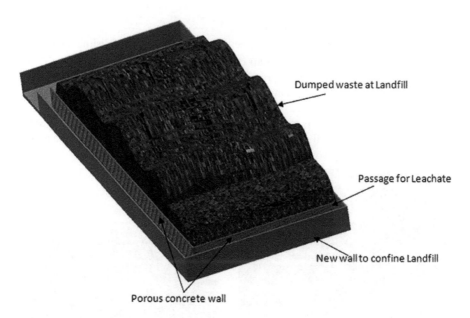

FIGURE 12.8   Reclaiming landfill.

required near landfill areas. It is worthy to note fines from legacy waste contain heavy metals, which are beneficial for concrete and the cement used for concrete. It helps to minimise the leaching of heavy metal entering underground water. The leachate thus can be innovatively augmented by creating a boundary of porous concrete.

Case Study

To understand the cost-effectiveness of the proposed iMSWDS, the landfills of Shankarpur (Durgapur City, West Bengal, India) were considered. Approximately 150 tonnes of mixed waste per day is dumped in Shankarpur Landfill site with the help of 25 waste dumping vehicles (14 dumpers, 4 tractors and other vehicles). The garbage is collected from around 450 concrete VAT, 125 single steel VAT and 220 double steel VAT (localised waste dumping sites). The transportation route of one of the dumping vehicles is shown. On average, the travel distance per trip is ~15 km. Approximately 60,000–80,000 L of diesel per year is required for transportation of the wastes. The approximate cost related to transportation of waste is listed in Table 12.2.

Two major drawbacks of dumping total solid in landfills are:

TABLE 12.2

Cost Associated with Transportation of Waste

Sl. No.	Description	Annual Expenditure (Rs)	Remarks
1	Cost of consumables	70 lakhs	Diesel consumption (monthly consumption: ~6,000 L)
2	Cost of maintenance	15 lakhs	No. of vehicles under operation – 30 (maintenance, registration, insurance, etc.)
3	Cost of manpower	100 lakhs	Manpower engaged in transportation (driver, helper and loader) – 60
Total expenditure		185 lakhs	

FIGURE 12.9   Waste dump at Chandidas Market, Durgapur.

- Waste collected from the cities appears to be at an advanced stage of degradation, as shown in Figure 12.9, which means a major portion of CO_2 and some amount of CH_4 is released before the garbage is transported to landfill.
- Transportation of the wastes to dumpsites consumes lots of fossil fuel, which in itself is an economic hindrance and polluting factor.

The situation demands the essentiality of processing the biodegradable waste at the local sites and promotes a decentralised model of waste processing, which might help in substantial curbing transportation costs and any chances of spillage/contamination as a result of the transportation of wastes. In addition, it saves the environment from GHG gases emissions.

There are two methods to effectively utilise biodegradable waste at a local level: first, by adopting biogas plant model and another by making biomass briquettes. Both the models are comprehensively sustainable (as illustrated in Table 12.3), since the generated biogas can be utilised for cooking rice, lentils, chapattis, etc. The biomass briquettes can also be utilised for cooking, heating, etc.

As per the data listed in Table 12.3, the average saving from de-centralised processing bio-waste/ward is around Rs. 1,500/day (5 lakhs per year) and the total profit generated from 25 wards will be more than 100 lakhs per year. In other words, the data provided in Table 12.3 clearly indicates that processing of 70% waste at local sites and dumping of 30% plastic+inert in landfills is self-sustainable (lesser pollution, processing of 250 TPD instead of 140 TPD, economic).

To experiment with the hypothesis, garbage from two markets, Benachity and Chandidas, was picked up to process the same. After the usage of the Mechanised Segregation System, the segregated output has been listed in Table 12.4. The economics of 100% waste processing is provided in Table 12.5, which clearly indicates that the processing of inert and plastic materials at a local site may not be sustainable.

TABLE 12.3

Decentralised Processing of Biodegradable Waste (10 ton/day ward)

Component	Qty (%)	Process	Qty (kg)	Output		Price (INR)	Expenditure		Costing (INR)
Bio-waste briquette	65-75	Biogas plant	6000	Biogas	300 cum	Rs. 7,500	Manpower	10 No	Rs. 6,000
		Dry biomass	1000	Briquette	1,000kg	Rs. 3,000	Electricity	200kW	Rs. 1,500
Recyclables	1-2	Resale	100			Rs. 1,000	Consumable	1 lot	Rs. 1,000
Plastics	15-20	Transport	1750				Transport (daily)	1 Truck	Rs. 500
Inert	10-15	Transport	1150				Maintenance (monthly)		Rs. 30,000
				Sub total		Rs. 11,500	Sub total		Rs. 10,000

TABLE 12.4

Waste Collected from Benachity and Chandidas Markets

Components	Chandidas Market (kg)				Benachity Market (kg)				Overall
	Day-1	Day-2	Total	%Avg	Day-1	Day-2	Total	%Avg	%Avg
Food, vegetable waste, green leaves	263	103	366	34.27	237	173	410	23.54	28.90
Banana stem, fruit stalk	16	28	44	4.12	48	68	116	6.66	5.39
Dry leaves	171	215	386	36.14	360	285	645	37.03	36.58
Plastics and paper	35	25	60	5.62	154	122	276	15.84	10.73
Rags and textiles	5	0	5	0.47	14	40	54	3.10	1.78
Rubber and leather	0	12	12	1.12	2	0	2	0.11	0.62
Fish wastes (scales and fins)	70	40	110	10.30	0	0	0	0.00	5.15
Inert (dust, earthen cups, stones)	40	45	85	7.96	114	125	239	13.72	10.84
	600	468	1068		929	813	1742		

TABLE 12.5

Economics of Processing of 2,800 kg

Revenue	Quantity	Price	Expenditure	Quantity	Cost
Biogas	50 cum Eqv. = 0.4*50 = 20 cum = 20 kg	Rs 1,500	Manpower Salary	04 persons (Rate: Rs 300/days) 4 days operation	Rs 4,800/-
Compost	550 kg	Rs 550	Electricity	35 kW/day 4 days operation	Rs 1,000/-
Briquette	100 kg	Rs 400	Consumables	Cement: 1 bag	Rs 400/-
Pyrolysis oil	100 L	Rs 1,500	Fuel	45 L (Rs 80/L)	Rs 3,600/-
Brick from inert	100 pcs	Rs 500			
Tipping fee	Rs 750/ton	Rs 2,250			
Recyclables	400 kg	Rs 3,000			
Total revenue		Rs 9,700/-	Total expenditure		Rs 9,800/-

Based on this case study, it can be deduced that the processing of biodegradable waste at local sites is a better option. It saves around 60% of the transportation cost and reduces the environmental pollution.

12.3 Conclusion

Technology-backed sustainable management of solid wastes has been presented. Since segregation of wastes at the point-of-origin is a challenge in the Indian Society, the segregation mechanism of the technology is ideal to handle mixed wastes. The segregation technology is essential for treating legacy waste. With such a technology, the urban land spaces will be relieved from the burgeoning pathogen-infested landfills and thus promoting healthy environs for communities. Innovative suggestions are also elaborated in the chapter in the form of providing a controlled passage for the leachate, through porous concrete, to prevent its leakage onto the adjacent water bodies.

The chapter is also a blueprint for integrating the participation of the local community by engaging them in the operations and maintenance of the system. The logistics associated with the transportation of wastes will be substantially reduced, thereby reducing expenditure on fossil fuels. The economic rationalisation achieved through the deployment of the technology is analysed and deducted through real-life data observed from field studies.

Acknowledgements

The author acknowledges the efforts of Dr. Amit Ganguly, Mr. Partha Das and Mr. Subho Samanta for their involvement in waste management projects, and the author also acknowledges the inputs provided by them for this article.

REFERENCES

Byun Y., Namkung W., Cho M., Chung J.W., Kim Y.S., Lee J.H., Lee C.R., Hwang S.M. (2010) Demonstration of thermal plasma gasification/vitrification for municipal solid waste treatment. *Environmental Science & Technology*, Vol. 44, pp. 6680–6684.

Byun Y., Cho M., Chung J.W., Namkung W., Lee H.D., Jang S.D., Kim Y.-S., Lee J.H., Lee C.R., Hwang S.M. (2011) Hydrogen recovery from the thermal plasma gasification of solid waste. *Journal of Hazardous Materials*, Vol. 190, pp. 317–323.

Byun Y., Cho M., Hwang S.M., Chung, J. (2012) Thermal plasma gasification of municipal solid waste (MSW). In Yun Y. (ed.) *Gasification for Practical Applications*. IntechOpen. DOI: 10.5772/48537.

CSIR-CMERI's Municipal Solid Waste Processing Facility, https://www.drishtiias.com/daily-updates/daily-news-analysis/csir-cmeri-s-municipal-solid-waste-processing-facility.

Hazra A., Das, S., Ganguly, A., Das, P., Chatterjee, P.K., Murmu, N.C. and Banerjee, P. (2019) Plasma arc technology: A potential solution toward waste to energy conversion and of GHGs mitigation. In: Ghosh S. (ed.) *Waste Valorisation and Recycling*. Springer, Singapore. DOI: 10.1007/978-981-13-2784-1_19.

Hossain M., Das S., Hossain M. (2014) Impact of landfill leachate on surface and ground water quality, *International journal of Environmental Science and Technology*, Vol. 7, Issue 6, pp. 337–346.

https://www.atsdr.cdc.gov/hac/landfill/html/ch2.html.

https://www.epa.gov/lmop/basic-information-about-landfill-gas.

https://www.hindustantimes.com/mumbai-news/navi-mumbai-municipal-corporation-targets-100-waste-segregation-by-year-end/story-U79dxOXbOWpq9eTgYXrg9L.html.

https://www.indiaspend.com/how-indore-became-indias-cleanest-city-and-how-others-can-follow/.

https://www.nmmc.gov.in/navimumbai/solid-waste-management-department-services.

https://www.smartcityindore.org/solid-waste/.

https://www.suratmunicipal.gov.in/Departments/SolidWasteManagementApproaches.

https://www.thehindu.com/news/cities/mumbai/swachh-survekshan-nmmc-to-conduct-mock-survey/article33711335.ece.

Kashyap R.K., Chugh P., Nandakumar T., (2016) Opportunities & challenges in capturing landfill gas from an active and un-scientifically managed land fill site: A case study, *Procedia Environmental Sciences*, Vol. 35, pp. 348–367.

Maiti S.K., De S., Hazra T., Debsarkar A., Dutta A. (2016) Characterization of leachate and its impact on surface and groundwater quality of a closed dumpsite: A case study at Dhapa, Kolkata, India, *Procedia Environmental Sciences*, Vol. 35, pp. 391–399.

Naveen B.P., Sumalatha J., Malik R.K. (2018) A study on contamination of ground and surface water bodies by leachate leakage from a landfill in Bangalore, India, *International Journal of Geo-Engineering*, Vol. 9, Issue 27, pp. 1–20.

Seo Y.-C., Alam M.T., Yang, Y.-C. (2018) Gasification of municipal solid waste. In Yun Y. (ed.) *Gasification for Low-grade Feedstock*. IntechOpen. DOI: 10.5772/intechopen.73685.

Singh, R. (2021) Municipal solid waste management in the city of Indore: A case study. *Journal of Civil Engineering and Environmental Sciences*, Vol. 7, Issue 1, pp. 8–17. DOI: 10.17352/2455-488X.000039.

Sustainable Processing of Municipal Solid Waste: 'Waste to Wealth', (2020) https://pib.gov.in/pressrelesedetailm.aspx?prid=1667099.

Swati, Thakur I.S., Vijay V.K., Ghosh P. (2018) Scenario of landfilling in India: Problems, challenges, and recommendations. In: Hussain C. (ed) *Handbook of Environmental Materials Management*. Springer, Cham. DOI: 10.1007/978-3-319-58538-3_167-1.

USEPA Office of Water (1994) EPA Drinking Water Glossary: A Dictionary of Technical and Legal Terms Related to Drinking Water, p. 17.

Section III

Use of Mathematical and Statistical Modelling in Setting Waste Management Strategies

13

Phenol Adsorption by Biological and Industrial Wastes and ANN Modeling

Ashanendu Mandal, Nirjhar Bar, and Sudip Kumar Das
University of Calcutta

CONTENTS

DOI: 10.1201/9781003231608-16

13.1 Introduction

Phenol is considered a severe pollutant that is produced as a by-product in many chemical industries, e.g., oil refineries, petrochemical plants, dye and paints manufacturing, rubber and plastic factories, coal tar processing, steel plants, pharmaceuticals, leather, and pesticides. (Dabrowski et al., 2005). The phenol content in industrial wastewaters is shown in Table 13.1. Besides industrial effluents, phenol is introduced into surface water from domestic wastewaters and agricultural run-off.

Phenol is soluble in water, and its solubility is 98 gm/L. It is colorless but becomes pink in the air due to easy oxidation. It melts at 181°C and forms a toxic liquid. It is also called carbolic acid with weak acidity. Its half-life ranges from 2 to 72 days.

Phenol's toxicity limits its acceptance in the environment. Its exposure occurs quickly through the nasal, skin, oral, and eye through many consumer products. The health hazards from phenol are given in Table 13.2. Phenol's impact on health depends on its quantity and duration of exposure. Its impact on the environment occurs through soil damage. It also affects the germination of seeds and the growth of plants. Due to the pollution impact, the phenol content in surface discharge is set at a maximum 1 mg/L by Environmental Protection Agency. The World Health Organization allows phenol content of up to 0.001 mg/L in drinking water (Saleh et al., 2018).

Phenol removal through adsorption has widely used activated carbon as an adsorbent (Atieh, 2014). However, the high costs, regeneration loss of 10%–15%, and disposal problems restrict the use of activated carbon in many countries (Ingole and Lataye, 2015). It has intensified the hunt for cheaper alternatives.

Many alternatives to activated carbon, e.g., carbon from corncob waste, eggshell, fly ash, wood and formaldehyde-pretreated Pinus pinaster bark, sewage sludge, etc. been attempted for phenol removal (Larous and Meniai, 2012). Many agricultural and industrial wastes are also attempted as adsorbents. These materials are cheap and readily available, and no pretreatment is required. The authors have undertaken multiple studies on phenol removal by agricultural and industrial wastes (Mandal and

TABLE 13.1

Concentration of Phenol in Industrial Effluents

Industry	Concentration of Phenol (mg/L)
Refineries	6–500
Coal processing	9–6,800
Coking	28–3,900
Petrochemicals manufacturing	2.8–1,220

TABLE 13.2

Health Hazards of Phenol

Exposure Mode	Effects Due to Exposure
Inhalation	Can irritate the nose, throat, and lungs. Higher exposures may cause a build-up of fluid in the lungs.
Oral	Ingestion of as little as 1 g can be fatal to humans.
Eyes	Can also cause severe eye damage, including blindness.
Skin	Is irritating and corrosive to the skin.
	Little or no pain may be felt on initial contact due to its local anesthetic effect.
	Skin contact may cause the skin to turn white, and later severe burns may develop.
	Rapidly absorbed through the skin, and toxic or fatal amounts can be absorbed through relatively small areas.
Chronic exposure	Repeated or prolonged exposure to phenol or its vapors may cause headache, nausea, dizziness, difficulty swallowing, diarrhea, or vomiting.
	Affect the central nervous system, liver, and kidneys.

Das, 2019a,b; Mandal et al., 2018, 2019a,b, 2020a–c). The results of these studies have been used for Artificial Neural Network (ANN) modeling in this research work.

The application of ANN has increased exponentially in the last few years in science and engineering. Better processor speed and increment in the RAM capacity of the computers and the availability of these components in the markets have revolutionized computer science. It has been used in adsorption for the last 15 years (Singha et al., 2014).

13.2 Material and Method

13.2.1 Chemicals and Equipment

Phenol, activated alumina, HCl, NaOH, NaCl, and 4-amino antipyrine from M/S Merck, India, were utilized in the study. UV-Spectrophotometer (Hach Germany), pH Meter (Hach Germany), X-ray Powder Diffractor (Bruker Germany), Scanning Electron Microscope (Hitachi Japan), Fourier-Transform Infrared Spectroscopy (ThermoFisher USA), and Brunauer–Emmett–Teller analyzer (NOVA USA) were utilized for characterization of the adsorbents. The B.O.D. (Biological Oxygen Demand) incubator shaker and the heating furnace, and the digital balance were also used for the experimentation. Origin Pro 8.0 software was used for graphical interpretations.

13.2.2 Adsorbate (Phenol Solution) Preparation

An aqueous solution of phenol was used in this study. The stock solution (1,000 mL) of 1,000 ppm (mg/L) was prepared, taking 1-gm phenol and 1,000-mL double-distilled water. The desired phenol concentration (5–300 mg/L) was then prepared by diluting more double-distilled water. The residual phenol concentration of the filtrate after filtration through Whatman (0.45 µm) filter paper was measured by UV-spectrophotometer, for which the calibration chart between absorbance and known phenol concentrations was already evaluated. The 4 mL of this 4-amino antipyrine solution was added to each 100 mL of residual phenol solution before measuring the absorbance in the spectrophotometer set at a wavelength of 460 nm (APHA, 1998). 4-amino antipyrine solution was prepared by mixing 0.376 gm of solid 4-amino antipyrine in 20 mL water. The solution pH was checked using the pH meter, and variations of pH (2–11) were carried out using NaOH and HCl.

13.2.3 Adsorbent Preparation

13.2.3.1 Guava Tree Bark

The guava tree bark arranged from Kolkata, India, was washed thoroughly with water to remove surface impurities. Barks were dried in sunlight for a week and then heated in the oven at around 70°C for 6 hours. The dry barks were ground with a small grinding machine and sieved with −44 +52 mesh sieve. The powdered fraction of guava barks were then stored in desiccators.

13.2.3.2 Rice Husk

The rice husk arranged from Kolkata, India, was washed repeatedly in water. It was then dried in sunlight for a week, powdered in a mixer-grinder, screened in the −44 +52 sieves, and kept in desiccators.

13.2.3.3 Neem Leaves

The neem (*Azadirachta indica*) leaves collected from a neem tree inside the University of Calcutta, Kolkata, India, were cleaned thoroughly with water to get rid of impurities and dust. They were then dehydrated in sunlight for a week, powdered in a mechanical grinder, placed in the oven at around 70°C for 10 hours, sieved through −44 +52 mesh, and preserved in desiccators.

13.2.3.4 Activated Carbon from Coconut Coir

This adsorbent was an offshoot of coconut and was primarily used as fuel in rural areas. The coconut coir was arranged from the sub-urban area near Kolkata, India. It was washed carefully in distilled water and was sun-dried for 7 days. Then it was processed for activation at 800°C for 1 hour, washed in distilled water, and dried again at 100°C for a day. It was then ground to powder, passed in −44 +52 sieves, and kept in desiccators.

13.2.3.5 Rice Husk Ash

This adsorbent was an offshoot of the rice crop and was generally disposed of as waste. Rice husk collected from Kolkata, India, was cleaned extensively in water. It was then dehydrated in sunlight for a week and burnt completely to form ash. The ash was screened in −44 +52 mesh sieves and kept in desiccators.

13.2.3.6 Red Mud

The red mud used was collected from a bauxite plant at Damanjodi, Odisha, India. It was mainly the red sludge of iron oxide. The red mud was cleaned in water thoroughly and heated in a furnace at 110°C for 8 hours. It was then ground to powder, screened in −44 +52 mesh, and kept in desiccators.

13.2.3.7 Clarified Sludge from Basic Oxygen Furnace

The clarified sludge collected from a steel plant, Rourkela, Odisha, India, was washed thoroughly in water. It was then dried in sunlight, powdered, meshed in −44 +52 sieves, and kept in desiccators.

13.2.3.8 Activated Alumina

Activated alumina powder (M/S Merck India) was arranged from Kolkata, India. Without any further treatment, this was kept in the desiccators for the experiments.

13.2.4 Adsorbent Characterization

The characterizations of the adsorbents are usually based on the surface characteristics like porosity, pore size, bulk density and surface area, etc. These characteristics are essential in any adsorption process because they control the adsorption capacity. The functional groups are also assumed to be necessary during the adsorption of ions onto the adsorbent surface. These characteristics are very significant in the kinetics, isotherms, and thermodynamics study of the adsorption process.

13.2.5 Adsorption Experiments—Batch Studies

The batch experiments were conducted to evaluate the phenol adsorption percentage with all the adsorbents. The studies were performed in various operating conditions, e.g., time, adsorbent dosage, pH, initial phenol content, and temperature.

The aqueous phenol solution (100 mL) with the desired initial phenol content was placed inside 250-mL conical flasks. The solution pH varied by NaOH solution and HCl solution. The quantity of the adsorbent dosage was varied as per the requirement, and the solution was shaken for different periods in a shaker at 120 strokes/minute at different temperatures. The solution after filtration was taken in a UV-spectrophotometer. This procedure was repeated thrice so that the average value could be taken for more precision.

The phenol removal percentage was evaluated from Equation (13.1).

$$\text{Phenol removal}(\%) = \frac{(C_0 - C_t)}{C_0} \times 100\% \tag{13.1}$$

where C_0 and C_t denote phenol concentrations (mg/L) at initial and any specific time, t (minutes), respectively. The adsorbed phenol quantity, q_t (mg/gm), was evaluated from Equation (13.2).

$$q_t = \frac{(C_0 - C_t)V}{W} \tag{13.2}$$

where W denotes adsorbent quantity (gm) and V denotes solution volume (L). The batch experimental results of phenol removal percentage versus the process variable, i.e., pH, time, dose, temperature, and phenol concentration, are plotted graphically to determine the optimum conditions. The batch experimental results are also tested for ANN modeling to find out the conformity of the prediction. The batch study process diagram is shown in Figure 13.1.

13.3 Results and Discussions

13.3.1 Adsorbent Characterizations

13.3.1.1 Scanning Electronic Microscopy (SEM)

The scanning electronic microscopy (SEM) results for the adsorbents showed that the adsorbents were spongy with numerous pores for all the adsorbents except for the adsorbent clarified sludge, which showed micrometric fibrous and light-colored marks.

13.3.1.2 X-Ray Diffraction (XRD)

The X-ray diffraction (XRD) analysis in Bragg angle 2θ ($10°–80°$) was performed for rice husk and rice husk ash. The result of the rice husk indicated the existence of crystalline silica at $22°$. The crystallinity

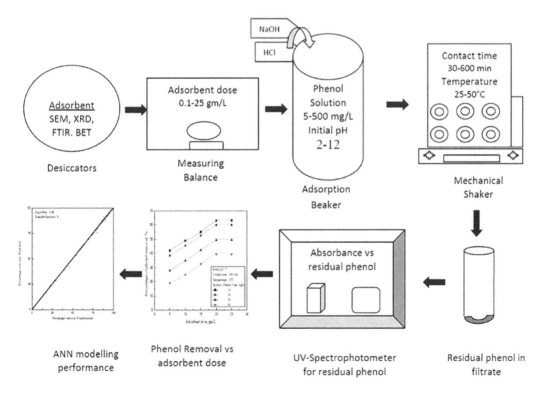

FIGURE 13.1 Batch study process diagram.

was high because of silica in the rice husk. The XRD result of rice husk has exhibited the existence of quartz at 22.85°, 26.63°, and 42.47°, the presence of cristobalite at 21.91°, 35.99°, and 69.50°and the presence of anorthite at 27.91° and 29.42°. The excess porosity of rice husk ash compared to rice husk was due to more dehydration, resulting in better adsorption for rice husk ash (Figure 13.2).

13.3.1.3 Fourier-Transform Infrared Spectrophotometer (FTIR)

The Fourier-transform infrared spectrophotometer (FTIR) results based on the wavenumbers determined the functional groups associated with the adsorbents. The FTIR results are shown in Table 13.3.

13.3.2 Physical Characteristics of Adsorbents

The physical characteristics of adsorbents, viz., bulk density, point of zero charge, and surface area are shown in Table 13.4.

13.3.3 Adsorption Experiments

13.3.3.1 Variation of pH

The tests were performed at different pH and different phenol concentrations, with constant dose, time, and temperature for all the adsorbents. Figure 13.2 shows the effects of pH on the removal of phenol for neem leaves and activated carbon from coconut coir. The results showed the rise of phenol removal with pH at the start. However, after a particular pH value, the removal percentage started decreasing. Similar graphs were also found for the other six adsorbents. The reason was possibly due to the amphoteric properties of adsorbents at the point of zero charge (Asmaly et al., 2015). The peak pH value for maximum phenol removal by guava tree bark, rice husk, neem leaves, coconut coir, rice husk ash, red mud, clarified sludge, and activated alumina were 5, 5, 3, 6, 9, 8, 7, and 3, respectively.

13.3.3.2 Variation of Time

The batch experiments were conducted at different times and with different phenol content but at constant pH, dose, and temperature for all the adsorbents. Figure 13.3 shows the effect of contact time on the phenol removal for rice husk and clarified sludge. The results showed that the phenol removal

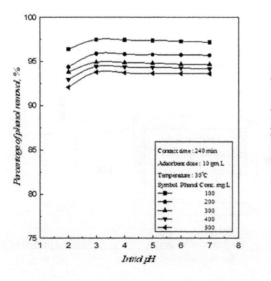
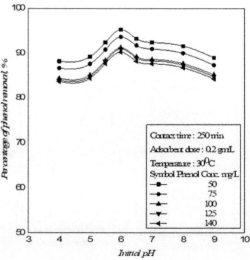

FIGURE 13.2   Effects of pH on the phenol removal: (a) neem leaves (b) activated carbon from coconut coir.

TABLE 13.3

FTIR Analysis of the Adsorbents

Adsorbent	Wave Number	Functional Group
Guava tree bark	551–887 cm^{-1}	Alkyl halides
	1,035 and 1,105 cm^{-1}	=C-O-C and -C-C-C
	1,318 and 1,622 cm^{-1}	Alkyl halides and alkenes
	2,923 and 3,428 cm^{-1}	Alkenes and phenols stretching
Rice husk	3,774 and 3,363 cm^{-1}	N-H and O-H bonds
	2,929 to 2,128 cm^{-1}	C-H stretching
	1,658 to 1,426 cm^{-1}	Aromatic C=C
	1,080 and 466 cm^{-1}	C-O and silica
Neem leaves	662 cm^{-1}	C-O-H twist broads
	1,104 cm^{-1}	C-C-C bonds
	1,384 and 1,531 cm^{-1}	CH$_3$ deformations and amide
	2,923 and 3,433 cm^{-1}	Alkenes and phenols stretching
Activated carbon from coconut coir	3,406 and 2,924 cm^{-1}	Alcohol O-H and alkane C-H
	1,621 and 1,515 cm^{-1}	N-H and N-O stretching
	1,383 and 1,266 cm^{-1}	C-H and C-O ester bending
	1,057 and 616 cm^{-1}	Alcohol and C-Cl bond
Rice husk ash	3,403 cm^{-1}	Alcohol O-H bond
	1,594 cm^{-1}	NH bond
	1,384 and 1,097 cm^{-1}	Alkane CH and ester CO bond
	797 and 466 cm^{-1}	C-Cl and C-I bond
Red mud	515 and 669 cm^{-1}	Alkyl bromide and =C-H bend
	780 and 1,036 cm^{-1}	Alkyl chloride and alcohol, ether, ester
	1,318–1,383 cm^{-1}	Alkyl fluoride
	1,622 and 2,924 cm^{-1}	N-H bend and C-H stretch
	3,429 cm^{-1}	O-H stretch
Clarified sludge from basic oxygen furnace	617 and 779 cm^{-1}	Alkyl bromide and alkyl chloride
	1,106 and 1,308 cm^{-1}	Alkyl fluoride
	1,625 and 3,427 cm^{-1}	N-H bond and N-H symmetric stretch
Activated alumina	3,436 and 1,648 cm^{-1}	OH and NH bond
	1,383 cm^{-1}	Alkane CH bond
	868 cm^{-1}	1-3 di-substituted C-H bond

TABLE 13.4

Physical Properties of the Adsorbents

Adsorbent	Bulk Density (gm/cm^3)	Surface Area (m^2/gm)	Point of Zero Charge
Guava tree bark	0.21	210	4.10
Rice husk	0.79	320	4.61
Neem leaves	0.46	370	2.7
Activated carbon from coconut coir	0.31	830	5.50
Rice husk ash	0.96	57.5	7.61
Red mud	3.00	300	7.65
Clarified sludge from basic oxygen furnace	1.98	78.54	9.80
Activated alumina	3.00	300	2.85

improved initially and then became constant. Similar results were obtained for other adsorbents also. The reason was that faster adsorption occurred initially due to empty adsorbent sites (Asmaly et al., 2015). The adsorbent's oxygenated surface facilitated the adsorption by hydrogen bonding with the hydroxyl surface of phenol. At equilibrium, the adsorption sites became saturated, and therefore no more adsorption

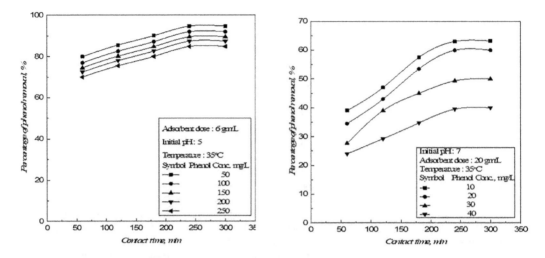

FIGURE 13.3 Impact of contact time variation for phenol removal: (a) rice husk (b) clarified sludge from basic oxygen furnace.

was possible after that. The contact time corresponding to maximum phenol removal by guava tree bark, rice husk, neem leaves, coconut coir, rice husk ash, red mud, clarified sludge, and activated alumina are 120, 120, 240, 250, 180, 480, 240, and 240 minutes, respectively.

13.3.3.3 Variation of Dose

The tests were performed at different doses and different phenol concentrations but at constant pH, time, and temperature for all the adsorbents. Figure 13.4 shows the variation of adsorbent dosage on the phenol removal for rice hush ash and red mud. The results showed the rise of removal initially but became almost constant after a specific quantum of dose. The reason was that the adsorption was faster initially due to empty adsorbent sites. The higher rate of adsorption continued because of the addition of surface area through the addition of adsorbent. However, afterward, the removal percentage might have become almost constant due to saturation of active spots and reduction of surface area through the overlapping of adsorbents

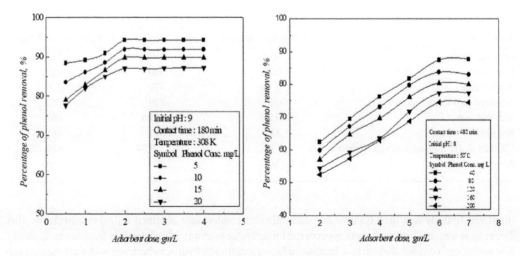

FIGURE 13.4 Impact of adsorbent dose variation for phenol removal: (a) rice husk ash (b) red mud.

(Asmaly et al., 2015). The observations for the rest of the adsorbents were also similar. The optimum adsorbent dose for maximum phenol removal by guava tree bark, rice husk, neem leaves, coconut coir, rice husk ash, red mud, clarified sludge, and activated alumina are 4, 6, 10, 0.2, 2, 6, 20, and 2 gm/L, respectively.

13.3.3.4 Variation of Temperature

The tests were performed at different temperatures and different phenol concentrations but at constant pH, dose, and time for all the adsorbents. Figure 13.5 shows the effect of temperature on phenol removal for red mud and activated alumina. The results showed the change in removal percentage with the increase in temperature. The result showed that phenol adsorption increased with temperature while the process was endothermic and decreased with temperature while the adsorption was exothermic (Ekpete et al., 2010). The results showed that the adsorption process for guava tree bark, rice husk, rice husk ash, red mud, and clarified sludge from basic oxygen furnace were endothermic, and the adsorption process for neem leaves, activated carbon from coconut coir, and activated alumina was exothermic. The adsorption temperature for maximum phenol removal by guava tree bark, rice husk, neem leaves, coconut coir, rice husk ash, red mud, clarified sludge, and activated alumina are 50°C, 35°C, 30°C, 30°C, 35°C, 50°C, 35°C, and 30°C, respectively.

13.3.3.5 Variation of Phenol Concentration

The tests were conducted at different phenol content with constant conditions of other process variables, e.g., pH, adsorption period, temperature, and adsorption dose. Figure 13.6 shows the variation of initial phenol concentration on phenol removal for guava bark and red mud. The results described that the phenol removal reduced with the rise of phenol concentration. Similar results were also found for the remaining six adsorbents. The probable reason was the insufficiency of active sites after phenol adsorption because each adsorbent had a restricted quantity of active spots, which was saturated quickly at upper phenol content. Further, the higher mass transfer rate at higher phenol content increased the adsorption uptake resulting in quicker saturation (Uddin et al., 2007). The maximum phenol removal percentage for all the adsorbents under optimum process conditions are summarized and shown in Table 13.5.

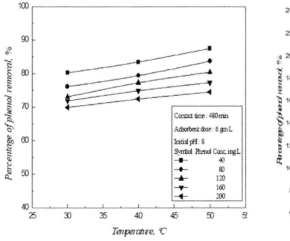
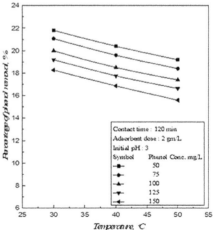

FIGURE 13.5 Impact of temperature variation for phenol removal (a) red mud (b) activated alumina.

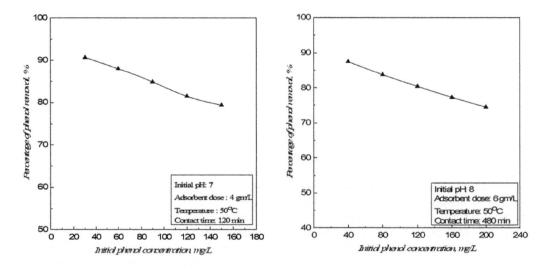

FIGURE 13.6   Impact of initial phenol concentration variation for phenol removal (a) guava tree bark (b) red mud.

TABLE 13.5

Maximum Phenol Removal Percentage for the Adsorbents

Adsorbent	Adsorbent Dose (gm/L)	Contact Time (min)	Temperature (°C)	Initial Phenol Conc. (mg/L)	pH	Maximum Phenol Removal (%)
Guava tree bark	4	120	50	30	5	90.67
Rice husk	6	240	35	50	5	94.74
Neem leaves	10	240	30	100	3	97.50
Activated carbon from coconut coir	0.2	250	30	50	6	95.20
Rice husk ash	2	180	35	5	9	93.33
Red mud	6	480	50	40	8	87.50
Clarified sludge from basic oxygen furnace	20	240	35	10	7	63.00
Activated alumina	5	240	35	50	3	21.80

13.4 ANN Modeling

ANN modeling has become extremely popular to use when the solution to a problem in data mining is not tractable from the mathematical point of view. This tool that was mainly used for problems related to computer science is now being used in all scientific areas related to engineering and technology. They have also been applied successfully in solving problems related to adsorption (Maiti et al., 2014, 2018; Jana et al. 2014; Singha et al. 2015; Das et al. 2015; Bar and Das, 2011; Bar et al., 2010a,b). From the literature survey, the application of ANN related to the adsorption of phenol is rare. Therefore, the modeling was attempted to predict the maximum percentage of phenol removal by biological and industrial waste adsorbents. A three-layer neural network (viz. input layer, hidden layer, and output layer) was

used. Hyperbolic tangent, $T_1 = \left[\dfrac{\left(e^{\beta x} - e^{-\beta x} \right)}{\left(e^{\beta x} + e^{-\beta x} \right)} \right]$ and sigmoid, $T_2 = \left[\dfrac{1}{\left(1 + e^{-\beta x} \right)} \right]$ transfer functions were used

in the hidden and output layer. β is the gain. The input layer consisted of adsorbent material, pH, dosage, phenol content, temperature, and time, whereas in the output layer, the removal percentage of phenol was

TABLE 13.6

Description of Data for ANN Modeling

Parameter	Range
Adsorbent number	1–8
pH	2–12
Adsorbent dose (g/L)	0.10–25
Initial concentration (mg/L)	5–500
Time (minutes)	30–600
Temperature (K)	298–323
% removal of phenol	8.67–97.50
Number of data	813

used. Table 13.6 gives the experimental data used to predict the proposed removal percentage of phenol through ANN modeling.

The schematic diagram of ANN is presented in Figure 13.7. The modeling optimization was carried out with the Levenberg–Marquardt (LM) and the Scaled Conjugate Gradient (SCG) algorithms. The performances of both algorithms with both the transfer functions are shown in Figures 13.8 and 13.9. The cross-validation, as well as the training, was done at the same time (simultaneous process). For the training with LM algorithm, the application of the stopping criteria was evident as the training and cross-validation processes stopped long before its maximum limit of 1,000 epochs.

Table 13.7 shows the values for training and cross-validation along with stopping criteria for both the algorithms. The maximum number of epochs are presented in Table 13.7 along with the number of epochs to allow for improvement in the standard deviation (σ), average absolute relative error (AARE), mean squared error (MSE), and correlation coefficient (CCC) concerning cross-validation (i.e., stopping criteria).

From Table 13.7, it is clear that the application of the LM algorithm is successful based on the χ^2 test. Figure 13.10 shows the visual demonstration of the best performing network corresponding to the LM algorithm using the experimental phenol removal percentage and the predicted phenol removal percentage. The graph suggested a close correlation of experimental and predicted parameters using ANN modeling. The successful ANN modeling of the research can thus avoid the repetition of the

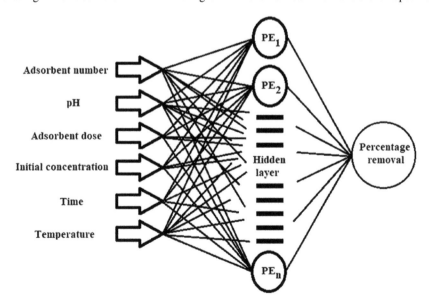

FIGURE 13.7 Schematic diagram of ANN.

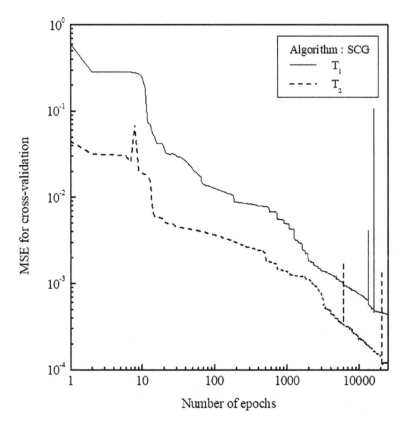

FIGURE 13.8 Cross-validation curve (SCG).

experiments before commercial applications yet can predict the final phenol removal percentage in any operational conditions.

13.5 Scale-Up Process Design

The isotherm models are used for the scale-up design in this adsorption process. For the process design, the equilibrium condition is considered, and therefore Equation (13.2) is converted to Equation (13.3).

$$\frac{W}{V} = \frac{(C_0 - C_t)}{q_t} = \frac{(C_0 - C_t)}{q_e} \tag{13.3}$$

where W denotes the adsorbent quantity (gm), V denotes the waste water volume (L), C_0 and C_t denote initial and final phenol concentrations (mg/L), and q_t and q_e signify phenol adsorbed per adsorbent quantum (mg/gm) in specific time and equilibrium (Figure 13.7). Replacing the value of q_e derived from the suitable isotherm model for all the adsorbents, the values of required adsorbent quantity (gm) to wastewater volume (L) considering 80% adsorption are shown in Table 13.8.

13.6 Safe Disposal of Spent Adsorbent

The adsorbents used for phenol removal from wastewater were readily available either naturally or from industrial wastes. The availability of such adsorbents is usually huge all over the world and mainly at a

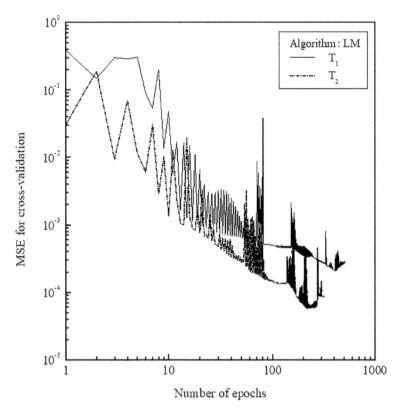

FIGURE 13.9 Cross-validation curve (LM).

meager cost. Therefore, their regeneration after phenol adsorption is not significant. The phenol desorption of the spent adsorbents with typically 30% (v/v) ethyl alcohol was tested in this research. It was found that the regeneration efficiency of all the adsorbents was within 50%–60%. So, the research does not insist on any desorption technique for the reuse of such low-cost adsorbents. The spent adsorbents are usually destroyed by incineration above 800°C, and the ash can be utilized for road repairing, land fillings, or brick manufacturing in many areas. The incineration destroys the phenol toxicity of the

TABLE 13.7

Performance of ANN for Prediction of Removal Percentage

Algorithm	Levenberg–Marquardt		Scaled Conjugate Gradient	
Maximum epochs	1,000		25,000	
Minimum epochs (stopping criteria)	100		5,000	
Transfer function	T_1	T_2	T_1	T_2
AARE	0.014575	0.013119	0.025782	0.027043
SD (σ)	0.014077	0.016076	0.031200	0.028166
MSE	1.839553	1.381446	4.963412	5.218691
CCC	0.998011	0.998583	0.994703	0.994441
χ^2	1.055376	0.957759	3.394655	3.295799
Minimum MSE for cross-validation	0.000208	0.000080	0.000428	0.000116
Optimum number of nodes in the hidden layer	20	17	22	15

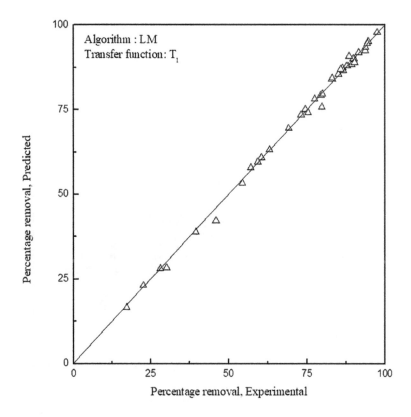

FIGURE 13.10 Final ANN prediction (LM algorithm).

adsorbents. The use of spent adsorbent is recommended for brick manufacturing, as the temperature required for this process needs 1,000°C and, therefore, can save the expenditure for separate incineration.

13.7 Conclusion

The phenol removal rates using various bio-sorbents and industrial waste materials were investigated in this research. The bio-adsorbents used were guava tree bark, rice husk, neem leaves, and activated carbon from coconut coir, and the industrial wastes used were rice husk ash, red mud, and clarified sludge from the basic oxygen furnace and activated alumina. The characterizations of the adsorbents were done

TABLE 13.8

Required Weight of Adsorbent for Different Volume of Wastewater

Volume of Waste Water (L)	Required Weight of Adsorbent (gm) Considering 80% Adsorption							
	Guava Tree Bark	Rice Husk	Neem Leaves	Activated Carbon from Coconut Coir	Rice Husk Ash	Red Mud	Clarified Sludge from Basic Oxygen Furnace	Activated Alumina
2	25.22	28.38	44.71	1.41	1.70	11.89	51.31	93.12
4	50.45	56.76	89.42	2.81	3.39	23.77	102.62	186.24
6	75.67	85.14	134.12	4.22	5.09	35.66	153.93	279.36
8	100.89	113.53	178.83	5.63	6.79	47.54	205.24	372.48
10	126.12	141.91	223.54	7.03	8.48	59.43	256.55	465.60

by SEM, BET, XRD, and FTIR. The phenol removal rate was tested using all the adsorbents at different phenol concentrations (5–500 mg/L), pH (2–12), dose (0.10–20 gm/L), temperature (25°C–50°C), and time (30–600 minutes). The maximum phenol removal percentage at optimum process conditions for guava tree bark, rice husk, neem leaves, activated carbon from coconut coir, rice husk ash, red mud, clarified sludge from basic oxygen furnace, and activated alumina was observed at 90.67%, 94.74%, 97.50%, 95.20%, 93.33%, 87.50%, 63.00%, and 21.80%, respectively. The experimental results of all the adsorbents in different process conditions were tested for ANN modeling using LM and SCG algorithms. The results showed that the training and cross-validation process for the LM method were in close harmony with experimental results as per the χ^2 test and, therefore, can suitably be used to predict maximum phenol removal percentage for all the adsorbents in any operating conditions. The pilot plant scale-up design conditions for all the adsorbents were deliberated using the appropriate isotherm model, and the requirement of adsorbent quantity for specific wastewater volume was calculated. The phenol desorption of spent adsorbents with 30% (v/v) ethyl alcohol showed 50%–60% regeneration efficiency. The safe dumping of used adsorbents was checked, and it was established that the used adsorbents could safely be used for landfilling or road repairing after incineration above 800°C. This investigation concludes that all the adsorbents are applicable for phenol removal, and the ANN modeling can successfully predict the phenol removal efficiency by an adsorbent at any process conditions.

Conflict of Interest

The authors have no conflicts of interest to disclose.

REFERENCES

APHA, AWWA, WEF. Standard methods for examination of water and wastewater. APHA, New York. 1998;28.

Asmaly, H. A., Abussaud, I. B., Saleh, T. A., Laoui, T., Gupta, V. K., Atieh, M. A. Adsorption of phenol on aluminum oxide impregnated fly ash. *Desal Wat Treat.* 2015;57(15):6801–6808.

Atieh, M. A. Removal of phenol from water different types of carbon: A comparative analysis. *APCBEE Procedia.* 2014;10:136–141.

Bar, N., Das, S. K. Comparative study of friction factor by prediction of frictional pressure drop per unit length using empirical correlation and ANN for gas-non Newtonian liquid flow through 180° circular bend. *Int Rev Chem Eng.* 2011;3(6):628–643.

Bar, N., Bandyopadhyay, T. K., Biswas, M. N., Das, S. K. Prediction of pressure drop using artificial neural network for non-newtonian liquid flow through piping components. *J Pet Sci Eng.* 2010a;71(2–3):187–194.

Bar, N., Biswas, M. N., Das, S. K. Prediction of pressure drop using artificial neural network for gas non-Newtonian liquid flow through piping components. *Ind Eng Chem Res.* 2010b;49(19):9423–9429.

Dabrowski, A., Podkoscielny, P., Hubicki, M., Barczak, M. Adsorption of phenolic compounds by activated carbon: A critical review. *Chemosphere.* 2005;58:1049–1070.

Das, B., Ganguly, U. P., Bar, N., Das, S. K., Holdup prediction in inverse fluidization using non-Newtonian pseudoplastic liquids: Empirical correlation and ANN Modeling. *Powder Technol.* 2015;273:83–90.

Ekpete, O. A., Horsfall, M., Tarawou, T. Potential of fluid and commercial activated carbons for phenol removal in aqueous systems. *ARPN J Eng App Sci.* 2010;5(9):39–47.

Ingole, R.S., Lataye, D. H. Adsorptive removal of phenol from aqueous solution using activated carbon prepared from babul sawdust. *J Haz Tox Radioact Waste.* 2015;19(4):04015002.

Jana, S. K., Biswas, A. K., Das, S. K., Pressure drop in tapered bubble columns using non-Newtonian pseudo plastic liquid: Experimental & ANN prediction. *Can J Chem Engg.* 2014;92(3):578–584.

Larous, S., Meniai, A. H. The use of sawdust as by product adsorbent of organic pollutant from wastewater: Adsorption of phenol. *Ener Procedia*, 2012;18:905–914.

Maiti, S. B., Let, S., Bar, N., Das, S. K. Non-spherical solid non-Newtonian liquid fluidization and ANN modeling: Minimum fluidization velocity. *Chem Eng Sci.* 2018;176:233–241.

Mandal, A., Das, S. K. Phenol adsorption from wastewater using clarified sludge from basic oxygen furnace. *J Environ Chem Engg.* 2019a;7(4):103259.

Mandal, A., Das, S. K. Adsorptive removal of phenol by activated alumina and activated carbon from coconut coir and rice husk ash. *Wat Cons Sci Eng.* 2019b;44(4):149–161.

Mandal, A., Mukhopadhyay, P., Das, S. K. Removal of phenol from aqueous solution using activated carbon from coconut coir. *IOSR J Eng.* 2018;8(12):41–55.

Mandal, A., Mukhopadhyay, P., Das, S. K. The study of adsorption efficiency of rice husk ash for removal of phenol from wastewater with low initial phenol concentration. *SN App Sci.* 2019a;1(2):192–204.

Mandal, A., Mukhopadhyay, P., Das, S. K. Efficiency analysis of rice husk as adsorbent for removal of phenol from wastewater. *J Environ Anal Toxicol.* 2019b;3:605–612.

Mandal, A., Bar, N., Das, S. K. Phenol removal from wastewater using low-cost natural bio-adsorbent neem (Azadirachta indica) leaves: Adsorption study and MLR modeling. *Sus Chem Pharm.* 2020a;17:100308.

Mandal, A., Dey, B. B., Das, S. K. Thermodynamics, kinetics and isotherms for phenol removal from wastewater using red mud. *Wat Prac Tech.* 2020b;15(3):705–722.

Mandal, A., Mukhopadhyay, P., Das, S. K. Adsorptive removal of phenol from wastewater using guava tree bark. *Environ Sci Poll Res.* 2020c;39(5):23937–23949.

Mitra, T., Singha, B., Bar, N., Das, S. K., Removal of Pb(II) ions from aqueous solution using water hyacinth root by fixed-bed column and ANN modeling, *J Hazard Mat.* 2014;273:94–103.

Saleh, T. A., Adio, S. O., Asif, M., Dafalla, H. Statistical analysis of phenols adsorption on diethylenetriamine-modified activated carbon. *J Clean Pro.* 2018;182:960–968.

Singha, B., Bar, N., Das, S. K., The use of artificial neural network (ANN) for modeling of Pb(II) adsorption in batch process. *J Mol Liquids* 2015;211:228–232.

Singha, B., Bar, N., Das, S.K. The use of artificial neural networks (ANN) formodeling of adsorption of Cr(VI) ions, Desalin. *WaterTreat.* 2014;52(1–3):415–425.

Uddin, M. T., Islam, M. S., Adedin, M. Z. Adsorption of phenol from aqueous solution by water hyacinth ash. *ARPN J Eng App Sci.* 2007;2(2):11–17.

14

Achieving Consistency of Judgment Matrix in Analytical Hierarchy Process

Bhupendra K. Sharma and Munish K. Chandel
Indian Institute of Technology Bombay

CONTENTS

14.1 Introduction

Analytic Hierarchy Process (AHP) was introduced by Saaty (1977), which is one of the most popular and widely applied multiple criteria decision-making (MCDM) methods (Peng et al., 2011). AHP is a systematic procedure, which decomposes a decision problem into a hierarchy that consists of an overall goal, a set of evaluation criteria and alternative options. The basic idea of AHP is to transform subjective judgments of relative importance to a score or weight (Saaty, 2008). It involves a large number of pair-wise comparisons among the evaluation criteria and alternatives. The alternatives are compared against each of the criteria and then the alternative having the highest score is selected as the best alternative. AHP consists of four steps (Islam and Abdullah, 2006): (i) decision-making problem is decomposed to a finite number of levels and then a hierarchy is constructed, which consists of the objective at the top followed by a set of criteria and alternatives; (ii) pair-wise comparison matrix (PCM) is constructed based on the experts' judgments for all the criteria and alternatives; (iii) weights or priority vectors of criteria and alternatives are determined through PCM; and (iv) ranking of alternatives is done with respect to the goal of the problem. In AHP, the elements are systematically compared and the judgment of relative importance of considered elements is given using Saaty's nine-point scale which is the most popular scale (Saaty, 1978, 2001; Benítez et al., 2011). In Saaty's scale, a score of 1 indicates that the elements are considered to be equally important, while 9 indicates that one element is extremely more important than the other. In AHP, the priority vectors/weights are derived from the PCM which is constructed based on the experts' judgments. For simplicity, let $A = \left(a_{ij} \right)_{n \times n}$ be a PCM with n criteria where $i, j \in (1, 2, \ldots, n)$, and then PCM has the following properties (Shiraishi et al., 1998; Benítez et al., 2011):

> **Definition 1:** A is said to be a positive reciprocal matrix if, $a_{ij} > 0, a_{ij} = 1$ and $a_{ij} = 1 / a_{ji}$ for all i, $j = 1, 2, \ldots, n$.

Definition 2: A positive reciprocal matrix is consistent if $a_{ik} \, a_{kj} = a_{ij}$ for all $i, j, k = 1, 2, ..., n$.

Definition 4: Reciprocal matrix is transitive if $A > B$ and $B > C$ then $A > C$.

Definition 5: PCM can pass the consistency test, if the consistency ratio $(\text{CR}) = \text{CI}/\text{ARI}$, where the consistency index $(\text{CI}) = \left[(\lambda_{\max} - n)/(n-1) \right]$; where ARI is an average random index (Table 14.1), and $\lambda_{\max}$ is the principal eigenvalue of a matrix having size of n (Saaty, 1977).

It is an important issue in AHP to derive reliable priority vectors from a PCM that is constructed from experts' judgments (Table 14.1). A number of prioritization methods to extract the priority vectors from these comparison matrices have been discussed in literature including the eigenvector method, weighted least squares method, logarithmic least squares method, correlation coefficient maximization method, etc. (Saaty, 1977; Crawford and Williams, 1985). However, the eigenvector method proposed by Saaty is the most robust and popular method for estimation of priority vectors particularly when the matrix is inconsistent (Saaty and Hu, 1998; Benítez et al., 2011; Kou and Lin, 2014). These priorities facilitate the decision-makers in ranking or selecting the most appropriate alternative to MCDM problems (Corrente et al., 2012).

According to Saaty, the judgment matrix having consistency ratio (CR) > 0.1 is acceptable. Therefore, the decision-makers need to adjust the inconsistent PCM as the consistency will affect the final ranking of results. However, it is difficult to build a judgment matrix satisfying Saaty's condition of consistency due to the complexity of the compared objects and the limited capability of human thinking (Khatwani and Kar, 2017). The way of dealing with such inconsistent matrices is returning such matrices to experts to construct new matrices based on their new decisions and following this procedure until the matrices satisfy the condition of consistency. This method is reliable and accurate but impracticable as it requires a huge amount of work for too long period (Zeshui, 2004). Therefore, some other techniques can be embraced to modify the judgment matrix legitimately in such a way that the revised matrix satisfies the condition of consistency and ensures that the revised matrix retains most of the information of the original matrix. There are two aims for the adjustment of consistency in AHP (Wu et al., 2017): (i) to make the PCM meet the consistency requirement and (ii) to minimize the deviation between the original and the revised PCM so that the revised PCM preserves most of the information of original matrix. Many studies have presented different ways for the improvement of the consistency of the matrix. A number of consistency improvement methods have been presented in literature including the Eigen value improvement (Dadkhah and Zahedi, 1993), convergent iterative algorithm (Zeshui and Cuiping, 1999), least square method (Chen and Triantaphyllou, 2001), triplet selection (Lamata and Peláez, 2002), heuristic algorithm (Peters and Zelewski, 2003), controlled error consistent matrix development (Ishizaka and Lusti, 2004), weak transitivity (Ma et al., 2006), Gower plot and linear programming (Li and Ma, 2007), auto generate consistent matrix (Cao et al., 2008), controlled linguistic preference deviation (Dong et al., 2008), adaptive AHP method (Lin et al., 2008), missing value multi-layer perceptron (Gomez-Ruiz et al., 2010), linearization technique (Benítez et al., 2011), improved AHP method (Li et al., 2013), integer programming (Bozóki et al., 2011), consistency and consensus improvement (Wu and Xu, 2012), consistency optimization (Benítez et al., 2012) and ordinal consistency improvement methods (Siraj et al., 2012), cosine maximization method (CMM) (Kou and Lin, 2014), induced bias matrix (IBM) model (Ergu et al., 2014), marginal optimization method (Xu and Xiong, 2017), and algorithm proposed by Jarek (2016), and optimization models (Negahban, 2018). As compared to the other methods, CMM provides a number of advantages. It provides an efficient and valid method for calculating the priority vector of PCM. In addition, it enables the derivation of a consistency index for the matrix and eradicates the need for statistical modeling. Kou and Lin (2014) developed a new consistency index related to CMM to measure the consistency level of a PCM, which is known as the cosine consistency index (CCI). CMM revises the entries of the matrix, which exhibits the maximum error on an iterative basis until CCI of the

TABLE 14.1

Average Random Index (Saaty and Hu, 1998)

n	1	2	3	4	5	6	7	8	9	10	11	12	13	14	15
ARI	0	0	0.58	0.9	1.12	1.24	1.32	1.41	1.45	1.49	1.51	1.54	1.56	1.57	1.58

matrix becomes >90 (CCI > 90) and ensures that the revised matrix maintains all the major information of the original matrix. According to the literature, PCM is acceptable for a CCI to be above 90, but anything below 90 is unacceptable (Kou and Lin, 2014; Khatwani and Kar, 2017). However, CMM does not extend to the incomplete and imprecise PCM and it does not follow Saaty's condition of consistency after even improving the CCI value of the matrix. Therefore, two prioritization methods, eigenvector method and CMM, can be combined to derive a better priority vector and improve the inconsistency of a PCM (Srdjevic, 2005). So far, no method considers both the conditions of consistency (CR and CCI) derived from these prioritization methods simultaneously. Therefore, there is a need to revise the judgment matrix in such a way that it follows both the conditions of consistency, which is also suggested by Khatwani and Kar (2017) in their future work.

This study aims to propose an approach that can be applied to improve the inconsistency of judgment matrix that utilizes the CCI and Saaty's condition of CR. The remainder of the chapter is structured as follows: in methodology, the algorithm used to improve the CCI and CR values of a matrix is described. Then, the application of the proposed approach is illustrated through some numerical examples and finally, the findings of the study are presented in the conclusion section.

14.2 Methodology

All the inconsistent judgment matrices can be made consistent using the algorithm developed by Khatwani and Kar (2017) and Saaty's condition of CR. A modified algorithm can be used to revise the elements of the matrix and to improve the CCI and CR value of matrix. Both the conditions of consistency can be calculated simultaneously through the modified algorithm. If for a matrix, CCI > 90 and CR < 0.1, then the matrix is acceptable; otherwise, this process is iterated until these conditions are followed. The judgment matrix can be revised in such a way that the revised matrix preserves all the major information of the original PCM. A MATLAB code script is prepared to satisfy both the consistency conditions. The algorithm used to improve the CCI and CR value is described below in the following section.

14.3 Algorithm: Improving Inconsistency of PCM

For $n \times n$ judgment matrix $A = (a_{ij})$, let k denote the k times of the iteration, and $\lambda \in (0,1)$. The degree of deviation of the revised matrix is reduced for a high value of λ. The algorithm consists of the following steps:

Step 1: Let $A^{(0)} = \left(a_{ij}^{(0)}\right) = \left(a_{ij}\right)$ and $k = 0$, for all $i, j \in (1, 2, ..., n)$;

Step 2: Compute normalized matrix by dividing every element in a given column by the sum of elements in that column (normalization);

Step 3: Calculate the priority vector of matrix by averaging the entries of each row of the normalized matrix;

Step 4: First multiply each row of matrix with the column of priority vector and then divide the calculated values by the corresponding priority vector. The obtained values are the eigenvalues;

Step 5: Calculate λ_{max} by averaging the eigenvalues computed in step 4;

Step 6: Calculate the optimal objective function value C^* and CI;

$$C^* = \sqrt{\sum_{i=1}^{n} \left(\sum_{j=1}^{n} b_{ij}\right)^2} \quad \text{and} \quad CI = \left[\left(\lambda_{max} - n\right) / \left(n - 1\right)\right]$$

where $b_{ij} = a_{ij} / \sqrt{\sum a_{ij}^2}$

Step 7: Calculate initial CCI (CCI = C^*/n) and CR (CR = CI/RI) for PCM;

Step 8: If CCI > 90 and CR < 0.1 then go to Step 11 otherwise continue to the next step;

Step 9: Calculate the numbers r and s, such that $\varepsilon_{rs} = \max_{i,j}\left\{a_{ij}^{(k)}\left(w_j^{(k)} / w_i^{(k)}\right)\right\}$, and let $A^{(k+1)} = \left(a_{ij}^{(k+1)}\right)$, where $\left(a_{ij}^{(k+1)}\right)$ can be obtained by using one of the following method. ε_{rs} is an entry that has the largest deviation in matrix A and the corresponding entry ε_{sr} is also modified in order to ensure that a positive reciprocal matrix is produced

i. The weighted arithmetic mean form

$$\left(a_{ij}^{(k+1)}\right) = \begin{cases} \lambda\left(a_{rs}^{(k)}\right) + (1-\lambda)\left(\dfrac{w_r^{(k)}}{w_s^{(k)}}\right), & (i,j) = (r,s); \\[3ex] \dfrac{1}{\lambda\left(a_{rs}^{(k)}\right) + (1-\lambda)\left(\dfrac{w_r^{(k)}}{w_s^{(k)}}\right)}, & (i,j) = (s,r); \\[3ex] a_{ij}^{(k)}, & (i,j) \neq (r,s),(s,r) \end{cases}$$

ii. The weighted geometric mean form

$$\left(a_{ij}^{(k+1)}\right) = \begin{cases} \left(a_{rs}^{(k)}\right)^{\lambda}\left(\dfrac{w_r^{(k)}}{w_s^{(k)}}\right)^{(1-\lambda)}, & (i,j) = (r,s); \\[3ex] \dfrac{1}{\left(a_{rs}^{(k)}\right)^{\lambda}\left(\dfrac{w_r^{(k)}}{w_s^{(k)}}\right)^{(1-\lambda)}}, & (i,j) = (s,r): \\[3ex] a_{ij}^{(k)}, & (i,j) \neq (r,s),(s,r) \end{cases}$$

Step 10: Let $k = k+1$ and return to step 2;

Step 11: Output k, $A^{(k)}, CCI^{(k)}, CR^{(k)}$ and $w^{(k)}$, then $A^{(k)}$ is the revised judgment matrix and $w^{(k)}$ is the priority vector.

Step 12: End.

14.4 Numerical Examples: Method Validation

In this section, the application of the approach described above is illustrated within three numerical examples in order to more fully understand the approach and highlight the advantages of CR and CCI improvement approaches. The approach was examined for different sizes of PCM to demonstrate the effectiveness of the proposed approach.

Example 14.1

The original matrix A (4 × 4) and its CCI, CR, and w are:

$$A = \begin{pmatrix} 1.00 & 2.00 & 0.50 & 9.00 \\ 0.50 & 1.00 & 3.00 & 7.00 \\ 2.00 & 0.33 & 1.00 & 6.00 \\ 0.11 & 0.14 & 0.16 & 1.00 \end{pmatrix}$$

CCI = 88.60, CR = 0.21, w = (0.34, 0.34, 0.28, 0.04)

The final transformed matrix $A^{(k)}$ and its $CCI^{(k)}$, $CR^{(k)}$ and $w^{(k)}$, by using the geometric method are (here $\lambda = 0.9$):

$$A^{(k)} = \begin{pmatrix} 1.00 & 2.00 & 0.72 & 9.00 \\ 0.50 & 1.00 & 2.18 & 7.00 \\ 1.37 & 0.45 & 1.00 & 6.00 \\ 0.11 & 0.14 & 0.16 & 1.00 \end{pmatrix}$$

$CCI^{(k)}$ = 93.37, $CR^{(k)}$ = 0.09, $w^{(k)}$ = (0.36, 0.32, 0.27, 0.04), $k = 9$

Example 14.2

The original matrix A (6×6) and its CCI, CR, and w are:

$$A = \begin{pmatrix} 1.00 & 2.00 & 5.00 & 4.00 & 6.00 & 8.00 \\ 0.50 & 1.00 & 5.00 & 3.00 & 5.00 & 7.00 \\ 0.20 & 0.20 & 1.00 & 3.00 & 5.00 & 7.00 \\ 0.25 & 0.33 & 0.33 & 1.00 & 3.00 & 8.00 \\ 0.17 & 0.20 & 0.20 & 0.33 & 1.00 & 7.00 \\ 0.13 & 0.14 & 0.14 & 0.13 & 0.14 & 1.00 \end{pmatrix}$$

CCI = 95.07, CR = 0.15, w = (0.37, 0.27, 0.15, 0.11, 0.07, 0.03)

The final transformed matrix $A^{(k)}$ and its $CCI^{(k)}$, $CR^{(k)}$ and $w^{(k)}$, by using the geometric method are (here $\lambda = 0.9$):

$$A^{(k)} = \begin{pmatrix} 1.00 & 2.00 & 4.65 & 4.00 & 6.00 & 8.00 \\ 0.50 & 1.00 & 3.17 & 3.00 & 5.00 & 7.00 \\ 0.22 & 0.32 & 1.00 & 2.78 & 5.00 & 7.00 \\ 0.25 & 0.33 & 0.36 & 1.00 & 3.00 & 8.00 \\ 0.17 & 0.20 & 0.20 & 0.33 & 1.00 & 4.26 \\ 0.13 & 0.14 & 0.14 & 0.13 & 0.24 & 1.00 \end{pmatrix}$$

$CCI^{(k)}$ = 96.32, $CR^{(k)}$ = 0.09, $w^{(k)}$ = (0.38, 0.25, 0.16, 0.12, 0.06, 0.03), $k = 13$

Example 14.3

The original matrix A (8×8) and its CCI, CR, and w are:

$$A = \begin{pmatrix} 1.00 & 5.00 & 3.00 & 7.00 & 4.00 & 6.00 & 0.50 & 0.25 \\ 0.20 & 1.00 & 8.00 & 5.00 & 3.00 & 3.00 & 0.20 & 0.14 \\ 0.33 & 0.13 & 1.00 & 6.00 & 3.00 & 4.00 & 6.00 & 5.00 \\ 0.14 & 0.20 & 0.17 & 1.00 & 0.33 & 0.25 & 0.14 & 0.13 \\ 0.25 & 0.33 & 0.33 & 3.00 & 1.00 & 0.50 & 0.20 & 0.20 \\ 0.17 & 0.33 & 0.25 & 4.00 & 2.00 & 1.00 & 0.20 & 0.25 \\ 2.00 & 5.00 & 0.17 & 7.00 & 0.50 & 0.50 & 1.00 & 0.50 \\ 4.00 & 7.00 & 0.20 & 8.00 & 0.50 & 4.00 & 2.00 & 1.00 \end{pmatrix}$$

CCI = 89.86, CR = 0.035, w = (0.18, 0.15, 0.24, 0.02, 0.04, 0.05, 0.1, 0.21)

The final transformed matrix $A^{(k)}$ and its $CCI^{(k)}$, $CR^{(k)}$ and $w^{(k)}$, by using the geometric method are (here $\lambda = 0.9$):

$$A^{(k)} = \begin{pmatrix} 1.00 & 4.28 & 2.04 & 7.00 & 4.00 & 6.00 & 1.27 & 0.65 \\ 0.23 & 1.00 & 0.96 & 5.00 & 3.00 & 3.00 & 0.59 & 0.31 \\ 0.49 & 1.04 & 1.00 & 6.00 & 3.00 & 4.00 & 4.73 & 2.44 \\ 0.14 & 0.20 & 0.17 & 1.00 & 0.33 & 0.25 & 0.14 & 0.13 \\ 0.25 & 0.33 & 0.33 & 3.00 & 1.00 & 0.50 & 0.20 & 0.20 \\ 0.17 & 0.33 & 0.25 & 4.00 & 2.00 & 1.00 & 0.20 & 0.25 \\ 0.79 & 1.69 & 0.21 & 7.00 & 0.50 & 0.50 & 1.00 & 0.50 \\ 1.54 & 3.20 & 0.41 & 8.00 & 0.50 & 4.00 & 2.00 & 1.00 \end{pmatrix}$$

$CCI^{(k)}$ = 90.07, $CR^{(k)}$ = 0.033, $w^{(k)}$ = (0.24, 0.11, 0.23, 0.02, 0.05, 0.06, 0.10, 0.19), k = 65

From the above examples, it is observed that the improvement of CCI and CR can be established using the proposed approach. Thus, the proposed approach can be used for the adjustment of the consistency of PCM.

14.5 Demonstration of Proposed Algorithm Using a Hypothetical Case Study

This section demonstrates the potential use of the proposed algorithm in AHP method for municipal solid waste (MSW) management through a hypothetical case study. The objective is to select the most suitable MSW treatment option using AHP method. Four MSW treatment options are considered: (i) composting, (ii) anaerobic digestion, (iii) incineration, and (iv) sanitary landfill. The sustainability of these options has been assessed using four criteria: environmental, economic, social, and technical. In this hypothetical study, the relative importance of each treatment option with respect to each criteria (Tables 14.2–14.5) and the relative importance of each criterion (Table 14.6) has been provided as per the authors' knowledge judgment using the Saaty scale from 1 to 9. As can be seen, the comparison

TABLE 14.2

Relative Priority of Alternatives against Environmental Criterion

Environmental	Composting	Anaerobic digestion	Incineration	Landfill	Priority/Weight
Composting	1.00	0.33	4.00	3.00	0.28
Anaerobic digestion	3.00	1.00	4.00	4.00	0.51
Incineration	0.25	0.25	1.00	2.00	0.12
Landfill	0.33	0.25	0.50	1.00	0.09

Note: CCI = 97.32 > 90, CR = 0.078 < 0.1.

TABLE 14.3

Relative Priority of Alternatives against Economic Criterion

Economic	Composting	Anaerobic digestion	Incineration	Landfill	Priority/Weight
Composting	1.00	3.00	7.00	3.00	0.51
Anaerobic digestion	0.33	1.00	5.00	3.00	0.27
Incineration	0.14	0.20	1.00	0.20	0.05
Landfill	0.33	0.33	5.00	1.00	0.17

Note: CCI = 96.95 > 90, CR = 0.086 < 0.1.

TABLE 14.4

Relative Priority of Alternatives against Technical Criterion

Technical	Composting	Anaerobic digestion	Incineration	Landfill	Priority/Weight
Composting	1.00	1.00	5.00	3.00	0.40
Anaerobic digestion	1.00	1.00	4.00	2.00	0.35
Incineration	0.20	0.25	1.00	2.00	0.13
Landfill	0.33	0.50	0.50	1.00	0.12

Note: CCI = 98.19 > 90, CR = 0.088 < 0.1.

TABLE 14.5

Relative Priority of Alternatives against Social Criterion

Social	Composting	Anaerobic digestion	Incineration	Landfill	Priority/Weight
Composting	1.00	3.00	6.00	4.00	0.52
Anaerobic digestion	0.33	1.00	5.00	3.00	0.27
Incineration	0.17	0.20	1.00	0.25	0.06
Landfill	0.25	0.33	4.00	1.00	0.15

Note: CCI = 97.59 > 90, CR = 0.078 < 0.1.

TABLE 14.6

Pair-wise Comparison Matrix of Criteria

Criteria	Environmental	Economic	Technical	Social	Priority/Weight
Environmental	1.00	2.00	0.50	9.00	0.34
Economic	0.50	1.00	3.00	3.00	0.34
Technical	2.00	0.33	1.00	6.00	0.28
Social	0.11	0.14	0.16	1.00	0.04

Note: CCI = 88.60 < 90, CR = 0.21 > 0.1.

TABLE 14.7

Transformed Matrix of Criteria

Criteria	Environmental	Economic	Technical	Social	Priority/Weight
Environmental	1.00	2.00	0.72	9.00	0.36
Economic	0.50	1.00	2.18	7.00	0.32
Technical	1.37	0.45	1.00	6.00	0.27
Social	0.11	0.14	0.16	1.00	0.04

Note: CCI = 93.37 > 90, CR = 0.09 < 0.1.

matrix of criteria (Table 14.6) does not follow both the conditions of consistency (CCI = 88.60 < 90 and CR = 0.21 > 0.1). Therefore, there is a need to improve the inconsistency of this matrix in order to achieve robust and reliable results. Table 14.7 shows the transformed matrix obtained through the proposed algorithm with CCI = 93.37 > 90 and CR = 0.07 < 0.1. Table 14.8 shows the score of MSW treatment options

TABLE 14.8

Score of MSW Treatment Options Trough AHP

Criteria	Score	Rank
Composting	0.3928	1
Anaerobic digestion	0.3753	2
Incineration	0.0967	4
Landfill	0.1252	3

obtained through AHP. Results show that composting is the top-ranked while incineration is the least sustainable option. The algorithm proposed in this work can successfully improve the inconsistency of PCM in AHP and provide robust priority order of MSW treatment options. It could play a vital in selecting an appropriate option and hence aid the decision-making process.

14.6 Conclusion

In this chapter, a modified approach based on the algorithm developed by Khatwani and Kar (2017), and Saaty's condition of consistency is proposed to improve the consistency of PCM. A MATLAB code is prepared to satisfy both the conditions of consistency. The judgment matrix is modified in such a way that the revised matrix preserves all the major information of the original PCM. The effectiveness of the proposed approach is demonstrated through the matrices of different orders and a hypothetical case study for MSW management. This approach has overcome the shortcomings of the existing methods, which pay attention to improve the consistency by negating the preservation of information of the original matrix. The approach presented in this chapter is intuitive, capable, and can be performed using computer programming. Thus, it may provide a new tool for the adjustment of inconsistent PCM in AHP.

REFERENCES

Benítez, J., Delgado-Galván, X., Izquierdo, J. and Pérez-García, R., 2011. Achieving matrix consistency in AHP through linearization. *Applied Mathematical Modelling*, 35(9), pp. 4449–4457.

Benítez, J., Delgado-Galván, X., Izquierdo, J. and Pérez-García, R., 2012. Improving consistency in AHP decision-making processes. *Applied Mathematics and Computation*, 219(5), pp. 2432–2441.

Bozóki, S., Fülöp, J. and Poesz, A., 2011. On pairwise comparison matrices that can be made consistent by the modification of a few elements. *Central European Journal of Operations Research*, 19(2), pp. 157–175.

Cao, D., Leung, L.C. and Law, J.S., 2008. Modifying inconsistent comparison matrix in analytic hierarchy process: A heuristic approach. *Decision Support Systems*, 44(4), pp. 944–953.

Chen, Q. and Triantaphyllou, E., 2001. Estimating data for multicriteria decision making problems: Optimization techniques: In: *Encyclopedia of Optimization*, pp. 567–576. Springer Science & Business Media: Berlin/Heidelberg.

Corrente, S., Greco, S. and SłOwińSki, R., 2012. Multiple criteria hierarchy process in robust ordinal regression. *Decision Support Systems*, 53(3), pp. 660–674.

Crawford, G. and Williams, C., 1985. A note on the analysis of subjective judgment matrices. *Journal of Mathematical Psychology*, 29(4), pp. 387–405.

Dadkhah, K.M. and Zahedi, F., 1993. A mathematical treatment of inconsistency in the analytic hierarchy process. *Mathematical and Computer Modelling*, 17(4–5), pp. 111–122.

Dong, Y., Xu, Y. and Li, H., 2008. On consistency measures of linguistic preference relations. *European Journal of Operational Research*, 189(2), pp. 430–444.

Ergu, D., Kou, G., Peng, Y. and Yang, X., 2014. Simulation experiments for improving the consistency ratio of reciprocal matrices. *International Journal of Computers Communications & Control*, 9(4), pp. 408–418.

Gomez-Ruiz, J.A., Karanik, M. and Peláez, J.I., 2010. Estimation of missing judgments in AHP pairwise matrices using a neural network-based model. *Applied Mathematics and Computation*, 216(10), pp. 2959–2975.

Ishizaka, A. and Lusti, M., 2004. An expert module to improve the consistency of AHP matrices. *International Transactions in Operational Research*, 11(1), pp. 97–105.

Islam, R. and Abdullah, N.A., 2006. Management decision-making by the analytic hierarchy process: A proposed modification for large-scale problems. *Journal for International Business and Entrepreneurship Development*, 3(1–2), pp. 18–40.

Jarek, S., 2016. Removing Inconsistency in pairwise comparisons matrix in the AHP. *Multiple Criteria Decision Making*, 11, pp. 63–76.

Khatwani, G. and Kar, A.K., 2017. Improving the cosine consistency index for the analytic hierarchy process for solving multi-criteria decision making problems. *Applied Computing and Informatics*, 13(2), pp. 118–129.

Kou, G. and Lin, C., 2014. A cosine maximization method for the priority vector derivation in AHP. *European Journal of Operational Research*, 235(1), pp. 225–232.

Lamata, M.T. and Peláez, J.I., 2002. A method for improving the consistency of judgements. *International Journal of Uncertainty, Fuzziness and Knowledge-Based Systems*, 10(6), pp. 677–686.

Li, H.L. and Ma, L.C., 2007. Detecting and adjusting ordinal and cardinal inconsistencies through a graphical and optimal approach in AHP models. *Computers & Operations Research*, 34(3), pp. 780–798.

Li, F., Phoon, K. K., Du, X., & Zhang, M., 2013. Improved AHP method and its application in risk identification. *Journal of Construction Engineering and Management*, 139(3), pp. 312–320.

Lin, C.C., Wang, W.C. and Yu, W.D., 2008. Improving AHP for construction with an adaptive AHP approach (A3). *Automation in Construction*, 17(2), pp. 180–187.

Ma, J., Fan, Z.P., Jiang, Y.P., Mao, J.Y. and Ma, L., 2006. A method for repairing the inconsistency of fuzzy preference relations. *Fuzzy Sets and Systems*, 157(1), pp. 20–33.

Negahban, A., 2018. Optimizing consistency improvement of positive reciprocal matrices with implications for Monte Carlo analytic hierarchy process. *Computers & Industrial Engineering*, 124, pp. 113–124.

Peng, Y., Kou, G., Wang, G., Wu, W. and Shi, Y., 2011. Ensemble of software defect predictors: an AHP-based evaluation method. *International Journal of Information Technology & Decision Making*, 10(01), pp. 187–206.

Peters, M. and Zelewski, S., 2003. A heuristic algorithm to improve the consistency of judgments in the Analytical Hierarchy Process (AHP). PIM.

Saaty, T.L., 1977. A scaling method for priorities in hierarchical structures. *Journal of Mathematical Psychology*, 15(3), pp. 234–281.

Saaty, T.L., 1978. Modeling unstructured decision problems: The theory of analytical hierarchies. *Mathematics and Computers in Simulation*, 20(3), pp. 147–158.

Saaty, T.L., 2001. Deriving the AHP 1–9 scale from first principles. *ISAHP 2001 Proceedings*, Bern, Switzerland.

Saaty, T.L., 2008. Decision making with the analytic hierarchy process. *International Journal of Services Sciences*, 1(1), pp. 83–98.

Saaty, T.L. and Hu, G., 1998. Ranking by eigenvector versus other methods in the analytic hierarchy process. *Applied Mathematics Letters*, 11(4), pp. 121–125.

Shiraishi, S., Obata, T. and Daigo, M., 1998. Properties of a positive reciprocal matrix and their application to AHP. *Journal of the Operations Research Society of Japan*, 41(3), pp. 404–414.

Siraj, S., Mikhailov, L. and Keane, J., 2012. A heuristic method to rectify intransitive judgments in pairwise comparison matrices. *European Journal of Operational Research*, 216(2), pp. 420–428.

Srdjevic, B., 2005. Combining different prioritization methods in the analytic hierarchy process synthesis. *Computers & Operations Research*, 32(7), pp. 1897–1919.

Wu, Z. and Xu, J., 2012. A consistency and consensus based decision support model for group decision making with multiplicative preference relations. *Decision Support Systems*, 52(3), pp. 757–767.

Wu, S., Xie, J., Liu, X., He, B., Yang, M. and Li, Z., 2017. Marginal optimization method to improve the inconsistent comparison matrix in the analytic hierarchy process. *Journal of Systems Engineering and Electronics*, 28(6), pp. 1141–1151.

Xu, Q. and Xiong, M., 2017. A method for improving consistency of judgment matrix in the AHP. *In 2017 2nd Asia-Pacific Conference on Intelligent Robot Systems (ACIRS)*, Wuhan, China, pp. 83–87, IEEE.

Zeshui, X.U., 2004. A practical method for improving consistency of judgement matrix in the AHP. *Journal of Systems Science and Complexity*, 17(2), pp. 169–175.

Zeshui, X. and Cuiping, W., 1999. A consistency improving method in the analytic hierarchy process. *European Journal of Operational Research*, 116(2), pp. 443–449.

15

Approach for Indexing Site Selection Criteria and Its Suitability for Treatment Storage and Disposal Facility

Swathy Sadala, Saikat Dutta, and B. Chakradhar
Ramky Enviro Services Private Limited (wholly owned subsidiary of REEL)

T. Vijaya Lakshmi
Jawaharlal Nehru Technological University Hyderabad

T.S. Sasi Jyothsna
Ramky Enviro Services Private Limited (wholly owned subsidiary of REEL)

CONTENTS

15.1 Introduction

The circular economy in India has grown in the past couple of years, but the potential of the sectors remains unrealised (Bain et al., 2010). Many industrial sectors in India are resource-intensive and are inadequately equipped with pollution control measures leaving a huge negative footprint on the environment, creating an urgent need in the reduction of these impacts. Waste management is the paramount approach to deal with such issues (Dubey and Lal, 2009; Sovacool and Brown, 2010; Aichele and Felbermayr, 2012). To congregate the ever-growing demands of the world population, every industry has to play a key role to adopt innovative practices to utilize, reprocess and recover resources, conserve power, and utilize waste (Lifset and Graedel, 2002; Joseph, 2006; Shen et al., 2010). Earlier, the informal disposal of hazardous and other wastes through open burning and dumping caused the emission of harmful compounds such as Furans and Dioxins, Mercury, heavy metals causing air pollution and associated health problems (Mukherjee et al., 2016). The generators of hazardous waste (HW) are the principal interface in the waste management framework. They must identify whether their waste is dangerous and must regulate the ultimate fate of the waste. Moreover, generators must assure and keep complete track of the HWs that they produce are easily recognized, managed and treated before reusing or transferring to any TSFD facility. In sustainable management of HW, Common Hazardous Waste Treatment, Storage and Disposal Facilities (CHWTSDFs) play a major role (Morrison III et al., 2018). To treat generated HW in India, there are only around 46 TSDFs, out of which 22 facilities have both Secured Land Fill (SLF) and incinerator, 10 facilities have only common incinerators and 14 facilities have TSDF with only common secured landfill (Karthikeyan et al., 2018; Sadala et al., 2019). The Major States in India generating HWs, in descending order of the number of wastes generated, are Gujarat, Maharashtra, Rajasthan, Jharkhand, Tamil Nadu, Andhra Pradesh and Telangana (Misra and Pandey, 2005; Babu and Ramakrishna, 2012; CPCB, 2019a). Disposing of HW can be at the captive treatment facility, which may be installed by the individual waste generator or at TSDFs. The common disposal facilities may have only secured landfill (SLF) or incinerator or a combination of both facilities. Before establishing TSDF, an appropriate site for a facility is one of the most important decisions in the HW management system (Abessi and Saeedi, 2010). Different sorts of risks from these wastes must be recognized regarding their impacts on man and the environment. It is essential to have the option to recognize these HW at the place of generation and to know about the associated hazards potential, to control these dangers adequately inside the model of waste management with proper planning and put it into reality.

15.2 Methodology

Qualitative and quantitative data has been gathered from available works of literature, guidelines and rules published by the Ministry of Environment, Forest and Climate Change (MoEF&CC), Central Pollution Control Board (CPCB), etc. and other respective units involved in handling and treating industrial wastes. For this analysis, the data was first interpreted to gain a general understanding of the site selection criteria, index and ratings for TSDF establishment. Integrated Common Hazardous Waste Treatment Storage and Disposal Facility (ICHWTSDF) may involve treating wastes and future strategies for improvement in HW management system in India. Some of the qualitative information was assigned numerical values and statistical analysis was conducted to generate quantitative data for assigning ranks on site selection criteria.

15.2.1 Site Selection Approach Evaluation and Its Criteria to Establish TSDF

The rapid growth of industries has completely changed the HW generation scenario in the nation. The quantity of HW generated has increased significantly and the nature of the waste generated has become critically complex. For the scientific and secured management of these wastes, MoEF&CC has promulgated Hazardous Waste (Management & Handling) Rules, 1989 and subsequently amended them in the year 2008 and 2016, as per need. According to the rules, HWs have to be managed safely and scientifically. For carrying out environmentally safe and systematic management of HW, integrated common facilities for handling, treatment and disposal of such wastes are more preferred than discrete waste management facilities, concerning industrial clusters and multi-industries spread all over the nation. Placement of these facilities becomes crucial in the management of HW from industries and hence the site selection for establishing such facilities has to be planned based on various need-based socio-environmental techno-economic aspects.

Selecting an appropriate site for a facility is one of the most important decisions in the HW management system. It requires systematic and technical screening procedures based on the geological, environmental, engineering, economic and regulatory considerations, as well as human values (Gorsevski et al., 2012; Sonkamble et al., 2013; Beskese et al., 2015). In an EIA report, a detailed site investigation was conducted considering various parameters for the selection of a potential TSDF site as per the regulatory guidelines (Envotech Waste Management Limited, 2015). The preliminary site investigation was conducted by San Envirotech Pvt. Ltd. (2015), followed the knockout criteria suggested by CPCB for establishing a TSDF (Figure 15.1). The project proponent should follow the environmental concerns during the site selection criteria with due respect to EIA process significantly, which further simplifies the environmental clearance process and subsequently leads to the passivity to the mitigation measures for the entire duration of the project life cycle (IL&FS, 2010). Babu et al. (2000) explained the mathematical models of site sensitivity indices for selection of TSDF sites and summarised that site selection criteria may vary from country to country but the purpose remains the same—select potential sites for HW management. Environmental protection and public health consideration should be the principal concern in site selection; some general criteria, e.g. parameters for site selection, are given in Table 15.1.

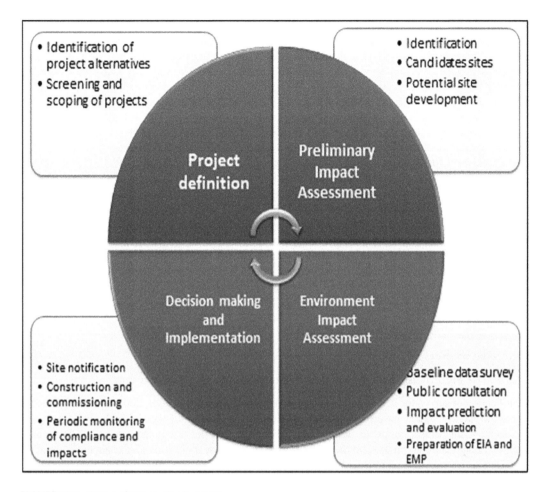

FIGURE 15.1 Stages of EIA studies for TSDF.

TABLE 15.1

Important Parameters Considered in Evaluations of TSDF Sites

Physical Parameters	Land Use
• Surface soil • Subsurface geology and aquifers • Topography, land stability • Surface water and streams, • Flooding • Seismic stability	• Agricultural value and land use type (residential, industrial, etc.) • Extractive industry • Mining • Water supply (surface and subsurface) Development potential • Transportation corridor or utility use
Ecological Parameters	**Climate Conditions**
• Flora and Fauna, habitat values • Conversation values	• Wind direction • Rainfall
Human Values	**Waste Disposal Suitability**
• Landscape • Historical, architectural and cultural values • Population density, employment opportunities	• Proximity to users • Transport access, availability to utilities and services, • Adjacent land use and zoning.

Source: CPCB (2003).

The important parameters for site selection that are to be excluded or rejected while selecting a site include (i) areas with unstable geological features (active fault plain); (ii) flood prone areas; (iii) wetlands, (iv) areas with unique archaeological, historical, paleontological and religious inserts; (v) area within 500 m from water supply zone; (vi) areas of groundwater recharge and extremely high water table zones; (vi) major natural hazard-prone zones like volcanic activity, seismic disturbance and landslides; and (vii) site evaluation criteria is prepared for ranking the selected site concerning six groups.

In this study, considering the above-mentioned cases, followed by the CPCB knockout criteria and relevant regulatory guidelines, each group is further subdivided into relevant categories based on the general information, hydrology, hydrogeology, geology, geotechnical and socio-economic/ecology conditions. A total of 100 points are divided among the six groups with a scoring of 25, 12.5, 12.5, 12.5, 12.5, and 25, respectively, based on the magnitude of the impacts. A total of 23 attributes are evaluated from six groups on five levels of sensitive scale marked as excellent to bad (100% to 20%). The aspects to be considered for attribute measurements are identified depending on the importance of the same. The relative values of the sensitive scale will be multiplied by the corresponding weightage of the given attribute score. In the same way, the score for all the attributes will be calculated and the final attributed score for the site is obtained. The total score out of 100 can thus be interpreted in terms of the sensitivity of site as follows. The detailed analysis index is presented in Table 15.2. In our study, the weightage against each attribute in the six criteria is calculated by assuming the relative values fall under the good category site suitability for TSDF.

15.3 Discussion

15.3.1 The Significant Role of TSDF in Hazardous Waste Management

Identification of site (s) for TSDF will be as per regulatory guidelines (CPCB, 2003; MoEF&CC, 2015) where the facility should have the following necessities: (i) location of the site with layout map; (ii) arrangement for transportation of waste from generators; (iii) the treatment processes like secured landfills, incineration and reuse and recycle facilities; (vi) leachate collection and treatment system; (v) NABL (National Accreditation Board of Testing and Calibration Laboratories) accredited laboratory; (vi) fire fighting systems; (vii) safe storage of the waste and storage capacity; and (viii) Environmental management plan including monitoring. Due to the complexity of waste characteristics and the fast increase in the rate of waste products along with other related issues, handling of HW requires an integrated and holistic approach to be implemented for the efficient and safe management of waste (Mrayyan and Hamdi, 2006).

The present trend in the treatment of HW is through ICHWTSDF, with its recent advances in technology and research implementation, which would provide a 'one-stop' solution for treatment and disposal of HW (CPCB, 2016a; Mmereki et al., 2017; Higgins, 2018). The ICHWTSDF includes secured landfill, waste treatment/stabilization unit, common incinerator, recycling facility for E-waste, spent solvent, used oil, used led acid batteries, alternative fuel and raw material, waste paper and plastic recycling and biomedical wastes facilities (CBWTDF, CPCB, 2016a). The toxic, corrosive and reactive emissions from these wastes that affect the health and environment further subject to scientific treatment to ensure the preservation of the ecosystem of the surroundings (Li et al., 2018; Sadala et al., 2019). Such facility contributes to creating a clean environment by reducing the indiscriminate dumping of waste in open sites that may pollute the air, land and water.

15.3.2 Present Scenario in the Sustainable Management of HW

The Hon' Supreme Court, during 2003, issued directives to all the states for creating an HW management facility in a time-bound manner and even appointed monitoring for overseeing the progress. India produces ~7.4 MTA of HW, which is generated from 43,936 industries (ASSOCHAM, 2017).

TABLE 15.2

TSDF Site Selection Criteria Index

S.N	Criteria (% Weightage)	Attributes	Units	5 — 100% — Excellent	4 — 80% — Ideal	3 — 60% — Good	2 — 40% — Poor	1 — 20% — Bad	Weightage	Overall Ranking	
1	General information (25%)	Transportation economy	km	0–5	5–10	10–20	20–40	>40	4.167	3.3336	0.8
		Slope (first scale)	%	1.5	1.5 to 1.2	1.2–0.75	0.75 to 0.5	<0.5	4.167	1.6668	
		Slope (second scale)	%	1.5	1.5–2.5	2.5–7.0	7.0–15	>15			0.4
		Topography	Shape	Convex				Concave	4.167	1.6668	0.4
		Flood prone							4.167	2.5002	0.6
		Optimum wind direction (downstream village)	km	>1	1–0.5	0.5 to 0.2	0.2 to 0.1	<0.1	4.166	3.3328	0.8
		Infrastructure (accessibility)		NH	SH	Local road		No road	2.083	1.2498	0.6
		Infrastructure (power supply)							2.083	0.8332	0.4
	A								**25**	**14.5832**	
2	Hydrology (12.5%)	Distance from surface water body/drinking water	km	>5	5 to 3	3 to 2	2 to 1	<1	6.25	1.25	0.2
		Annual rainfall	cm/year	<25	25–80	80–150	150–250	>250	6.25	5	0.8
	B								**12.5**	**6.25**	
3	Hydrogeology (12.5%)	Groundwater depth post monsoon	m	>15	15 to 10	10 to 5	5 to 1	<1	3.125	1.875	0.6
		Groundwater flow direction (distance to D/S village)	km	>5	5 to 3	3 to 1	1 to 0.5	<0.5	3.125	1.875	0.6
		Groundwater quality		Bad quality				Good quality	3.125	1.25	0.4
		Groundwater gradient	m/km	<5	5–10	10–20	20–50	>50	3.125	1.875	0.6
	C								**12.5**	**6.875**	
4	Geology (12.5%)	Subsidence		Settled soil				Filled up soil	4.167	3.3336	0.8
		Depth of bedrock	m	>15	15 to 10	10 to 5	5 to 1	<1	4.167	2.5002	0.6
		Seismic conditions	Intensity	V	VI	VII	VIII	IX	4.167	3.3336	0.8

(Continued)

TABLE 15.2 *(Continued)*

TSDF Site Selection Criteria Index

S.N	Criteria (% Weightage)	Attributes	Units	Relative Values					Weightage	Overall Ranking	
				5	4	3	2	1			
				100%	80%	60%	40%	20%			
				Excellent	Ideal	Good	Poor	Bad			
	D								**12.5**	**9.1674**	
5	Geotechnical (12.5%)	Permeability	(1×10-6 cm/s)	<0.1	0.1–1	1–10	10–100	>100	6.25	3.75	0.6
		Engineering property (MA, Pl, Sheer)							6.25	5	0.8
	E								**12.5**	**8.75**	
6	Socio-economic /ecological (25%)	Demography	km	>5	5 to 2.5	2.5 to 1.0	1.0 to 0.2	<0.2	6.25	5	0.8
		Land use pattern		Waste land/ saline	Grazing /fallow	Single crop/non irrigated	Double crop/ irrigated	Plantation	6.25	3.75	0.6
		Transportation impacts						Passing populated areas	6.25	5	0.8
		Special ecological features							6.25	6.25	1
	F								**25**	**20**	

A + B + C + D + E + F = 65.62.

Source: CPCB (2003).

The interim report of the monitoring committee on the management of HW, CPCB, 2019b, remarks on Indian states generating higher quantities of HW – Gujarat (39.20%), Rajasthan (10.10%), Odisha (8.30%), Jharkhand (8.07%), Tamil Nadu (5.34%), Maharashtra (5.32%), Karnataka (4.70%), Andhra Pradesh (3.94%), Telangana (3.86%) and Uttar Pradesh (2.60%), which together contributes around 91% of total HW generated in the country (CPCB, 2019b) as presented in Figure 15.2.

The same report briefs that among the Indian states generating HW, Gujarat tops the list with ~2811723 (MTA) HW of which around 1648699 (MTA) of waste (59%) is treated through various treatment technologies. The 708270 (MT) of HW is sent to common sanitary landfill (43%), 425746 (MT) belongs to schedule IV (26%) which is recycled, about 289276 (MT) is co-processed in cement kilns (18%), while 140467 (MT) waste under Rule 9 and hence has the potential for power/energy generation (9%) and 84940 (MT) is incinerated (5%). Here, Schedule IV refers to commonly recyclable HW, for example, paint and ink sludge/residues and Rule 9 refers to Utilization of HW and other wastes as a resource or after pre-processing either for co-processing, etc. example HW is used in cement kiln as a fuel.

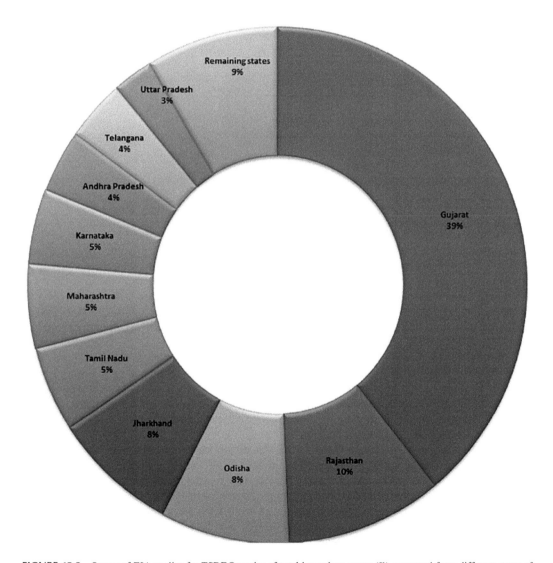

FIGURE 15.2 Stages of EIA studies for TSDF.Quantity of total hazardous waste (%) generated from different states of India. Pie chart presentation of major HW (%) generating states in India. SLF, Secured landfill; In, Incineration; CK, Cement kiln; Rule 9, wastes supplementary resource or for energy recovery; SchIV, Wastes comes under recyclable HW (Hazardous and Other Wastes (Management and Transboundary Movement) Rules, 2016); AT, Actual treated HW. (CPCB (2019b).)

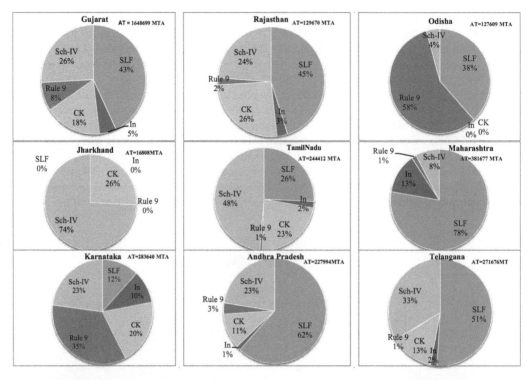

Figure 15.3 Pie chart presentation of major HW (%) generating states in India. SLF, Secured landfill; In, Incineration; CK, Cement kiln; Rule 9, wastes supplementary resource or for energy recovery; SchIV, Wastes comes under recyclable HW (Hazardous and Other Wastes (Management and Transboundary Movement) Rules, 2016); AT, Actual treated HW.

Rajasthan generates 724449 (MTA) HW of which around 129670 (MTA) of waste (18%) is treated through various treatment technologies. Around 58374 (MT) of HW is sent to common sanitary landfill (45%), 30868.32 (MT) belongs to Schedule IV* (24%) which is recycled, about 33656 (MT) is co-processed in cement kilns (26%), while 2656 (MT) is the waste under Rule 9 and hence has the potential for power/energy generation (2%) and 4117 (MT) is incinerated (5%). Odisha generates 595339 (MTA) and the actual volume of treated HW is 127609 (MTA) which is (21%) of the total volume and is treated through various treatment technologies. Around 48692.9 (MT) of HW is sent to common sanitary landfill (38%), 5514.88 (MT) belongs to Schedule IV* (4%) which is recycled, about 31.42 (MT) co-processed waste in cement kilns (0.02%), while 73370.15 (MT) (57%) is the waste under Rule 9 and hence has the potential for power/energy generation. Likewise, for the remaining states, the volume of waste generated and the numbers of treated wastes are presented in Figure 15.3. Noting down such a huge quantum of HW generation and the way they are addressed under various treatments suggests the need for more numbers of ICHWTSDFs as a feasible outcome.

The TSDFs or its further edition as ICHWTSDFs will not only help in treating HW but then emphasize on extracting out the reusable and recyclable products which then serve the purpose of waste to valorisation. During 2018–2019, the central pollution control board has introduced standard operating procedures for the utilization of HW under Rule 9 of the Hazardous and Other Wastes (Management and Transboundary Movement) Rules, 2016, and accordingly, guidelines were prepared based on the types of HW, their source of generation and anticipated recovery/product from the waste. Among the mentioned wastes, spent sulphuric acid, spent aluminium carbonate, tarry residues come under schedule I and aluminium dross residues, spent aluminium chloride come under schedule-II. Here schedule I indicates processes generating HW, while schedule-II indicates waste constituents with concentration limits and recovers the products from these HWs as mentioned in Table 15.3.

TABLE 15.3

Utilization of Hazardous Wastes as a Recovery Product

Schedule	Type of HW	Source of Generation	Recovery/Product
I	Spent sulphuric acid	Generated from dyes and dyes intermediates industries	As a supplementary resource in the manufacturing of dyes and dye intermediates
	Spent sulphuric acid	Generated from dyes and dyes intermediates industries chemical manufacturing industries	As a neutralized agent in ETP/CETP
	Spent sulphuric Acid	Generated during manufacturing of G-salt (dye and dye intermediate sector)	R-complex and gamma acid
	Tarry residue waste	Coal gasifier units	For the production of creosote oils and coal tar pitch
	Spent aluminium dross residue/ rejects	Generated during aluminium smelting process	Calcium aluminate (synthetic slag)
	Spent ammonium carbonate	During manufacturing of copper phthalocyanine blue (dye and dye intermediate)	Metallic carbonates (i.e. Zinc, Cu, Mn, Mg and Ferrous)
II	Aluminium dross residues	Generated from separation of metal from aluminium dross/aluminium dross reprocessing units	As a supplementary resource for manufacturing of alum
	Spent aluminium chloride	During the production of CPC green and 2,4,6-trimethylbenzoyl chloride.	As a supplementary resource to manufacturing liquid aluminium hydroxide chloride/poly aluminium chloride for further use in ETP
	Spent solvent	Industrial uses of solvents production or industrial use of synthetic dyes, dye intermediates and pigments; production and formulation of pesticides including stock piles.	Recovered solvent or mixture of solvent containing: acetone, toluene, benzene, xylene, cyclohexane, methyl iso butyl ketone, methanol, isopropyl alcohol, methylene dichloride, tetra hydro furan, ethyl acetate, dimethyl acetate, etc.

Source: CPCB (2016b).

15.4 Conclusion

Site selection criteria and evaluation of the site features play a significant role in the establishment of TSDF. Disposal of HW on land is the most convenient and economical way of disposing of waste, but this leads to environmental degradation of natural resources like groundwater contamination and land degradation. It is necessary to isolate hazardous contaminants of the waste from the environment by using proper technology. A total of 23 attributes were evaluated from six environmental parameters on five levels sensitive scale measured from excellent to bad (100% to 20%) to select the ideal site that does not disturb the surrounding environmental features. In the theoretical study presented in this chapter, weightages for chosen attributes were assigned to measure the environmental conditions that indicate good suitability of site under the site selection criteria, with a value of 65%. Such exercise of assigning weightages for the management of waste through the integrated facility is very effective, as it can suggest suitable sites that can handle different types of wastes generated from various sources, which in turn supports sustainable waste management, cost-effectiveness, time and land. Based on the information gathered on major HW generating states of India, it is suggested that these states should follow the detailed analysis and indexing for TSDFs site selection for the establishment of new integrated common HW TSDFs with appropriate treatment, majorly focusing on reuse recycling technologies for effective waste management.

Conflict of Interest

The authors state no conflict of interest for this work.

Acknowledgement

The authors would like to thank Ramky Group, India for providing knowledge resources for the article. The authors express gratitude to the Director, Jawaharlal Nehru technological University Hyderabad for support and providing the facility to carry out the work.

REFERENCES

Abessi, O. and Saeedi, M., 2010. Hazardous waste landfill siting using GIS technique and analytical hierarchy process. *Environment Asia*, 3(2), pp. 47–53.

Aichele, R. and Felbermayr, G., 2012. Kyoto and the carbon footprint of nations. *Journal of Environmental Economics and Management*, 63(3), pp. 336–354.

Babu, B.V. and Ramakrishna, V., 2000. Extended studies on mathematical modeling of site sensitivity indices in the site selection criteria for hazardous waste treatment, storage and disposal facility. *Journal of the Institution of Public Health Engineers India*.

Babu, B.V. and Ramakrishna, V., 2012. Hazardous waste management in India. *Birla Institute of Technology and Science Pilani*, 333, 31.

Bain, A., Shenoy, M., Ashton, W. and Chertow, M., 2010. Industrial symbiosis and waste recovery in an Indian industrial area. *Resources, Conservation and Recycling*, 54(12), pp. 1278–1287.

Beskese, A., Demir, H.H., Ozcan, H.K. and Okten, H.E., 2015. Landfill site selection using fuzzy AHP and fuzzy TOPSIS: a case study for Istanbul. *Environmental Earth Sciences*, 73(7), pp. 3513–3521.

CPCB, 2016a. Revised guidelines for common bio-medical waste treatment and disposal facilities.

CPCB, 2016b. Standard operating procedure for schedule-I&II (CPCB, 2018-2019) for hazardous waste under rule 9 of the HW rule.

CPCB, 2003. Guidelines for conducting environmental impact assessment: Site selection for common hazardous waste management facility, document series HAZWAMS/25/2002-2003.

CPCB, 2019a. Guidelines for preparation of inventories on hazardous and other waste generation and their management, New Delhi, India.

CPCB, 2019b. Interim report of monitoring committee on management of hazardous waste, New Delhi, India.

Dubey, A. and Lal, R., 2009. Carbon footprint and sustainability of agricultural production systems in Punjab, India, and Ohio, USA. *Journal of Crop Improvement*, 23(4), pp. 332–350.

Envotech Waste Management Limited. 2015. Environment impact assessment report of integrated waste management facility comprising of TSDF.

Gorsevski, P.V., Donevska, K.R., Mitrovski, C.D. and Frizado, J.P., 2012. Integrating multi-criteria evaluation techniques with geographic information systems for landfill site selection: A case study using ordered weighted average. *Waste Management*, 32(2), pp. 287–296.

Higgins, T.E., 2018. *Hazardous Waste Minimization Handbook*. CRC Press: Boca Raton, FL.

IL&FS, 2010. Technical EIA guidance manual for common hazardous waste treatment storage and disposal facilities, Hyderabad.

Joseph, K., 2006. Stakeholder participation for sustainable waste management. *Habitat International*, 30(4), pp. 863–871.

Karthikeyan, L., Suresh, V.M., Krishnan, V., Tudor, T. and Varshini, V., 2018. The management of hazardous solid waste in India: An overview. *Environments*, 5(9), p. 103.

Li, M., Xu, J. and Li, B., 2018. Analysis of development of hazardous waste disposal technology in China. In *IOP Conference Series: Earth and Environmental Science* (Vol. 178, No. 1, p. 012027). IOP Publishing.

Lifset, R. and Graedel, T. E., 2002. Industrial ecology: Goals and definitions. In: *A Handbook of Industrial Ecology*, 3–15. Edward Elgar Publishing: Northampton, MA.

Misra, V. and Pandey, S.D., 2005. Hazardous waste, impact on health and environment for development of better waste management strategies in future in India. *Environment International*, 31(3), pp. 417–431.

Mmereki, D., Baldwin, A., Li, B. and Liu, M., 2017. Healthcare waste management in Botswana: Storage, collection, treatment and disposal system. *Journal of Material Cycles and Waste Management*, 19(1), pp. 351–365.

MoEF&CC. 2015. Hazardous and other wastes (management and trans boundary movement) rules, New Delhi, India.

Morrison III, D.R., Stern, P.M.C. and Osorio-Amado, P.C.H., 2018. Handle hazardous waste safely. *Chemical Engineering Progress*, 114(4), pp. 42–48.

Mrayyan, B. and Hamdi, M.R. 2006. Management approaches to integrated solid waste in industrialized zones in Jordan: A case of Zarqa City. *Waste Management*, 26(2), pp. 195–205.

Mukherjee, A., Debnath, B. and Ghosh, S.K., 2016. A review on technologies of removal of dioxins and furans from incinerator flue gas. *Procedia Environmental Sciences*, 35, pp. 528–540.

Sadala, S., Dutta, S., Raghava, R., Jyothsna, T.S., Chakradhar, B. and Ghosh, S.K., 2019. Resource recovery as alternative fuel and raw material from hazardous waste. *Waste Management & Research*, 37(11), pp. 1063–1076.

San Envirotech Pvt. Ltd. 2015. Environmental impact assessment report for M/s. Cluster Enviro Pvt. Ltd. for the proposed TSDF.

Shen, L.Y., Tam, V.W., Tam, L. and Ji, Y.B., 2010. Project feasibility study: The key to successful implementation of sustainable and socially responsible construction management practice. *Journal of Cleaner Production*, 18(3), pp. 254–259.

Sonkamble, S., Sethurama, S., Krishnakumar, K., Dhunde, P., Amarender, B. and Kumar, V.S., 2013. Role of geophysical and hydrogeological techniques in EIA studies to identify TSDF site for industrial waste management. *Journal of the Geological Society of India*, 81(4), pp. 472–480.

Sovacool, B.K. and Brown, M.A., 2010. Twelve metropolitan carbon footprints: A preliminary comparative global assessment. *Energy Policy*, 38(9), pp. 4856–4869.

The Associated Chambers of Commerce of India (ASSOCHAM), 2017. PricewaterhouseCoopers (PwC). Waste Management in India-Shifting Gears, New Delhi, India.

16

Using a Composite Index to Assess the Sustainable Urban Access to Basic Services and Solid-Waste Management

Sudha Panda and Sukanya Dasgupta
KIIT University

CONTENTS

16.1 Background

India has experienced phenomenal growth of its cities. The rate of urbanization has grown from 26% in the 1990–2000 decade to 30% in the 2000–2010 decade. Projecting this growth rate by 2030, the urban population will be 40% of the total projected population of 1,470 million. Any economic development in the cities is bound to have a cascading effect on its surrounding rural areas. This growth of urbanization is a double-edged sword and together with the economic potential it poses, it could strain the already tattered infrastructure and services. Since reactive solutions are costly, India should anticipate demand rather than play catch up. The first step towards managing this urbanization is its measurement through a composite index for Basic Infrastructure services.

DOI: 10.1201/9781003231608-19

TABLE 16.1

Benchmarking Indian Cities with Global Cities

Groups	City Prosperity Index	Cities	Productivity Index	Quality of Life Index	Infrastructure Index	Environment Index	Equity Index
A	0.925	Vienna	0.939	0.882	0.996	0.932	0.883
	0.906	Tokyo	0.925	0.931	0.989	0.936	0.828
B	0.709	Mexico	0.743	0.764	0.900	0.866	0.405
	0.793	Moscow	0.806	0.813	0.960	0.908	0.550
C	0.694	Mumbai	0.645	0.739	0.745	0.632	0.715
	0.636	New Delhi	0.596	0.690	0.786	0.448	0.712

Source: UN-Habitat (2016).

16.1.1 Benchmarking Indian Cities on Basic Urban Infrastructure

So how do Indian cities compare with global cities? Indian cities like Mumbai and New Delhi perform averagely as seen by the City Prosperity Index, a global initiative by UN-Habitat, which maps 50 cities across the world on Urban Sustainability with its 5 thematic indices of Productivity, Quality of Life, Infrastructure, Equity, and Environment.

An excerpt of 6 cities (State of World's Cities 2016–2017) classified into 3 groups based on their ranking (Table 16.1) shows that the Indian cities perform weakly on the Infrastructure Indices, whereas the top rung cities of Vienna and Tokyo show balanced high score on all five indices.

16.1.2 Nature of Urban Crisis in India

The following section highlights some of the urban policy crisis in India which implies this research. This research deals with mid-sized cities with population <10 lakhs, which have difficulty in securing funds as they find it difficult to prove their creditworthiness to international agencies that look for quick returns.

The official vision of sustainable urban development is limited to seeing this as an ecological issue, which is subsequently linked to the development infrastructure through independent funding. Hence, it is a very fragmented approach wherein poverty alleviation is visualized independently of infrastructure programmes; the decentralization of governance is not linked with financing of urban development, and so on. In Tamil Nadu, the municipal administration has contracted out solid-waste collection to the women's groups (Rural Development and Panchayati Raj Department Report, 2020). This is a comprehensive approach whereby local communities and government are participating to tackle the environment pollution and poverty alleviation issues jointly.

16.1.2.1 Privatization and Commercialization of Infrastructure

In the name of decentralization and popular participation, there is withdrawal of the state from welfare responsibilities. There is privatization and commercialization of infrastructure including the social sector. 'Sustainable Cities Program' (SCP) has been launched as a tool for implementing Agenda 21. It has been implemented in Chennai, Hyderabad, Bangalore, Delhi, and Calcutta. Infrastructure projects are usually funded by international loans. At the city level, multilateral and bilateral agencies like Financial Institutions Reform and Expansion (FIRE) in Hyderabad and Chennai have opened up new business avenues in India under the guise of improving the urban environment.

16.1.2.2 Urban Finance Restricted to Large Cities

Preliminary observations indicate that megacities are exalted by policymakers by virtue of the contribution they make to the national and state economy. Indeed, the larger the city, the more anti-poor it seems to become and this adversely impacts women, children, and the elderly (Jha and Udas-Mankikar, 2019).

For example, the benefits of Healthy Cities Programme (HCP) and FIRE project will benefit only the large cities. In spite of the minor amount invested by the international agencies compared to the internal resources, they have a large say on official projects. Cities have to invest in projects that give fast returns to make them commercially viable and basic services projects that are not as revenue fetching get sidelined. Hence the poor continue to live in deplorable conditions, while the debt-ridden cities are diverting the funds into commercially viable projects like flyovers and other infrastructure projects. Only metros and large cities have been the main recipients of these loans as they are able to prove their creditworthiness.

16.1.2.3 Basic Services
16.1.2.3.1 Inadequate Network and Coverage

Basic services access is considered to be the single step to improving the urban environment. For example, water supply, and sanitation infrastructure reduce water pollution, and these projects are usually funded by international loans. Since only large cities can prove that they are creditworthy, they have been the main recipients of these loans.

According to official statistics, 50% of the urban population has safe drinking water network coverage, but the average availability is less than four hours per day; however, and in some areas, water is supplied only for one hour on alternate days (Service Level Benchmarking Report, 2011). Many people are forced to draw water from unsafe sources, which leads to widespread waterborne diseases.

Only 69% of the urban population has sanitation facilities and the rest use either dry latrines or defecate in the open. Out of 1,400 towns in the country with municipalities and municipal corporations, only 12.2% have partial sewerage network facilities and 5.3% have some form of primary treatment facilities before final disposal (Service Level Benchmarking Report, 2011).

16.1.2.3.2 Inadequate Collection and Processing of Solid Waste

Not more than 60% of the solid waste generated daily in the urban centres is collected daily, which leads to the accumulation and decomposition of waste in public places, which has adverse effects on public health. There is no arrangement for waste processing except in a few cities where composting is done on a limited scale.

16.1.2.3.3 Small and Medium Sized Cities Worst Hit

Access to basic services is best in the largest cities; it steadily worsens with declining city size. Access to sewerage and drainage facilities especially worsens as city size decreases. Access to drinking water is a bigger problem in smaller cities than in their larger counterparts (Indian Institute of Human Settlements, 2016). Access to network services such as piped water is concentrated in the core of the city, with access levels gradually decreasing towards the periphery. The benefits of urbanization are, therefore, not spilling beyond the core. No major city in India provides more than a small percentage of its population with a continuous water supply.

16.2 Research Methodology

1. To form a framework of Basic Urban Infrastructure, it has to be tailor-made for India by

   a. Mapping theoretical definitions and global practices to Indian policy regulations
   b. Making a gap analysis of India's performance vis-à-vis the global aggregate performance-based indicators.

2. To benchmark the city's performance with the state's (Odisha) urban performance and urban India's performance on Basic Urban Infrastructure and identify the indicators in which the city underperforms.

16.2.1 Obtaining the Model

It is also important that there is a detailed study to identify areas where urban India lags behind the urban world. Special attention needs to be paid to these sectors and these indicators have to be necessarily included in the framework. An analysis to obtain a customized framework of indicators, which will evaluate the typical problems pertaining to the Indian situation, is necessary.

To form an index for Access to Basic Services, in Table 16.2, a mapping was done for each theme between theory, policy and practice, and a gap analysis was done on the resultant indicators to show the potential weak zones where India falls behind the world aggregates and hence those indicators had necessarily to be included in the model. The theoretical basis was obtained from the relevant literature study. If there was mapping on three or more areas, it justified the inclusion in the resultant framework. The criteria for selection of indicators were that it should follow the Bellagio practices (Hass et al., 2002) of being policy-relevant, simple, understandable, and valid. There should be availability of time series data that is cost-effective, i.e. good-quality affordable data. All data for comparison were of 2011. In the absence of a national sustainable strategy, the Report of the Sub Committee on Development of Sustainable Habitat Parameters in the Field of Urban Planning by Town and Country Planning Organization, Government of India, Ministry of Urban Development (Govt of India, 2011) formed the basis of sustainable urban policy guidelines. The global initiatives taken for mapping were the Social Progress Index (Stern et al., 2014), Global Urban Indicator database (UN-Habitat, 2000), Millennium Development Goals (UN-Habitat, 2009), City Data Book (Asian Development Bank, 2001). It was important that the framework was not indicator rich and information poor (OECD, 2008). While referring to Table 16.2, the discussion will be limited to the final five indicators finally chosen because they address the typical problems of India. India's problems are very specific and hence cannot be addressed by anyone of the global frameworks.

16.2.1.1 Access to Basic Services

Access to reliable supplies of safe drinking water and sanitary disposal of sewage are two of the most important means of improving human health. Referring to Table 16.1, the final indicators under the access to basic services theme were as follows:

1. Network coverage (water supply) and households with tap water from treated source: Both are lower than the world average and needs improvement.
2. Percentage of households with toilet facilities and improved sanitation: This is much lower than the world average.
3. Percentage of households with sewerage network: This is also lower compared to the world.
4. Percentage of households with electrical connections: It has risen from 89% in 2000 to 93% in 2011, which is a positive trend and is higher compared to the world.
5. Percentage of households with landline and mobile telephone connections: The percentage of the population with telephone connections has gone up from 0.60% in 91% to 64% in 2011, which is a huge jump in connectivity.

16.3 Data Collection

16.3.1 Sampling Universe

The sampling universe consists of mid-sized cities (population between one to ten lakhs) in India because they constitute two-thirds of the total number of cities having a Municipal Corporation. These mid-sized cities are unable to prove their creditworthiness for international loans, and hence the constraints to their economic sustainability are very severe.

TABLE 16.2

Selection of Indicators for the Framework of the Index

Indicators	Indicators of Access to Basic Services								Gap Analysis	
	S P I	GU I D	MDG	C D B	Global City Indicator	Policy Relevance with National Sustainable Strategy	Theoretical Base	Resultant Framework	Urban India Average	Urban World Average
Percentage households with piped water connections (percent)	√	√	√	√	√	√	√	√	52% network coverage 62% treated supply	63% network coverage 91% treated supply
Percentage households with electricity connections (percent)	√	√		√		√	√	√	92.68	76.7
Percentage households with sewerage network (percent)	√	√	√	√		√	√	√	32.68	61
Percentage households with telephone connections (percent)	√	√		√		√	√	√	64	81
Government expenditure per capita on each service				√		√			*	*
Water consumption				√	√	√			*	*
Price of water						√			*	*
Price of electricity						√			*	*
Quality of electrical supply							√		*	*
Percentage households with toilet connections (percent)	√			√		√	√	√	74	80
Percentage households with solid-waste collections (percent)	√			√		√	√	√	38.23	*

* Data unavailable.

Source for theoretical base: Access to water and sanitation facilities (Global City Indicators, 2008), authorized electrical connections and interruptions ,nutrition (Stern et al., 2014) quality of service (Kaufmann et al., 2005) provision and access to civic amenities and a clean safe and healthy living environment for all (Mahadevia, 2001) safe and sufficient drinking water (Mahadevia, 2001).

Source for data: World Development Indicators (2011), India Human Development Report (2011), World Health Organization (2011), and Service Level Benchmarking Report (2011).

16.3.2 State of Odisha

Odisha has been intentionally chosen for testing the framework. Almost half of Odisha is below the poverty line. It has lagged behind the national average in terms of literacy rate, life expectancy rate, infant mortality rate, and child mortality rate (Government of Odisha, 2004). This high incidence of poverty and relatively low level of human development feed into each other creating unsustainable conditions. Although it has always been one of the least urbanized states in India, since 2011, Odisha has seen rapid rates of urbanization, as seen by the decadal growth rate of 30.28, almost in line with the national average of 32.60. The population of the state has grown by 14% between 2001 and 2011, but the urban population has grown by double the rate. This provides the best context for the sustainability impact to be studied in terms of whether the infrastructure services match the population growth.

The sampling frame is the cities having a Municipal Corporation. Out of the five cities having municipal corporations, four cities have been studied, with Bhubaneswar and Cuttack following a similar demographical pattern and Berhampur and Sambalpur following another pattern. While Bhubaneswar being the capital is a new and planned city, Cuttack is a heritage city with a 1,000-year-old history. Bhubaneswar has extensive scope for growth because of its location and topography but the city of Cuttack is limited between two rivers with a saucer-shaped topography. While Berhampur is a commercial centre, Sambalpur acts as a residential hub for the rapidly industrialized area of Jharsguda.

16.4 Data Analysis

Most of the data for the cities has been obtained from the relevant municipalities (Comprehensive Development Plans), NUIS (National Urban Information System), and census data as of 2011.

16.4.1 Normalization

In real-life situations, indicator values have different measurement units (income in local currencies, electricity in KWh, etc.). For developing composite indicators, it is essential to transform the values of all these indicators into some standard form. Thus, for each of the indicators included in the analysis, a relative indicator is estimated using the actual, minimum and maximum sustainability threshold values. For comparing best case and worst case scenario, the four cities (Siliguri, Asansol, Raipur, and Amritsar) of comparable population, character, and regional setting have been taken (Tables 16.3–16.7).

The relative indicator is developed using a scaling technique where the minimum value is set to 0 and the maximum to 1. The equation used for this is (Table 16.8):

$$\text{Relative indicator} = \frac{\text{Actual value} - \text{Minimum threshold value}}{\text{Maximum threshold value} - \text{Minimum threshold value}}$$

TABLE 16.3

Percentage of Households with Piped Water Connections

	Percentage Households with Treated Piped Water	Network Coverage (%)	Per Capita Supply
Bhubaneswar	47	45% (911 km covers 55% of 1,600 km road length)	150
Cuttack	68	53%	150
Berhampur	64	70%	72.38
Sambalpur	76	80%	135

Source: Census (2011).

TABLE 16.4

Percentage Households with Toilet Facilities

	Percentage of Households with Toilet Facilities	Percentage of Households with Piped Sewer System	Percentage of Households with Septic Tank	Percentage of Households With Other
Bhubaneswar	75	24	41	3
Cuttack	84	17	61	1.26
Berhampur	85	11	63	2.5
Sambalpur	65.63	7	53	2.33

Source: Census (2011).

TABLE 16.5

Percentage Households with Solid-Waste Collections

	Percentage of Households with Solid-Waste Collection (Door to Door)	Percentage of Household Level Coverage	Efficiency of Collection
Bhubaneswar	28.2	75	74.5
Cuttack	0%	39	67
Berhampur	2.6	35	81.6
Sambalpur	0%	62	65

Source: Census (2011).

TABLE 16.6

Percentage Households with Electrical Connections

	Percentage Households with Electricity
Bhubaneswar	86.22
Cuttack	94.36
Berhampur	94.40
Sambalpur	95.08

Source: Census (2011).

TABLE 16.7

Percentage Households with Landline/Mobile Telephone Connections

	Percentage Households with Landline	Percentage Households with Mobile	Percentage Households with Both
Bhubaneswar	4.78	64.78	13.18
Cuttack	4.91	66.39	12.97
Berhampur	4.29	71.70	8.56
Sambalpur	4.22	59.31	11.34

Source: Census (2011).

16.4.2 Linear Aggregation

Weights to the indicators were obtained by doing an expert survey with a questionnaire using a constant sum rating scale. The weights so obtained almost matched the weights obtained from factor analysis confirming the validity of the expert survey and choice of indicators (Table 16.9).

The composite thematic value can be found by

TABLE 16.8

Access to Basic Needs (Relative Indicators)

	Indicators	Unit	A	B	C	D	E	F	G	H	Urban Odisha Average	Urban India Average	Weights
1a	a. Coverage connections- water supply	%	0.72	0.68	0.40	0.53	0.31	0.00	0.23	1.00	0.2	0.7	0.32
1b	b. Percentage households with tap water from treated source	%	0.00	0.56	0.45	0.78	1.00	0.29	0.18	0.65	0.9	0.4	
2	Percentage households with toilet facilities	%	0.31	0.61	0.64	0.00	0.45	1.00	0.28	0.96	0.0	0.3	0.21
3	Percentage households with solid- waste collection	%	0.46	0.03	0.00	0.00	0.00	0.00	0.03	1.00	0.1	0.4	0.08
4	Percentage households with electrical connections	%	0.00	0.65	0.66	0.71	0.14	0.60	0.86	1.00	0.3	0.5	0.13
5a	a. Percentage households with landline telephone connections	%	0.46	0.50	0.32	0.30	0.24	0.00	0.56	1.00	0.1	0.8	0.24
5b	b. Percentage households with mobile telephone connections	%	0.53	0.64	1.00	0.15	0.65	0.90	0.00	0.46	0.3	0.5	

Source: OECD (2008).

A, Bhubaneswar; B, Cuttack; C, Berhampur; D, Sambalpur; E, Asansol; F, Siliguri; G, Raipur; H, Amritsar.

TABLE 16.9

Index Formation for Four Odisha Cities

Theme	Bhubaneswar	Cuttack	Berhampur	Sambalpur
Access to Basic Services	0.34	0.46	0.39	0.36

$$\text{Ti} = \sum\nolimits_{k=1} w_k x_{ik} + w_1 s_{i1} + w_2 s_{i2} + \ldots + w_m s_{il} \tag{16.1}$$

Where Ti the overall score of theme i and x_{ik} the relative indicator value i for criterion j of which w_j is the weight.

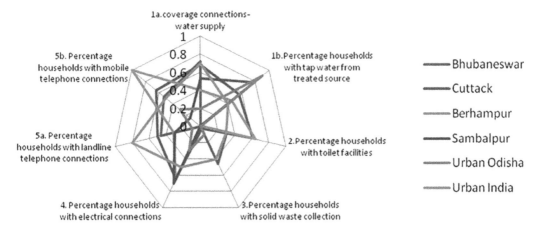

FIGURE 16.1   Indicators of access to basic services for the four cities of Odisha.

16.5 Results Analysis and Discussion

16.5.1 Diagnosis of Weak Sectors for the Cities

The percentage of the city population served by a potable water supply and its area coverage is an indicator of city health, cleanliness and quality of life. Percentage of households with toilet facilities and sewerage connections have a lot of significance on sanitation facilities which has an impact on health also, especially in the Swachh Bharat Mahan context. Percentage households with electrical connections are an indicator of lawful provision of a basic urban service. The number of telephone connections is an indicator of communication technology connectivity.

As can be seen in Table 16.9, all four cities perform badly on the Access to Basic Services Index but Bhubaneswar fares worst with an index value of 0.34. Bhubaneswar should work on increasing the percentage of household with potable water (presently 47%) because it is a basic necessity and Cuttack on having a sewerage network (presently at a dismal rate of 2%). Except Bhubaneswar's solid-waste collection which is 28% of households, all other cities are extremely poor at 0%–2% of households.

In general, the spider charts of all the cities are skewed without a balanced development on all indicators (compared to the urban India figures) (Figure 16.1).

16.6 Conclusion and Practical Significance of the Model

This model has done mapping of the theory, practices, and policy besides identifying the critical problems confronting the Indian cities. The purpose is to relate the indicators to the goals of sustainability and the crisis confronting the cities. If the indicators are goal-related, they can be used for policy fine-tuning. For example, there is a limited approach to urban poverty. Medium and small-sized cities have a larger share of the poor and a larger share of the slum population in India than the million-plus population cities. Hence, the sampling universe consists of 89 mid-sized Indian cities (population between one to ten lakhs because they represent two-thirds of the total number of cities having a Municipal Corporation).

The model with its case study in Odisha is standardized and replicable across India, with 5 indicators. The data for a majority of the indicators were taken from the census which ensured that cost-effective, reliable and good-quality data is available in time series so that the progress in social sustainability can be charted. Since all themes and indicators are interconnected, the linear aggregation model is ideal to capture the trade-offs between different indicators. It can be used for projecting what-if scenarios according to different policy scenarios.

Linear additive model is a simple model which can be easily understood by decision-makers and can be projected to reflect different what-if scenarios. It can be used as a diagnostic tool and for target setting. The research has shown that it can be used for prediction and for rewarding good performance. It can give a scientific basis for resource allocation and understand the parameters that lead to sustainable city planning. It can be used to identify the growth pattern of the urban centres in a time frame of a decade. This can be used at a broader level of regional planning to plan the growth matrix of cities in a particular region.

One of the main purposes of forming a composite social sustainability index is the allocation of resources amongst the various indicators of the theme. There are so many cross-currents between urban growth and policy reality, that these indicators not only help to inform and empower policymakers, citizens, researchers, and activists to measure and monitor sustainable development but also for policy-making and for prioritization of budget allocation between various sectors. To achieve faster economic growth, there is a need to enhance human capabilities. The challenge of urbanization is to ensure planning for these basic services delivery at the minimum standards.

The integrative process for measurement of sustainability has been done through an integrative conceptual framework, integration by the selection of specific indicators and aggregation of the indicators. A very good example of this integrative mechanism in policymaking is seen in Curitiba, Brazil. Here garbage collection by slum dwellers is encouraged by the exchange of 6 bags of trash with 1 bag of grocery, instead of cash which fosters health and economic well-being so that productive workforce increases.

Recently, Bhubaneswar has been ranked the premier city of India among the 28 smart cities; however, the sustainability in the access to basic services as seen by the research leaves much to be desired. The solid-waste collection process in Bhubaneswar is weak in terms of manpower and vehicle availability. Bin capacity provided is inadequate and locations are also inappropriate. Untreated solid waste is dumped on open land after collection. In comparison to other cities in the country, the budget allocation is also inadequate. Besides political will, technological capability and management competence in all functional areas need intensive development to achieve the status of a smart and clean city (Nanda and Panda, 2019).

A scorecard of the cities based on this model reveals the actual sustainability position of the city and can be used to make politically informed judgements. It encourages public accountability and would go a long way in evaluating our cities.

REFERENCES

Asian Development Bank (2001) *Urban Indicator for Managing Cities*. Chapter 2 from Cities Data Book. Retrieved on 26 July, 2015 from: https://www.adb.org/sites/default/files/publication/30020/urban-indicators-managing-cities.pdf.

Census (2011) Percentage of households to total households amenities and assets. Census of India. Retrieved from: http://censusindia.gov.in/.

Global City Indicators (2008) *Global City Indicators Program Report - Part of a Program to Assist Cities in Developing an Integrated Approach for Measuring City Performance*. Retrieved from: http://www.cityindicators.org/ Deliverables/ Final % 20 Indicators %20Report% 203_21_08_4-23-2008-924597.pdf.

Government of Odisha (2004) Human development report: 2004. Planning and Coordination Department. Retrieved from: www.in.undp.org//human_development_report_2004_orissa_full_report.pdf.

Govt of India (2011) Report of the Sub Committee on Development of Sustainable Habitat Parameters in the Field of Urban Planning by Town and Country Planning Organisation Government of India Ministry of Urban Development.

Hass J.L., Brunvoll F., and Hoie H. (2002). Overview of sustainable development indicators used by National and International Agencies. OECD Statistics Working Papers. DOI: 10.1787/838562874641, Retrieved from: http://www.oecd-ilibrary.org/economics/overview-of-sustainable-development-indicators-used-by-national-and-international-agencies.

India Human Development Report (2011) Govt of India, Planning Commission.

Indian Institute for Human Settlements (2016) How urban is India? IIHS Case No 1-0002. Bangalore.

Jha, R., Udas-Mankikar, S. (2019) India's urban challenges: Recommendations for the new government (2019-2024), ORF Special Report No. 90, Observer Research Foundation.

Kaufmann, D., Léautier, F., Mastruzzi, M. (2005) Governance and the city: An empirical exploration into global determinants of urban performance. Policy Research Working Papers, World Bank Institute. Retrieved from: http://elibrary.worldbank.org/doi/pdf/10.1596/1813-9450-3712.

Mahadevia, D. (2001). Sustainable urban development in India: An inclusive perspective. *Development in Practice*, 11(2–3), 242–259. doi: 10.1080/09614520120056388.

Nanda, S., Panda, H. (2019). Solid waste management in Bhubaneswar: Practices and challenges. *International Journal of Management, Technology and Engineering*, 9, 821–838.

NUIS data (2011) Retrieved from: www.tcpomud.gov.in.

OECD (2008) *Handbook on Constructing Composite Indicators: Methodology and User Guide.* Retrieved from: www.oecd.org.

Rural Development and Panchayati Raj Department Report (2020) Govt of Tamil Nadu. Retrieved from: https://tnrd.gov.in/policynotes.html.

Service Level Benchmarking Report (2011) Ministry of Urban Development, Government of India.

Stern, S., Porter, E.M., Green, M. (2014) Social progress index 2014 methodological report. Social progress imperative. Retrieved from: http://www.socialprogressimperative.org/wp-content/uploads/2016/06/Social-Progress-Index-2014-Report-e-2.pdf.

UN-Habitat (2000) Global urban indicators database (version 2) by global urban observatory, United Nations Human Settlements programme.

UN-Habitat (2016) State of the world's cities report 2016/2017: Prosperity of cities. Retrieved from: www.unhabitat.org.

UN-Habitat Report (2009) Retrieved from: www.unhabitat.org.

Urban Indicator Guidelines (2009) "Better Information, Better Cities" Monitoring the Habitat Agenda and Millennium Development Goals-Slum targets.

World Development Indicators (2011) World Bank. Retrieved from: www.worldbank.org.

World Health Organization (2011) *Healthy Cities Indicators: Analysis of data from cities across Europe.* World Health Organization Regional Office for Europe (Copenhagen). Retrieved on July 17, 2015 from: http://glocalhealth consultants.com/wp-content/uploads/2014/11/globalAndLocalHealth.pdf.

17

Adsorption of Ammonia Using Cost-Effective Waste Camellia sinensis: Comparison of Linear and Nonlinear Regression Methods by Error Analysis

Baisali Rajbansi and Rajdeep Mullick
University of Calcutta

CONTENTS

17.1 Introduction

Nowadays, the removal of nitrogenous organic compounds present in wastewaters has attracted a good attention because they can be degraded to ammonia. Due to large quantities of industrial and municipal wastewater discharge, NH_4^+ concentration in certain surface waters is much higher than the permissible level [1]. This reduces the availability of safe drinking water [2] and thus indirectly affects human being. Also due to depletion of dissolved oxygen and toxicity, it is unsafe to fish and other aquatic organisms. So, nowadays in many countries, the removal of NH_4^+ from wastewater is getting more attention, as the consent levels for the amount of ammonium contained in the final effluent.

The release of ammonia in the atmosphere is mainly caused by many natural sources, such as animal wastes and microorganisms in soil and anthropogenic sources like fertilizer production, manufacturing industry (of ammonia and coke), sewage treatment plants, as well as refrigeration systems. Its toxicity affects the environment in two ways via photochemical smog formation and eutrophication of the surface waters and human health. Ammonia is colourless but has a pungent odour and corrosiveness, which can cause irritation, and its toxicity can harm the respiratory system or can leading cause to death. It is known to be harmful to aquatic organisms at concentration above 25 µg/L [3]. Due to its high solubility, it can easily affect skin, eyes and the respiratory system to form ammonium hydroxide via an exothermic process that is responsible for thermal injuries. Considering all these, the National Research Council (NRC) has identified ammonia emissions as a prime air quality concern up to a global level.

DOI: 10.1201/9781003231608-20

Adsorption is supposed as one of the best methods due to its inexpensiveness, universal nature and ease of operation among various wastewater treatment techniques. Usually, chemisorption gives much stronger interactions than physisorption [4] and enables a better selective adsorption process. In reactive adsorption, the surface chemistry mostly influences the transformations of the adsorbate. The interactions will differ depending on the chemical nature of the adsorbate molecules and the strength of functional groups present on the surface of the adsorbent. Effective adsorption of ammonia must involve some important factors like hydrogen bonding, acid–base reactions and complex formation (ammonia and common ligand).

The traditional processes mostly used to lessen ammonia from wastewater are biological nitrification–denitrification, air stripping and chemical precipitation. A methodical biological treatment is a very expensive system that would also require a relatively long time. However, in recent times, great efforts have been made to identify new economically feasible and environment friendly treatment alternatives, of which adsorption has received a big attention. Adsorption is one of the effective ways to remove colour, odour, and organic and inorganic pollutants from wastewater, as it is useful for the adsorbents to selectively adsorb a certain mass of components onto their surfaces.

Tea is basically the processed dried leaves of only one species of plant, *Camellia sinensis* [5]. It is one of the most popular beverages, and every year about 3.5 million tons of tea is consumed in the world [5]. So, a huge amount of spent tea leaf or its residue was disposed of to the environment through daily tea drinking. This is a domestic waste that is available in enormous amount all over in most of the country. This is an oxygen-demanding pollutant and takes a long time for biodegradation. The processed waste tea is not a pollutant, rather it could be used as an adsorbent. After extraction of tea from tea leaves, generally, makers face a great problem in disposing of the spent tea leaves. Therefore, utilization of the STL is very much desirable.

Activated carbon is undoubtedly considered universal adsorbent for the removal of different types of pollutants from wastewater. However, it is sometimes restricted due to its higher costs. So, the use of spent tea leaves as low-cost adsorbents will be attractive due to their contribution to the reduction of costs and contributing to environmental protection.

This study aims to produce low-cost adsorbent from spent tea leaves that were chemically activated. In this research spent tea leaves adsorbent (STLA) has been used for the removal of ammonia from an aqueous solution.

Over the last few years, linear regression has been developed as a major option in designing adsorption systems [6]. A number of error functions (sum of the squares of errors, ERRSQ; hybrid fractional error function, HYBRID; Marquardt's percent standard deviation, MPSD; average relative error, ARE; Sum of the absolute errors, EABS; chi-square function, χ^2; AICc, etc.) were highlighted and discussed for the comparisons of linear and nonlinear isotherm models [7–12]. According to recent research, the development of nonlinear isotherms represents a potentially viable and powerful tool, which also leads to greater improvement in the area of adsorption science [13]. However, according to Foo and Hameed [7], more explorations in this area are recommended. Moreover, these error estimating functions do not take into account the number of parameters in the models. Hence, they cannot be used for the selection of model(s) appropriateness and determination of its/their magnitude. The literature reports that researches are mostly done using the Akaike information criterion (AIC) for model selection and ranking in various field [14–16].

In this context, the scope of the present study was to investigate the applicability of the prepared modified adsorbent from waste tea leaves as an alternative "eco-friendly" adsorbent for ammonia removal from aqueous phase at equilibrium conditions. Linear and nonlinear regression analysis was performed to determine model parameters. Six error functions were applied to evaluate, compare and rank the feasibility of the four applied isotherm models (Langmuir, Freundlich, Temkin and Dubinin–Radushkevich).

17.2 Material and Methods

17.2.1 Adsorbent Preparation

The spent tea leaves taken from the nearest shop were washed several times with hot distilled water to remove surface impurities, colour and odour and dried in the sunlight for about 20 days. This naturally dried material is then kept in the hot air oven for 12 hours at 50°C. Then it was crushed in a grinder and passed through the 100 mesh sieve plate to produce an adsorbent of uniform size. This powdered mass was digested with dilute H_2SO_4 solution and kept at 30°C for 12 hours. Then it was washed 20–22 times using hot distilled water and kept for drying 24 hours at 30°C. Finally, dried mass was taken in a silica crucible and kept at 500°C in a muffle furnace for 15 minutes. This chemically modified mass was collected in air-tight packets and kept in desiccators to keep it moisture free.

17.2.2 Equilibrium Studies

The study of ammonia adsorption experiments was conducted with a batch system. A known amount of adsorbent and aqueous solution in a series of 250-mL conical flasks was mixed under a magnetic stirrer. The effects of the stirring time, initial ammonia concentration, adsorbent dose and pH on the adsorption of ammonia were investigated. The total ammonia in aqueous solutions was measured using the UV-VIS-NIR spectrophotometer (Make-U-4100 HITACHI). The absorbency of the solution was monitored by visible spectroscopy at the maximum wavelength of 570 nm (λ_{max}). The colour of the solution was developed using a certain amount of Nessler's reagent as reference and the corresponding concentration of total ammonia in aqueous solution is determined by using the standard curve method [17]. All calibrations were performed using a standard curve over the ammonia range of 0.5–10.0 mg/L and the obtained standard curve had linear correlation coefficients (R^2) of 0.996. The quantity of adsorbed ammonia (q, mg/g) was calculated.

$$\text{Removal percentage} = \frac{(C_0 - C)}{C_0} \times 100 \tag{17.1}$$

The adsorption capacity (mg/g) was calculated with respect to equation

$$q = \frac{(C_0 - C) \times V}{M} \times 100 \tag{17.2}$$

where q is the ammonia adsorption capacity (mg/g), C_0 is the initial ammonia concentrations in the solution (mg/L), C is the ammonia concentration in the solution at time t (mg/L), V is the solution volume (L) and M is the mass of adsorbent (g).

17.2.3 Isotherm Modelling

In order to optimize the design of a specific adsorbate/adsorbent system, it is important to establish the most appropriate correlation for the experimental equilibrium data [18]. The adsorption behaviour of ammonia on prepared adsorbent in the present research was modelled by the Langmuir, Freundlich, Temkin and Dubinin–Radushkevich isotherm equations (Table 17.1).

TABLE 17.1

Mathematical Equations of the Applied Isotherm Models

Isotherm Model	Nonlinear Form	Linear Expression	Plot
Langmuir	$q_e = \dfrac{q_{max}K_L C_e}{1+K_L C_e}$	$\dfrac{1}{q_e} = \dfrac{1}{q_{max}} + \dfrac{1}{q_{max}K_L C_e}$	$\dfrac{1}{q_e}$ versus $\dfrac{1}{c_e}$
Freundlich	$q_e = K_F\, C_e^{1/n}$	$\ln q_e = \ln K_f + \dfrac{1}{n}\ln C_e$	$\ln q_e$ versus $\ln C_e$
Temkin	$q_e = \dfrac{RT}{b}\ln\!\left(A_T C_e\right)$	$q_e = \dfrac{RT}{b_T}\ln A_T + \dfrac{RT}{b_T}\ln C_e$	q_e versus $\ln C_e$
Dubinin–Radushkevich	$q_e = q_s e^{\left(-k_{ad}\epsilon^2\right)}$	$\ln q_e = \ln q_s - K_{ad}\epsilon^2$	$\ln q_e$ versus ϵ^2

17.2.4 Error Analysis

In the present research, linear and nonlinear regression analysis was performed to determine the values of the isotherm model parameters. Six different error functions (ERRSQ, HYBRID, ARE, χ^2, MPSD, and R^2) were examined to evaluate the applicability of each model isotherm equation to the experimental data (Table 17.2) using the solver add-in functions of Microsoft Excel software.

17.3 Result and Discussion

In this section, among the four studied isotherms models, the best-fitting one was determined by the use of six well-known error functions to calculate the error deviation between experimental and predicted equilibrium adsorption data, after both linear and nonlinear analysis. Among all of the error methods, it was assumed that both the liquid and the solid phase concentration contribute equally to weighting the error criterion for the procedure of model solution. Alternative isotherm parameters were also determined by nonlinear regression using error functions (ERRSQ, HYBRID, MPSD, ARE, and χ^2) and the corrected AIC$_c$. The values of the four model constants and the isotherm error deviation data are presented in Table 17.3.

The experimental data points of ammonia adsorption on STLA are shown in Figures 17.1–17.4. It shows a gradual increase in adsorption rate from lower to higher concentration. So, it could be interpreted that due to the presence of a number of vacant active sites, as well as a larger continuance of the adsorption process in the later stages adsorption capacity increases.

The Langmuir equation predicted a maximum monolayer adsorption capacity Q_m 52.63 mg g^{-1} (Table 17.3), while the values of this parameter calculated by applying the nonlinear approach ranged between Q_m 49.96 and 76.33 mg g^{-1} (Table 17.4). The comparative analysis between the values of the error functions, obtained through the linear approach, showed the two-parameter Langmuir model as the one with the highest R^2 0.9661 (Table 17.3) and also characterized by the lowest ERRSQ, χ^2, HYBRID, ARE and MPSD error values showed in bold.

Apparently, the Langmuir isotherm presented a better correlation with experimental equilibrium data than the other three models and could be identified as the most suitable for modelling the equilibrium sorption behaviour of ammonia on the prepared adsorbent.

Considering the theoretical bases of the Langmuir model, the studied separation process could be described as monolayer adsorption onto a surface containing a finite number of identical sites and also represents the equilibrium distribution of ammonia among the solid and liquid phases.

Statistically, it is expected that the higher the number of parameters in a model equation, the closer the theoretical estimates should be to the empirical data. Moreover, the error functions HYBRID and MPSD could be accepted as the most indicative, adequate and essentially meaningful when determining the best-fit isotherm model, as the number of the isotherm parameters is accounted only by them [18].

TABLE 17.2

Error Functions Used to Discriminate between Models

Error Function	Abbreviation	Definition/Expression
Sum squares error	ERRSQ	$\sum_{i=1}^{n}\left(q_{e,calc}-q_{e,meas}\right)_i^2$
Hybrid fractional error function	HYBRID	$\dfrac{100}{n-p}\sum_{i=1}^{n}\left[\dfrac{q_{e,meas}-q_{e,calc}}{q_{e,meas}}\right]_i$
Average relative error	ARE	$\dfrac{100}{n}\sum_{i=1}^{n}\left|\dfrac{q_{e,meas}-q_{e,calc}}{q_{e,meas}}\right|_i$
Chi square	χ^2	$\sum_{i=1}^{n}\left|\dfrac{\left(q_{e,calc}-q_{e,meas}\right)^2}{q_{e,meas}}\right|_i$
Marquardt's percentage standard deviation	MPSD	$100\sqrt{\dfrac{1}{n-p}\sum_{i=1}^{n}\left[\dfrac{q_{e,meas}-q_{e,calc}}{q_{e,meas}}\right]_i^2}$
The coefficient of determination	R^2	$\dfrac{\left(q_{e,meas}-\overline{q_{e,calc}}\right)^2}{\sum\left(q_{e,meas}-\overline{q_{e,calc}}\right)^2+\left(q_{e,meas}-q_{e,calc}\right)^2}$
Corrected Akaike information criterion	AIC_C	$AIC+\left(\dfrac{2p(p+1)}{n-p-1}\right)$, where, $AIC=2p-n\left[\ln\left(\dfrac{SSR}{(n-p)}\right)\right]$
Akaike weight for the *i*th model	(λ_i)	$\lambda_i=\dfrac{\exp\left(-\left(\tfrac{1}{2}\right)\Delta_i\right)}{\sum_{i=1}^{R}\exp\left(-\left(\tfrac{1}{2}\right)\Delta_i\right)}$, where $\Delta_i=AIC_{Ci}-AIC_C$

To prove the observations, the modes of the experimental and predicted model isotherms obtained on the basis of the linear and nonlinear approaches were also compared (Figures 17.1–17.4). The Langmuir isotherms presented in Figure 17.1 correlate the experimental equilibrium data satisfactorily. It was also observed that the ERRSQ and Chi-square methods yielded the best fit in the low concentration range, while the LTFM (linear transform model) and other nonlinear isotherm could be applied for modelling the high concentration region.

Among the four applied isotherm equations, the Langmuir (Figure 17.1) isotherm could be identified as the most suitable for modelling the equilibrium sorption behaviour of ammonia in prepared adsorbent covering the entire concentration range. Besides, the individual model curves, derived on the basis of the studied error functions, practically coincided in both cases.

FTIR spectral analysis was done on the adsorbents to explore the presence of important functional groups that might be relevant to the interaction with ammonia. In adsorption, the most important functional groups are carboxyl, amino, hydroxyl and thiol groups due to their high nucleophilicity.

However, in the case of FTIR spectra of acid-treated spent tea leaves shown in Figure 17.5, there is a sharp shift in positions of O–H group [19]. The peak $1,398\,cm^{-1}$ indicates carboxylic acid group [20].

Complexation of ammonium ions may be formed with ionized O–H group of "free" hydroxyl groups and bonded O–H bands of carboxylic acids in the intermolecular and intra-molecular hydrogen bonding

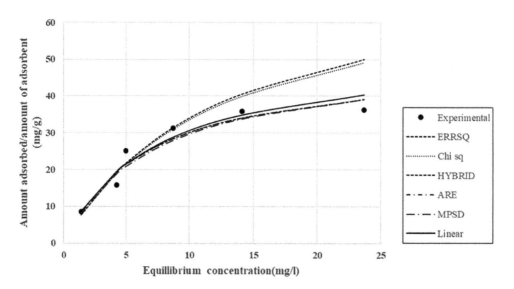

FIGURE 17.1 Experimental equilibrium data and Langmuir isotherms of ammonia sorption on the prepared adsorbent.

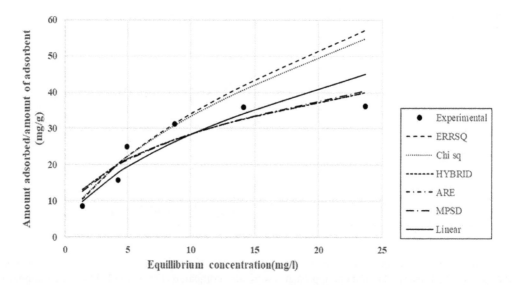

FIGURE 17.2 Experimental equilibrium data and Freundlich isotherms of ammonia sorption on the prepared adsorbent.

of formed new compounds [19,20]. It could be postulated that most of the hydroxyl is forming part of carboxyl groups that have a high affinity towards pollutants [21,22].

Hence, there is a possibility of chemical interaction between this positive charge and negative charge existent and delocalized in the anionic structure of the adsorbent.

The comparison between the values of the six error estimating functions and the modes of experimental versus model isotherms outlined the Langmuir (Figure 17.1) equation as the most adequate model describing ammonia adsorption in the prepared modified adsorbent.

In this work, the corrected AIC_C was used to verify the conclusions for the best-fitting model withdrawn on the bases of error analysis, so to rank the four isotherm models. The AIC developed by Akaike is a methodology for model selection in a situation where more than one model has been fitted to experimental data and screening of the best models is crucial to the objectives of the research work.

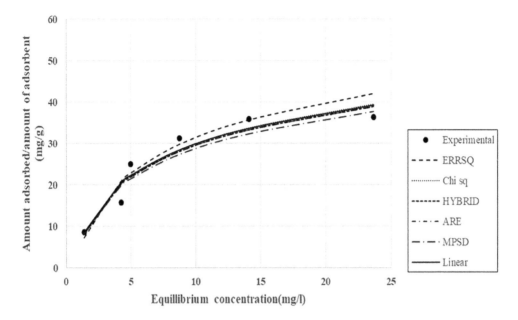

FIGURE 17.3 Experimental equilibrium data and Temkin isotherms of ammonia sorption on the prepared adsorbent.

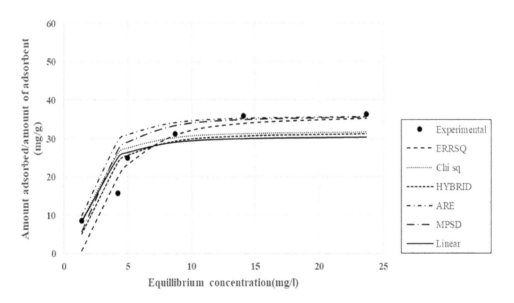

FIGURE 17.4 Experimental equilibrium data and Dubinin–Radushkevich isotherms of ammonia sorption on the prepared adsorbent.

Akaike's general approach not only allows the best model to be identified but also allows the ranking of the rest of the models under consideration [23,24]. The data in Table 17.4 show that the AIC_C for the Langmuir model has the minimum AIC_C value (18.8673). This implies that the theoretical data obtained from this model fits the experimental results better than the other three equilibrium isotherm equations. The ranking of the isotherm equations was made on the basis of the Relative Akaike Weight (RAW) (Table 17.2), the values of λ_i (Table 17.4) which clearly distinguished the appropriateness of the four isotherm equations ranking two-parameter Langmuir model as the ($\lambda_i - 0.4988$) best-fitted ones.

TABLE 17.3

Isotherm Error Deviation Data Related to the Adsorption of Ammonia on Prepared Activated Carbon Using Six Commonly Applied Error Functions—Linear Regression Analysis

	Freundlich		Langmuir		Temkin		Dubinin–Radushkevich	
	K_F	N	q_{max}	K_L	A_T	b_T	q_s	K_{ad}
Parameter	8.2	1.856	52.63	0.139	1.57	10.86	30.63	7×10^{-7}
R^2	0.8911		**0.9661**		0.918		0.8238	
ERRSQ	78.96323512		**32.22623847**		38.47540738		163.3267235	
χ^2	4.52203478		**1.565684411**		2.01182632		7.664610451	
HYBRID	113.0508695		**39.14211029**		50.29565801		191.6152613	
ARE	75.36724633		**26.09474019**		33.53043867		127.7435075	
MPSD	32.07128824		**15.40565192**		17.54140167		32.03278169	

Note: The highest R^2 and the lowest ERRSQ, HYBRID, MPSD, ARE, and χ^2 are in bold.

TABLE 17.4

Values of Langmuir, Freundlich, Temkin, and Dubinin–Radushkevich Model Parameters for the Adsorption of Ammonia Obtained through Nonlinear Regression Analysis and Error Deviation Data Using Five Error Functions

	ERRSQ	χ^2	HYBRID	ARE	MPSD	AIC_C	λ_i
Freundlich Model							
K_F	8.4666035	8.785380816	11.39068593	11.39068592	11.05186641		
N	1.65519968	1.726364667	2.522155716	2.522155716	2.43531009		
Error	28.47277146	1.663338471	75.59030408	50.39353606	18.3475447	19.77592329	0.316735802
Langmuir Model							
K_L	0.080465475	0.081564877	0.150693052	0.150693052	0.139954901		
q_{max}	76.3382003	74.63433256	49.96752083	49.96752096	50.85678942		
Error	**24.47166509**	**1.327556689**	**49.13162128**	**32.75441419**	**14.92810011**	**18.86732936**	**0.498879688**
Temkin Model							
A_T	1.292889394	1.453073998	1.510371881	1.510371868	1.574607344		
B_T	12.26886012	11.176263	10.88355159	10.88355163	10.43456922		
Error	34.80844029	1.9614815	58.77382484	39.1825499	16.88865886	20.9813932	0.173353531
Dubinin–Radushkevich Model							
K_{AD}	2.07E-06	7.00E-07	9.88E-07	7.00E-07	9.96E-07		
q_{max}	36.08265003	32	31.582844	36	36.08265003		
Error	87.18814567	8.341446006	178.0538125	36.0000007	40.71431718	26.4906441	0.011030979

Note: The lowest ERRSQ, HYBRID, MPSD, ARE, χ^2, and AICc are in bold.

17.4 Conclusion

The adsorption of ammonia on prepared activated carbon from aqueous medium was investigated on the basis of equilibrium studies. The experimental data was modelled and evaluated using four isotherm models and seven optimization and error functions, including the corrected Akaike information criterion (AIC_C) as a statistical estimating and ranking tool. The LTFM provided the highest R^2 regression coefficient for the case of Langmuir isotherm. The analysis of the other error functions for linear/nonlinear optimization and AIC_C showed that the conventional two-parameter Langmuir equation is the best-fitting model among the other two-parameter isotherms applied. The laboratory studies proved a good

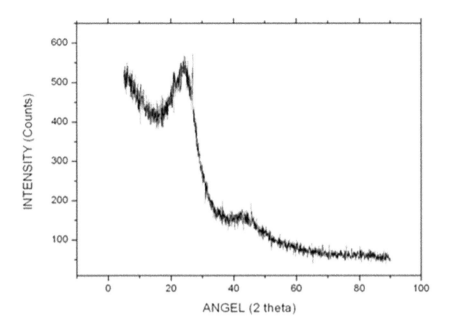

FIGURE 17.5 XRD analysis of acid activated tea waste.

efficiency of prepared modified adsorbent from waste tea leaves as an alternative inexpensive adsorbent for removal of the ammonia from aqueous solutions and outlined its future applicability in large-scale wastewater treatment technologies.

REFERENCES

1. A.H. Azhar, C.A. AimiShaza, Ammonia removal from an aqueous solution using chemical surface: Modified sand, *Health Environ. J.* 3(2) (2012) 17–24.
2. A. Dimirkou, M.K. Doula, Use of clinoptilolite and an Fe-over exchanged clinoptilolite in Zn^{2+} and Mn^{2+} removal from drinking water, *Desalination.* 224 (2008) 280–292.
3. W. Kerstin, T. Mayr, I. Klimant, Fluorescence sensors for trace monitoring of dissolved ammonia, *Talanta.* 77 (2008) 66–72.
4. R.I. Mase, *Principles of Adsorption and Reaction on Solid Surfaces.* Hoboken, NJ: John Wiley & Sons (1996).
5. N.S. Mokgalaka, R.I. McCrindle, B.M. Botha, Multielement analysis of tea leaves by inductively coupled plasma optical emission spectrometry using slurry nebulization, *J. Anal. At. Spectrom.* 19 (2004) 1375–1378.
6. O. Hamdaoui, E. Naffrechoux, Modeling of adsorption isotherms of phenol and chlorophenols onto granular activated carbon. Part II. Models with more than two parameters, *J. Hazard. Mater.* 147 (2007) 401–411.
7. K.Y. Foo, B.H. Hameed, Insights into the modeling of adsorption isotherm systems, *Chem. Eng. J.* 156(1) (2010) 2–10.
8. S.S.A. Alkurdi, R.A. Al-Juboori, J. Bundschuh, L. Bowtell, A. Marchuk. Inorganic arsenic species removal from water using bone char: A detailed study on adsorption kinetic and isotherm models using error functions analysis, *J. Hazard. Mater.* 405 (2021) 124112.
9. D.M. Pavlović, L. Ćurković, D. Blažek, J. Țupan. The sorption of sulfamethazine on soil samples: Isotherms and error analysis. *Sci. Total Environ.* 497 (2014) 543–552.
10. L. Zheng, Y. Yang, P. Meng, D. Peng, Absorption of cadmium (II) via sulfur-chelating based cellulose: Characterization, isotherm models and their error analysis, *Carbohydr. Polym.* 209 (2019) 38–50.
11. R. Kamaraj, A. Pandiarajan, S. Jayakiruba, M. Naushad, Vasudevan, Kinetics, thermodynamics and isotherm modeling for removal of nitrate from liquids by facile one-pot electro synthesizednano zinc hydroxide. *J. Mol. Liq.* 215 (2016) 204–211.

12. K.V. Kumar, K. Porkodi, F. Rocha, Isotherms and thermodynamics by linear and non-linear regression analysis for the sorption of methylene blue onto activated carbon: Comparison of various error functions, *J. Hazard. Mater.* 151 (2008) 794–804.
13. A. Saravanan, S. Karishma, S. Jeevanantham, S. Jeyasri, A.R. Kiruthika, P.S. Kumar, P.R. Yaashika, Optimization and modeling of reactive yellow adsorption by surface modified Delonixregia seed: Study of nonlinear isotherm and kinetic parameters, *Surf. Interfaces* 20 (2020) 100520.
14. C.P. McDonald, N.R. Urban, Using a model selection criterion to identify appropriate complexity in aquatic biogeochemical models, *Ecol Modell.* 221(3) (2010) 428–432.
15. Y. Li, A.M. Rauth, X.Y. Wu, Prediction of kinetics of doxorubicin release from sulfopropyl dextran ion-exchange microspheres using artifcial neural networks, *Eur. J. Pharm. Sci.* 24(5) (2005) 401–410.
16. R. Lancelot, M. Lesnoff, J.J. McDermott, Use of Akaike information criteria for model selection and inference. An application to assess prevention of gastrointestinal parasitism and respiratory mortality of Guinean goats in Kolda, Senegal, *Preventive Veterinary Medicine*, 55(4) (2002) 217–240.
17. APHA, *Standard Methods for the Examination of Water and Wastewater*, 19 ed. New York: American Public Health Association (1995).
18. M.C. Ncibi, Applicability of some statistical tools to predict optimum adsorption isotherm after linear and non-linear regression analysis, *J. Hazard. Mater.* 153 (2008) 207–212.
19. M.A. Wahab, S. Jellali, N. Jedidi, Ammonium biosorption onto sawdust: FTIR analysis, kinetics and adsorption isotherms modelling, *Bioresour. Technol.* 101 (2010) 5070–5075.
20. A. Kausar, K. Naeem, T. Hussain, Z. Nazli, H.N. Bhatti, F. Jubeen, A. Nazir, M. Iqbal, Preparation and characterization of chitosan/clay composite for direct Rose FRN dye removal from aqueous media: Comparison of linear and non-linear regression methods, *J. Mater. Res. Technol.* 8 (2019) 1161–1174.
21. R. Christie. *Colour Chemistry*. London: Royal Society of Chemistry (2001).
22. M. Rudolph. *Química Orgánica Simplificada*. New York: Chemical Publishing Co., Inc., (1986).
23. F.M. Mutua, The use of the Akaike information criterion in the identification of an optimum flood frequency model, *Hydrologic. Sci. J.* 39 (1994) 235–244.
24. H. Alrobei, M.K. Prashanth, C.R. Manjunatha, C.B. Pradeep Kumar, C.P. Chitrabanu, P.D. Shivaramu, K.Y. Kumar, M.S. Raghu, Adsorption of anionic dye on eco-friendly synthesised reduced grapheme oxide anchored with lanthanum aluminate: Isotherms, *Kinetics and Statistical Error Analysis*, 47 (2021) 10322–1033.

18

A Decision Support System for Environment-Friendly Sustainable Model of IT Service Sector Industries

Ipsita Saha
Jadavpur University

Amit Kundu
Techno India Group

Sadhan Kumar Ghosh
Jadavpur University

CONTENTS

18.1 Introduction

The competitive marketing environment forces organizations to think of environmental sustainability. In order to achieve this, each business model should ensure that there is no gap between supply of product and stakeholder's demand; product quality and customer needs. It has been observed that many organizations use ERP systems as the backbone of their business model. ERP can integrate and automate all operations within one module. ERP can combine all the functional modules within a core system to provide service to all departmental segments (Shen et al., 2016). The industries that use decision support system (DSS) include ERP system as their backbone. IT industries maintain DSS to control overall process flow starting from manufacturing, inventory management, ordering and shipping from the warehouse. The robust decision-making system helps IT sectors to implement the effective planning of productivity and employee performance improvement. Sustainable business model proposes a strategic tool (Ahlawat & Punam, 2011) to deliver an environmentally sustainable product. Here in this chapter, some IT sectors have been identified and their business models were analyzed. Analysis suggests that some IT sectors are suffering due to unsustainable work models (Dissanayake et al., 2017). We emphasized on some factors that can lead the organizations toward sustainability.

DOI: 10.1201/9781003231608-21

18.2 Literature Survey

Our country India is called a global hub in service sector industries. Gaurav Nayyar (2009) clearly explained that in India the growth of service sector shows rapid steps made by experienced professionals. IT, Telecom, ITES and BPO industries made revolution in the Indian share market and the Government of India has taken initiative to promote these sectors of India in front of the world. Robert S. Sullivan (2002) mentioned that the United States was the first country that was turned into postindustrial society. According to Teece (2010), a business model is such that where the correlation of factors (Byrne, 2010) can be clearly stated and delivered within the organization. Richardson (2008) stated that a business model should consist of design of products to be delivered to the customers. A sustainable business model can also be defined as the best business practices considered within the organization and can be considered as initiatives taken for the betterment of society (Høgevold et al., 2015).

The factors that are responsible for delivering a successful product to the consumers are called critical success factors (CSFs). According to Jiwat et al. (2013), the factors that are under the attention of managers and are liable for enhancing organizational performance are called CSFs (Rockart et al., 1996). Others have said that CSFs are the factors that can help managers to achieve the designated goal (Saade & Nijher, 2016). To study the implementation of critical factors, researchers across the globe developed different system models (Bancroft et al., 1998; Markus & Tanis, 1999; Ross, 1998), but any of them did not consider the CSFs. In other studies, it has been found that organizational business models are based on CSFs (Akkermans & van Helden, 2002; King & Burgess, 2005). Many of them had agreed upon that the simulation model was done based on data analysis and this model can be treated as DSS for implementing business models. These models help the organization to strengthen decision-making ability. IT service sector industries follow DSS to optimize the functional modules of their ERP system (Esteves, 2009). Designing the DSS model lean paradigm can be considered within the supply chain to enhance the effectiveness of the organization (Afonso & Cabrita, 2015). This technology can act as a key for developing sustainable business model innovation. Roome and Louche (2016), Schaltegger et al. (2016), Yang et al. (2016), Geissdoerfer et al. (2016, 2018), and Evans et al. (2017) mentioned a thorough review regarding previous work done on business model innovation. Basically, ERP software implementation is different compared to other customized system (Holland & Light, 1999). Authors have agreed on how sustainable models can help the organizations like IT service sector to meet their sustainability goals. Even Schaltegger et al. (2016) and Geissdoerfer et al. (2018) mentioned that the definition sustainable business model has got evolved.

18.3 Research Gap

Though many researchers worked on DSS for the overall sustainability of IT service sector, many organizations are still ahead in establishing environment-friendly model. So, the objective of the chapter is to propose a DSS to ensure sustainability of IT service sector organizations. In order to achieve that, certain parameters have been identified in environmental domain. Based on the survey report, latent factors were estimated through statistical analysis. Structured equation modeling was done to design the DSS.

18.4 Research Objective

To understand the intention of the service sector of India, several adaptation techniques have been considered in the present study for the implementation of a successful ERP system. Hence, the identification and enumeration of parameters that are responsible for developing a business model should be done. To establish a sustainable business model, the first objective involves the collection of parameters that are associated with the environmental perspective of the service sector organization. The second objective involves finding out factor loading concerning the above parameter. From empirical data using the exploratory factor analysis process, construct's validity has been measured. Before concluding sustainable business model, construct's validity is required to be tested.

18.5 Research Methodology

18.5.1 Questionnaire Design

The questionnaire was primarily framed based on a literature survey to ask for ratings on an ordinal Likert type five-point scale to get the perception of environmental sustainability. For this business modeling method (Montilva & Barrios, 2004) questionnaire was validated after a long discussion among the two academicians and one corporate manager who had in-depth knowledge and experience in schema design and implementation of ERP systems. Enhancing the acceptability of the questionnaire and validity of constructs is the basic objective of the testing. After the pilot tests, their suggestions were taken into consideration and modifications were done accordingly. The identified parameters in each mentioned perspective with their code are listed in Table 18.1.

18.5.2 Methodology of Analyses

Data was collected through the survey from IT specialists (Nah et al., 2001) of service sector and exploratory factor analysis (EFA) has been done on those collected data. Such opinion surveys often help in ascertaining the impact of ERP systems. The data collection was done after personal interviews from service sector only. Factor analysis has been used to estimate the constructs at operational levels. Since the testing for validity of construct is a necessary step to measure theoretical structure using observed variables (Cronbach & Meehl, 1955), exploratory factor analysis (EFA) was used.

To achieve that, the present study has taken into consideration the service sector specifically IT industries. The sample size was 152, and we got Cronbach's alpha=0.932 that indicates constructs' satisfactory internal consistency (Cheung & Rensvold, 2002).

18.6 Findings

18.6.1 Exploratory Factor Analysis: Environmental Perspective

Exploratory factor analysis is used not only to reduce data to a lesser set of variables but also to categorize the structure of the relationship. The scree plot is shown in Figure 18.1, and the result of EFA is shown in Table 18.2. The result exhibited Kaiser–Meyer–Olkin (KMO) of the sampling adequacy as 0.672.

18.6.2 Output Factor Analysis-Environmental Perspective

- The factor names and their corresponding variables are listed in Table 18.3. The findings of the factor analysis indicated that two factors, namely, decision regarding stakeholder's opinion and process flow (F_1) and environmental sustainability (F_2), have emerged as significant ones for establishing sustainable supply chain management.

TABLE 18.1

Identified Parameters Associated with Environmental Sustainability Issues

Parameters	Variable's Code	Variable Description
Environmental	BP 401	Level of sharing ecological information
	BP 402	Clients' review
	BP 403	Degree of processing information
	BP 404	Degree of waste generation in cloud storage
	BP 405	Management's role toward sustainability
	BP 406	Degree of selection of suppliers
	BP 407	Degree of awareness toward the elimination of hazardous waste

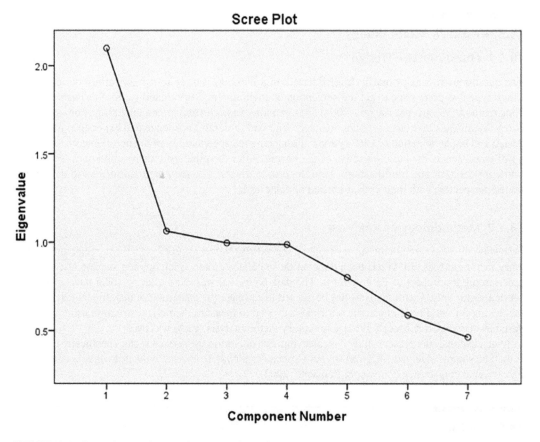

FIGURE 18.1   Scree plot—exploratory factor analysis (environmental) using SPSS 21.0.

TABLE 18.2

Results of Exploratory Factor Analysis (Environmental) Using SPSS 21.0

	Total Variance Explained								
	Initial Eigenvalues			**Extraction Sums of Squared Loadings**			**Rotation Sums of Squared Loadings**		
Component	**Total**	**% of Variance**	**Cumulative %**	**Total**	**% of Variance**	**Cumulative %**	**Total**	**% of Variance**	**Cumulative %**
1	2.099	29.988	29.988	2.099	29.988	29.988	2.058	29.398	29.398
2	1.063	15.190	45.178	1.063	15.190	45.178	1.105	15.780	45.178
3	0.996	14.231	59.409						
4	0.988	14.111	73.520						
5	0.802	11.458	84.978						
6	0.588	8.393	93.371						
7	0.464	6.629	100.000						

Extraction Method: Principal Component Analysis.

TABLE 18.3

Construct and Factor Loading of Environmental Perspective

Factor Name	Factor Definition	Variables	Variable Description	Factor Loading F1	F2
Decision regarding stakeholder's opinion and process flow	The factor is defined as the decision strategy regarding stakeholder's opinion and process flow to run the business	BP 402	Clients' review	0.578	
		BP 403	Degree of processing information	0.791	
		BP 404	Degree of waste generation in cloud storage	0.733	
		BP 405	Management's role toward sustainability	0.743	
Environmental sustainability	This factor is defined as the environmental information related to ecology, supplier, and waste generation of the organization for the purpose to achieve high performance	BP 401	Level of sharing ecological information		0.734
		BP 406	Degree of selection of suppliers		0.686
		BP 407	Degree of awareness toward the elimination of hazardous waste		0.225

18.7 Conclusion

The study identifies how the organizations can consider their system implementations to make them successful. ERP that can handle the DSS can be considered for implementing a successful venture. This research has attempted to identify the most significant factors that are responsible for the development of a sustainable business model through the Enterprise Resource Planning (ERP) system. The factors, namely, decision regarding stakeholder's opinion and process flow (Factor 1) and environmental sustainability (Factor 2) indicate the environmental sustainability of IT sector. Service sector industries have understood that some factors are responsible for the establishment of a business model. Future studies may append certain parameters to the model to give the best output to the market. Here also lies the shortcoming of the present study is only IT service sector have been considered. Until and unless, management understands the challenges of inclusion of environmental factors, the success of the business model will be far away.

REFERENCES

Afonso, H., & Cabrita, M. D. R. (2015). Developing a lean supply chain performance framework in a SME: A perspective based on the balanced scorecard. *Procedia Engineering*, 131, 270–279. doi: 10.1016/j.proeng.2015.12.389.

Akkermans, H.A., & van Helden, K. (2002). Vicious and virtuous cycles in ERP implementation: a case study of interrelations between critical success factors. *European Journal of Information Systems* 11, 35–46.

Bancroft, N., Seip, H., & Sprengel, A. (1998). *Implementing SAP R/3*, 2nd edn. Manning Publications, Greenwich.

Byrne, B. M. (2010). *Structural Equation Modeling with AMOS: Basic Concepts, Applications, and Programming*. Routledge Taylor & Francis Group, London.

Cronbach, L. J., & Meehl, P. E. (1955). Construct validity in psychological tests. *Psychological Bulletin*, 52(1), 281–302.

Daniel F. S. (2002). The variables of positive change. doi: 10.1002/he.75.

David J. T. (2010). Business models, business strategy and innovation. *Long Range Planning*, 43(2–3), 172–194. doi: 10.1016/j.lrp.2009.07.003.

Dissanayake, D. G. K., Perera, S. & Wanniarachchi, T. (2017). Sustainable and ethical manufacturing: A case study from handloom industry. *Text Cloth Sustain* 3, 2. doi: 10.1186/s40689-016-0024-3.

Esteves, J. (2009). A benefit realization road-map framework for ERP usage in small and medium enterprise. *Journal of Enterprise Information*, 22(1/2), 25–35.

Evans, S., Vladimirova, D., Holgado, M., Van Fossen, K., Yang, M., Silva, E., & Barlow, C. (2017). Business model innovation for sustainability: Towards a unified perspective for creation of sustainable business models. *Business Strategy and the Environment*. doi: 10.1002/bse.1939.

Geissdoerfer, M., Bocken, N.M.P., & Hultink, E.J. (2016). Design thinking to enhance the sustainable business modelling process. *Journal of Cleaner Production*, 135, 1218e1232. doi: 10.1016/j.jclepro.2016.07.020.

Geissdoerfer, M., Vladimirova, D., & Evans, S. (2018). Sustainable business model innovation: A review. *Journal of Cleaner Production*, 198. doi: 10.1016/j.jclepro.2018.06.240.

Cheung, G.W. & Rensvold, R.B. (2002). Evaluating goodness-of-fit indexes for testing measurement invariance. *Structural Equation Modeling*, 9(2), 233–255.

Høgevold, N. M., Svensson, G., & Padin, C. (2015). A sustainable business model in services: An assessment and validation. *International Journal of Quality and Service Sciences*, 7(1), 17–33.

Holland, C. & Light, B. (1999). A critical success factors model for ERP implementation. *IEEE Software*, 16(1), 30–36.

Ahlawat, J., & Punam (2011). ERP systems and business performance. *International Economics & Finance Journal*, 6(2), 273–286.

Jiwat, R., Corkindale, D., & Wu, M. (2013). Implementation critical success factors (CSFs) for ERP: Do they contribute to implementation success and post-implementation performance. *International Journal of Production Economics*, 144(1), 157–174.

King, S.F., & Burgess, T.F. (2005). Beyond critical success factors: A dynamic model of enterprise system innovation. *International Journal of Information Management*, 26, 59–69.

Markus, M.L., & Tanis, C. (1999). *The Enterprise Systems Experience: From Adoption to Success*. Claremont Graduate University, California.

Montilva, J. C., & Barrios, J. A. (2004). BMM: A business modeling method for information systems development. *The Clei Electronic Journal*, 7(2), 23–44.

Nah, F. F. H., Lau, J. L. S., & Kuang, J. (2001). Critical factors for successful implementation of enterprise systems. *Business Process Management Journal*, 7(3), 285–296.

Nayyar, G. (2009). *The Role of Services in Development: The Indian Experience*. Oxford University, Oxford.

Richardson, J. (2008). The business model: an integrative framework for strategy execution. *Strategic Change*, 17 (5e6), 133e144.

Rockart, J.F., Earl, M.J., Ross, J.W. (1996). The new IT organization: Eight imperatives. Retrieved: https://dspace.mit.edu/bitstream/handle/1721.1/2623/SWP-3902-40987801-CISR-292.pdf.

Ross, J. W. (1998). The ERP revolution: Surviving versus thriving. Centre for Information Systems Research, Sloan School of Management.

Roome, N., & Louche, C., 2016. Journeying toward business models for sustainability: A conceptual model found inside the black box of organisational transformation. *Organization & Environment* 29, 11e35. doi: 10.1177/1086026615595084.

Saade, R.G., Nijher, H. (2016). Critical success factors in enterprise resource planning implementation: A review of case studies. *Journal of Enterprise Information Management*, 29(1), 72–96. doi: 10.1108/JEIM-03-2014-0028.

Schaltegger, S., Hansen, E.G., & Lüdeke-Freund, F., 2016. Business models for sustainability: Origins, present research, and future avenues. *Organization & Environment*, 29, 3e10. doi: 10.1177/1086026615599806.

Yang, M., Evans, S., Vladimirova, D., & Rana, P., 2016. Value uncaptured perspective for sustainable business model innovation. *Journal of Cleaner Production*. doi: 10.1016/j.jclepro.2016.07.102.

Shen, Y.-C., Chen, P.-S., & Wang, C.-H. (2016) A study of enterprise resource planning (ERP) system performance measurement using the quantitative balanced scorecard approach. *Computers in Industry*, 75, 127–139.

Section IV

Case Studies on WM, Sanitation and Hygiene

19

Synergistic Interaction among Supplementary Cementitious Materials (SCMs) for Sustainable Solid-Waste Management

Kaveti Anithamma, N. Niranjan Kumar, S.V. Satyanarayana, and Dilip Kumar Behara
JNT University

CONTENTS

19.1 Introduction

The enormous increase in production of industrial wastes and agriculture wastes not only create severe disposal problems but also have adverse effects on environment. For example, several hectares of valuable land is required by thermal power plants for the disposal of fly ash (FA), which is light in weight and easily flies thereby creating severe health problems such as asthma, bronchitis, etc. (Patankar et al., 2013; Pandey et al., 2009). Similarly, mass production of sugarcane bagasse ash (SCBA), a by-product of sugar industries and coconut coir (CC) from coconut trees is occurring every year (Bahurudeen & Santhanam, 2015; Bhatnagar et al., 2016). Therefore, it is essential to develop alternative methodologies for better utilization of supplementary cementitious materials (SCMs) in a more economic way in perspective of environmental pollution control and solid-waste management. In light of these, several researchers were focused to use these materials as additive or partial replacement in cement concrete. As the main constituent species in these SCMs are carbon, silicon, calcium, alumina etc., which give pozzolonic properties to cement for better performance enhancement. For example, FA is widely used in blended cements as it improves the workability at lowers replacement levels (<40%) and at less water demand due to more spherically shaped particles (Oner et al., 2005;

Nochaiya et al., 2010). Further, studies containing SCMs with proposed activation energy function for strength calculations were also reported (Han et al., 2003). Also, some researchers have linked the compressive strength to the SCM's fineness and the amount of CaO, the ratio of potassium to alumina and loss of ignition (Das, 2006). Several studies also focused on the optimum utilization of SCMs for maximum compressive strength (Oner & Akyuz, 2007; Chindaprasirt & Rukzon, 2008; Jidrada et al., 2016; Guo et al., 2010).

Similarly, sugarcane bagasse ash material has been described to be a suitable SCM for use in concrete in previous research studies (Cordeiro et al., 2009a,b; Bahurudeen et al., 2015, 2016; Cordeiro et al., 2011). The high value of amorphous silica content which is an important property of sugarcane bagasse results in good pozzolonic properties (Cordeiro et al., 2009a; Chusilp et al., 2009; Frías et al., 2011). The sugarcane bagasse ash is mainly composed of main silica (60–75%), CaO, K_2O, minor oxides Al_2O_3, Fe_2O_3, SO_3 and loss of ignition (Cordeiro et al., 2009b; Morales et al., 2009). It is well reported in literature that the performance of bagasse ash in concrete will be maximum when it is directly burnt to 650°C for 4 hours and ground to 5.4 µm mean particle size (Ganesan et al., 2007). Further, coconut fibres are commercially available in three forms: namely bristle (long fibres), mattress (relatively short) and decorticated (mixed fibres). The different types of fibres have different uses depending upon the requirement. Brown coir fibres are mostly used in engineering (Aggarwal, 1992; Agopyan et al., 2005; Li et al., 2007). It is reported in literature that the optimum quantity and length of fibres is a key factor in achieving the maximum strength of composite (Ramaswamy et al., 1983; Verma et al., 2013; Verma & Gope, 2015), wherein plant fibres were used as alternative or additive fibres in composites such as concrete, cement paste and mortar. In several investigations, it is proved that coconut fibre is the toughest fibre amongst the all-natural fibres, and the strength will be increased in fibre reinforced mortar. However, a decrease in the strength of mortars is seen with an increase in the fibre content (Ali et al., 2012).

Herein, we propose a methodology to use these waste SCMs (sugar cane bagasse ash, coconut coir and FA) as additive/partial replacement in cement and concrete for potential environmental disposal problems. Several review articles have already reported the utilization of SCMs for performance improvement of concrete due to their consistent performance due to good pozzolanic properties (Aprianti 2017; Juenger et al., 2011; Johari et al., 2011; Elahi et al., 2010; Juenger & Siddique, 2015; Lothenbach et al., 2011). The individual, as well as composite materials, were prepared and characterized using SEM, EDX, and XRD. The morphological studies infer that the surface is rough and is more heterogeneous. Further, the elemental analysis (EDX) confirms the presence of silica (SiO_2), alumina (Al_2O_3) and cement (Ca-Al-Si) in the specimens. The XRD confirms the crystalline nature of FA, sugar cane ash and blended cement. The experimental tests/results signify that the concrete contained up to 30% of FA, SCBA and CC and showed a good compressive strength and split tensile strength in comparison with control concrete. The synergistic interaction studies reveal that among the three SCMs, sugarcane bagasse ash act as inhibitive but not promotive towards strength characteristics. The central outcomes of the present work will give directions to understand better material combinations for the overall performance improvement of concrete.

19.2 Materials and Methods

19.2.1 Materials

Ordinary Portland cement type I (OPC) grade 53 was used in the mixtures as control concrete. Locally available natural sand with 4.75-mm sieve passing maximum size was used as a fine aggregate and crushed stone with 20 mm maximum size was used as a coarse aggregate. FA (FA) collected from Rayalaseema Thermal Power Station (RTPP), Kadapa District was used in this study. The sugarcane bagasse ash (SCBA) and coconut coir fibre (CC) were collected from the Anantapuramu local area and are used in the present study. Further, coconut coir fibre cut into small pieces with an approximate length of 2 cm was used in this study.

19.2.2 Characterization Methods

19.2.2.1 Scanning Electron Microscopy (SEM)-Energy-Dispersive X-Ray Spectroscopy (EDX)

Scanning electron microscopy (SEM) has been performed to determine the surface morphology as well as to identify the surface roughness after testing specimens of all SCM materials. Further, Energy-dispersive X-ray spectroscopy (EDX) has been performed to confirm the presence of elements and their composition.

19.2.2.2 X-Ray Diffraction

X-ray diffraction (XRD) has been performed to understand the amorphous/crystalline nature of specimens as well as the residual stress generated after the application of load.

19.2.3 Mechanical Testing of Specimens

19.2.3.1 Compressive Strength Test

Concrete cubes of size $150 \times 150 \times 150 \, \text{mm}^3$ were tested in a 2,000 kN capacity compression testing machine (UTM). The compressive strength of concrete is determined by applying the load at the rate of $140 \, \text{kg/cm}^2$/minutes till the specimen fails.

19.2.3.2 Split Tensile Strength Test

The split tensile test of the concrete was executed as per ASTM C496-90 in a universal test machine. Concrete cylinders of 300 mm height and 150 mm diameter across were taken for measurements.

19.3 Results and Discussion

The SEM observation of coal FA has shown a series of spherical and irregular particles of different sizes (Figure 19.1a) which are heterogeneous in nature. The observed particles may be of silicates, calcium carbonate, quartz, etc. Similar morphologies were observed by Guo et al. (2010). SEM image of sugarcane bagasse ash (Figure 19.1b) shows various shapes/sizes with a roughened surface and similar morphologies observed by Cordeiro et al. (2009b). Further, SEM observations of blended cement (mixture of FA, SCBA, and CC) shown in Figure 19.1c depicts a change of microstructure with the incorporation of SCM materials. Few clusters were also observed due to the aggregation of particles during the moulding and curing steps. In order to confirm the presence of materials in specimens, EDX was performed. The EDX spectrum confirms the presence of the elements of O, Si and Al (inset image of Figure 19.1c). From EDX, it can be claimed that silica (SiO_2), alumina (Al_2O_3) and cement (Ca-Al-Si) are present in the specimens (Table inset image of Figure 19.1c).

XRD analysis was performed to identify the crystalline (or) amorphous nature of materials. The XRD pattern of FA, sugar cane bagasse ash and blended cement is shown in Figure 19.1d. The main crystalline phases detected for FA were quartz (Q), muscovite (M), calcite (C), and anorthite (A) matching with existing literature (Modolo et al., 2018). The XRD pattern of sugarcane bagasse ash shows two major crystalline phases i.e. quartz (Q) and crystobalite (C) matching with existing literature (Bahurudeen et al., 2016). These peaks primarily confirm the presence of silica in both SCBA and FA. The XRD pattern of blended cement shows peaks of both SCBA as well as FA along with normal cement characteristic peaks. The main crystalline phases of cement were Quartz (Q), Alite (A), and Celite (C), matching with existing literature (Ye et al., 2015). The sharp crystalline peaks obtained from XRD spectra of blended cement, FA, SCBA can also infer that few residual stresses are left in the specimen sample after 28 days.

TABLE 19.1

Compressive Strength and Split Tensile Strength at 28 Days Results Comparison of the Individual, Combination of Two Materials and Combination of All Waste Materials (FA, SCBA and CC)

S. No.	Materials	Compressive Strength (N/mm²) 28 days
1	Control concrete	19.99
2	Fly ash (FA)	33.88
3	Sugarcane bagasse ash (SCBA)	32.66
4	Coconut coir (CC)	33.72
5	Fly ash (FA) + sugarcane bagasse ash (SCBA)	33.22
6	Fly ash (FA) + coconut coir (CC)	34.00
7	Sugarcane bagasse ash (SCBA) + coconut coir (CC)	33.66
8	Combination of materials (CM)	36.44

The characterized specimen cube samples were tested for water absorption test (Figure 19.1) and the data values are shown in Table 19.1. The water absorption will increase with an increase in the percentage composition of additive. This result indicates the ingredients absorb more water for settling to form a homogenous mixture. Figure 19.2a shows the compressive strength of FA mortars, sugarcane bagasse

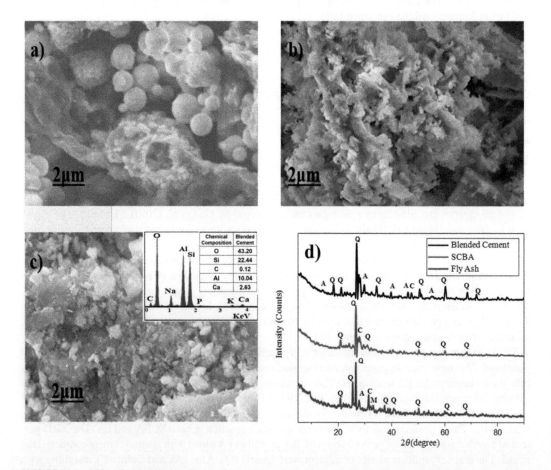

FIGURE 19.1 SEM micrograph of (a) fly ash, (b) sugarcane bagasse ash, (c) blended cement (inset image shows EDAX for blended cement with a table showing the composition (wt%.) of elements), and (d) XRD analysis of fly ash, sugarcane bagasse ash and blended cement.

ash and coconut coir cured for 7 days (FA 7, SCBA 7 and CC 7) and 28 days (FA 28, SCBA 28 and CC 28) with cement partial/additive replacement of 5%, 10%, 15%, 20%, 25% and 30%, respectively. For the FA, a concrete mixture with a replacement of 15% shows higher compressive strength than the control concrete. For, sugarcane bagasse ash, the concrete mortars with a replacement of 20% shows higher compressive strength than the control concrete. Water plays an important role to dissolute the FA to form a uniform mixture with added materials. The probable reason for decrease in compressive strength of concrete with increase of ash content and/or sugarcane bagasse ash may be due to insufficient water content in specimen. Therefore, there may be trade-off between powder volumes of FA and/or SCBA with water content in order to achieve good compressive strength. Further, coconut coir concrete mortars with additive of 25% resulted in higher compressive strength than the control concrete.

In order to study the interaction of materials on the compressive strength of concrete, it is proposed to conduct experiments with a mixture of SCM materials to cement concrete for 7 and 28 days curing. The specimen samples were tested for water absorption tests before measurements. The water absorption will increase with an increase in percentage composition of two materials (Figures 19.1 and 19.2). The same reasoning of water absorption for settling to form homogenous mixture is applicable. Figure 19.2b shows the compressive strength characteristics of cement blended with FA and sugarcane bagasse for 7 days (FA + SCBA 7) and 28 days (FA + SCBA 28). It is observed that the compressive strength of blended material increases for all compositions in comparison to control concrete and the highest compressive strength was observed for 20% composition. Figure 19.2c shows the compressive strength characteristics of cement blended with FA and coconut coir for 7 days (FA + CC 7) and 28 days (FA + CC 28). It is observed that the compressive strength of blended material increases for all compositions in comparison to control concrete, and the highest compressive strength was observed for 15% composition. Figure 19.2d shows the compressive strength characteristics of cement blended with sugar cane bagasse and coconut coir for 7 days (SCBA + CC 7) and 28 days (SCBA + CC 28). It is observed that the compressive strength of blended material increases for all compositions in comparison to control concrete and the highest compressive strength was observed for 25% composition. The same reasoning of water scarcity with an increase in powder volume of added SCM is applicable for all these cases. Further, it may be possible that there might be inhomogeneity of mortar with the addition of materials to the existing cement which demonstrates an understanding of the synergistic interaction among the components.

Figure 19.3 shows the compressive strength characteristics of the combination of all three waste materials (FA, SCBA and CC) for 7 days (CM 7) and 28 days (CM 28). It is observed that the compressive strength of the combination of materials increases for all compositions in comparison to control concrete and the highest compressive strength was observed for 20% composition. In similar arguments, the added three materials may segregate into different phases for above 20% composition and leading to a decrease in compressive strength. Therefore, it is expected to understand the interactions with materials at micro-level for control/design of materials towards better solid-waste management.

Before conducting split tensile strength measurements for prepared cylinders, water absorption tests were conducted for individual as well as a combination of materials (Figures 19.3 and 19.4). The water absorption will increase with an increase in percentage composition of additive. This result indicates the ingredients absorb more water for settling to form a homogenous mixture. Figure 19.4a shows the split tensile strength of FA mortars, sugarcane bagasse ash and coconut coir cured for 7 days (FA 7, SCBA 7 and CC 7) and 28 days (FA 28, SCBA 28 and CC 28) with cement partial/additive replacement of 5%, 10%, 15%, 20%, 25% and 30%, respectively. For the FA, a concrete mixture with a replacement of 15% shows higher split tensile strength than the control concrete. For, sugarcane bagasse ash, the concrete mortars with a replacement of 20% show higher split tensile strength than the control concrete. Water plays an important role to dissolute the FA to form a uniform mixture with added materials. The probable reason for the decrease in split tensile strength of concrete with an increase of ash content and/or sugarcane bagasse ash may be due to insufficient water content in specimen. Therefore, there may be trade-off between powder volumes of FA and/or SCBA with water content in order to achieve good split tensile strength. Further, coconut coir concrete mortars with the additive of 20% resulted in higher split tensile strength than the control concrete. The same reasoning of water scarcity with an increase in powder volume of added SCM is applicable for all these cases.

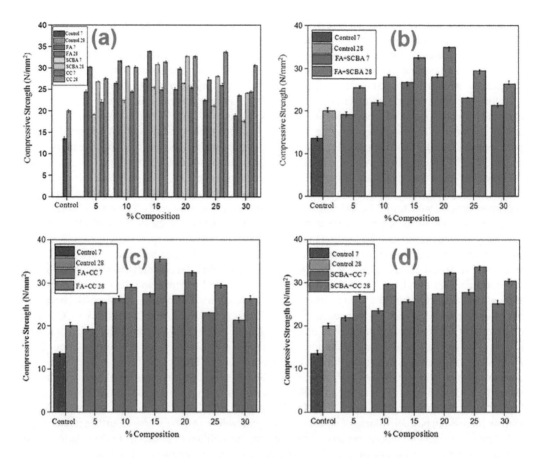

FIGURE 19.2 Compressive strength measured for 7 and 28 days of cubes prepared from (a) individual waste materials fly ash, sugarcane bagasse ash and coconut coir (FA, SCBA and CC) with cement concrete, (b) combination of two waste materials fly ash (FA) and sugarcane bagasse ash (SCBA), (c) combination of two waste materials fly ash (FA) and coconut coir (CC), and (d) combination of two waste materials sugarcane bagasse ash (SCBA) and coconut coir (CC).

In order to study the interaction of materials on the split tensile strength of concrete, it is proposed to conduct experiments with a mixture of SCM materials to cement concrete for 7 and 28 days. Figure 19.4b shows the split tensile strength characteristics of cement blended with FA and sugar cane bagasse ash for 7 days (FA + SCBA 7) and 28 days (FA + SCBA 28). It is observed that the compressive strength of blended material increases for all compositions in comparison to control concrete and the highest split tensile strength was observed for 15% composition. Figure 19.4c shows the split tensile strength characteristics of cement blended with FA and coconut coir for 7 days (FA + CC 7) and 28 days (FA + CC 28). It is observed that the split tensile strength of blended material increases for all compositions in comparison to control concrete and the highest split tensile strength was observed for 15% composition. Figure 19.4d shows the split tensile strength characteristics of cement blended with sugar cane bagasse and coconut coir for 7 days (SCBA + CC 7) and 28 days (SCBA + CC 28). It is observed that the split tensile strength of blended material increases for all compositions in comparison to control concrete and the highest split tensile strength was observed for 25% composition. The same reasoning of water scarcity with an increase in powder volume of added SCM is applicable for all these cases. Further, it may be possible that there might be inhomogeneity of mortar with the addition of materials to the existing cement which demonstrates the understanding of the synergistic interaction among the components.

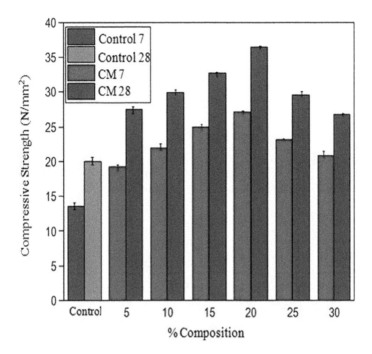

FIGURE 19.3 Compressive strength of the combination of waste materials fly ash (FA), sugarcane bagasse ash (SCBA) and coconut coir (CC) with cement concrete at 7 and 28 days.

Figure 19.5 shows the split tensile strength characteristics of the combination of all three waste materials (FA, SCBA and CC) for 7 days (CM 7) and 28 days (CM 28). It is observed that the split tensile of the combination of materials increases for all compositions in comparison to control concrete and the highest split tensile strength was observed for 15% composition. In similar arguments, the added three materials may segregate into different phases for above 15% composition and leading to decrease in split tensile strength. Therefore, it is expected to understand the interactions with materials at micro-level for control/design of materials towards better solid-waste management.

In order to understand the synergistic interaction among materials, the compressive strength and split tensile strength characteristics of all individuals and combinations of materials were compared (Tables 19.1, 19.3, and 19.4), and the synergistic interaction of materials is shown in Figure 19.5. It is clearly evident from the table that the addition of SCM materials (all three SCBA, CC, and FA) to concrete will result in an increase in both compressive strength and split tensile strength values. However, the addition of two materials will give mixed results of compressive strength and split tensile strength in comparison to individual materials. For example, FA alone will contribute a compressive strength of 33.88 N/mm²; however, the addition of sugarcane bagasse ash to FA will result in a slight decrement of compressive strength (33.22 N/mm²) in comparison to FA. Similarly, sugarcane bagasse ash added to coconut coir will also result in a slight decrement. This result indicates sugarcane bagasse ash acts as inhibitive but not promotive towards strength characteristics. However, this is not the case with coconut coir. These results provoke that the homogenous addition of waste materials can enhance the properties of cement. Further, it also ignites to study of the various interacting parameters comprising at micro-level that facilitates to increase the compressive/split tensile strength of concrete.

It can be observed from the compressive strength characteristics, that either addition or partial replacement of FA, SCBA, CC materials 15%, 20%, and 25% respectively with concrete mixture will result in better performance. For a combination of two and three materials, it is observed that the compressive strength of blended material increases for all compositions in comparison to control concrete

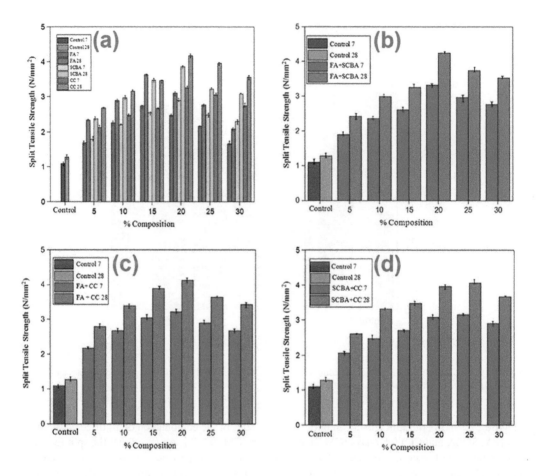

FIGURE 19.4 Split tensile strength measured for 7 and 28 days of cylinders prepared from (a) individual waste materials fly ash, sugarcane bagasse ash and coconut coir (FA, SCBA and CC) with Cement Concrete, (b) a combination of two waste materials fly ash (FA) and sugarcane bagasse ash (SCBA), (c) combination of two waste materials fly ash (FA) and coconut coir (CC), and (d) combination of two waste materials sugarcane bagasse ash (SCBA) and coconut coir (CC).

and highest compressive strength was observed for 20% composition in the case of FA + SCBA; 15% composition in case of FA + CC; 25% composition in case of SCBA + CC and 20% composition for in case of all three, i.e. SCBA + FA + CC. The split tensile strength characteristics reveal that as individual materials, FA, SCBA, CC have shown better performance results when replaced/partially added with 15%, 20%, and 20%, respectively, with concrete mixture. For a combination of two and three materials, it is observed that the split tensile strength of blended material increases for all compositions in comparison to control concrete and highest compressive strength was observed for 15% composition in the case of FA + SCBA; 15% composition in case of FA + CC; 25% composition in case of SCBA + CC 15% composition in case of all three materials, i.e. FA + SCBA + CC. The results of these experimental findings confirm the synergistic interaction among SCMs in giving better performance in comparison to individual materials. The outcomes of the present chapter will not only give insights into the better utilization of SCMs for energy conversion/production but also pave the path towards the development of alternative methodologies for better utilization of solid-waste materials for sustainable solid-waste management.

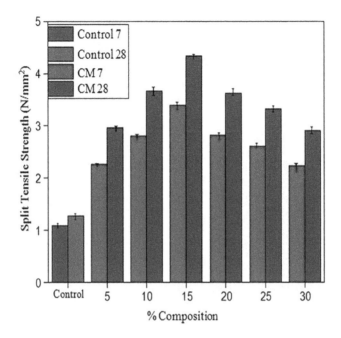

FIGURE 19.5   Split tensile strength of combination waste materials fly ash sugarcane bagasse ash and coconut coir with cement concrete at 7 and 28 days.

19.4 Conclusions

The present work envisages a methodology to use solid waste/SCMs as additive/partial replacement in cement and concrete for potential environmental disposal problems. The main objective of the present work is to study the synergistic interaction among these materials for better strength improvement of concrete. The water absorption test results of specimens (cubes and cylinders) for both individuals as well as the combination of materials indicate that with increase in percentage composition of additive, the water absorption increases indicating the ingredients absorb more water for settling to form a homogenous mixture. The compressive strength characteristics reveal that as individual materials, FA, SCBA, CC have shown better performance results when replaced/partially added with 15%, 20%, and 25%, respectively, with concrete mixture. For a combination of two and three materials, it is observed that the compressive strength of blended material increases for all compositions in comparison to control concrete, and highest compressive strength was observed for 20% composition in the case of FA+SCBA; 15% composition in the case of FA + CC; 25% composition in the case of SCBA + CC; and 20% composition in the case of FA + SCBA + CC. The split tensile strength characteristics reveal that as individual materials, FA, SCBA, CC have shown better performance results when replaced/partially added with 15%, 20%, and 20%, respectively, with concrete mixture. For a combination of two and three materials, it is observed that the split tensile strength of blended material increases for all compositions in comparison to control concrete, and highest compressive strength was observed for 15% composition in the case of FA + SCBA; 15% composition in the case of FA + CC; 25% composition in the case of SCBA + CC; and 15% composition in the case of FA + SCBA + CC. The results signify that the concrete contained up to 30% of FA, SCBA and CC showed good compressive strength and split tensile strength in comparison with control concrete. The synergistic interaction studies reveal that among the three materials, sugarcane bagasse ash act as inhibitive but not promotive towards compressive strength characteristics.

Acknowledgements

The authors gratefully acknowledge the Department of Chemical Engineering, JNTUA College of Engineering, Ananthapuramu, Andhra Pradesh for supporting research and providing experimental facilities. Further, the authors acknowledge Prof. P. R. Bhanu Murthy, Professor of Civil Engineering, JNTUA College of Engineering, Ananthapuramu, Andhra Pradesh for permitting to avail few facilities to carryout compressive strength/split tensile strength characteristics at Civil Engineering Department, JNTUACEA.

Research Funding

This work was supported partly from IEI, R&D grant in aid (2016–2017) and UGC XII plan funds of JNT University, Ananthapur (AP).

Disclosure Statement

No potential conflict of interest was reported by the authors.

REFERENCES

Aggarwal, L.K., 1992. Studies on cement-bonded coir fibre boards. *Cement and Concrete Composites*, 14(1), pp. 63–69.

Agopyan, V., Savastano Jr, H., John, V.M. and Cincotto, M.A., 2005. Developments on vegetable fibre–cement based materials in São Paulo, Brazil: An overview. *Cement and Concrete Composites*, 27(5), pp. 527–536.

Ali, M., Liu, A., Sou, H. and Chouw, N., 2012. Mechanical and dynamic properties of coconut fibre reinforced concrete. *Construction and Building Materials*, 30, pp. 814–825.

Aprianti, E., 2017. A huge number of artificial waste material can be supplementary cementitious material (SCM) for concrete production: A review part II. *Journal of Cleaner Production*, 142, pp. 4178–4194.

Bahurudeen, A. and Santhanam, M., 2015. Influence of different processing methods on the pozzolanic performance of sugarcane bagasse ash. *Cement and Concrete Composites*, 56, pp. 32–45.

Bahurudeen, A., Kanraj, D., Dev, V.G. and Santhanam, M., 2015. Performance evaluation of sugarcane bagasse ash blended cement in concrete. *Cement and Concrete Composites*, 59, pp. 77–88.

Bahurudeen, A., Wani, K., Basit, M.A. and Santhanam, M., 2016. Assesment of pozzolanic performance of sugarcane bagasse ash. *Journal of Materials in Civil Engineering*, 28(2), p. 04015095.

Bhatnagar, A., Kesari, K.K. and Shurpali, N., 2016. Multidisciplinary approaches to handling wastes in sugar industries. *Water, Air, & Soil Pollution*, 227(1), p. 11.

Chindaprasirt, P. and Rukzon, S., 2008. Strength, porosity and corrosion resistance of ternary blend Portland cement, rice husk ash and fly ash mortar. *Construction and Building Materials*, 22(8), pp. 1601–1606.

Chusilp, N., Jaturapitakkul, C. and Kiattikomol, K., 2009. Utilization of bagasse ash as a pozzolanic material in concrete. *Construction and Building Materials*, 23(11), pp. 3352–3358.

Cordeiro, G.C., Toledo Filho, R.D. and Fairbairn, E.M.R., 2009a. Effect of calcination temperature on the pozzolanic activity of sugar cane bagasse ash. *Construction and Building Materials*, 23(10), pp. 3301–3303.

Cordeiro, G.C., Toledo Filho, R.D., Tavares, L.M. and Fairbairn, E.D.M.R., 2009b. Ultrafine grinding of sugar cane bagasse ash for application as pozzolanic admixture in concrete. *Cement and Concrete Research*, 39(2), pp. 110–115.

Cordeiro, G.C., Toledo Filho, R.D., Tavares, L.M., Fairbairn, E.D.M.R. and Hempel, S., 2011. Influence of particle size and specific surface area on the pozzolanic activity of residual rice husk ash. *Cement and Concrete Composites*, 33(5), pp. 529–534.

Das, S.K., 2006. Geotechnical properties of low calcium and high calcium fly ash. *Geotechnical & Geological Engineering*, 24(2), pp. 249–263.

Elahi, A., Basheer, P.A.M., Nanukuttan, S.V. and Khan, Q.U.Z., 2010. Mechanical and durability properties of high performance concretes containing supplementary cementitious materials. *Construction and Building Materials*, 24(3), pp. 292–299.

Frías, M., Villar, E. and Savastano, H., 2011. Brazilian sugar cane bagasse ashes from the cogeneration industry as active pozzolans for cement manufacture. *Cement and Concrete Composites*, 33(4), pp. 490–496.

Ganesan, K., Rajagopal, K. and Thangavel, K., 2007. Evaluation of bagasse ash as supplementary cementitious material. *Cement and Concrete Composites*, 29(6), pp. 515–524.

Guo, X., Shi, H. and Dick, W.A., 2010. Compressive strength and microstructural characteristics of class C fly ash geopolymer. *Cement and Concrete Composites*, 32(2), pp. 142–147.

Han, S.H., Kim, J.K. and Park, Y.D., 2003. Prediction of compressive strength of fly ash concrete by new apparent activation energy function. *Cement and Concrete Research*, 33(7), pp. 965–971.

Jidrada, P., Sua-iam, G., Chatveera, B. and Makul, N., 2016. Recycling of combined coal-biomass ash from electric power plant waste as a cementitious material: characteristics and improvement. *Journal of Material Cycles and Waste Management*, 18(3), pp. 527–540.

Johari, M.M., Brooks, J.J., Kabir, S. and Rivard, P., 2011. Influence of supplementary cementitious materials on engineering properties of high strength concrete. *Construction and Building Materials*, 25(5), pp. 2639–2648.

Juenger, M.C. and Siddique, R., 2015. Recent advances in understanding the role of supplementary cementitious materials in concrete. *Cement and Concrete Research*, 78, pp. 71–80.

Juenger, M.C.G., Winnefeld, F., Provis, J.L. and Ideker, J.H., 2011. Advances in alternative cementitious binders. *Cement and Concrete Research*, 41(12), pp. 1232–1243.

Li, Z., Wang, L. and Ai Wang, X., 2007. Cement composites reinforced with surface modified coir fibers. *Journal of Composite Materials*, 41(12), pp. 1445–1457.

Lothenbach, B., Scrivener, K. and Hooton, R.D., 2011. Supplementary cementitious materials. *Cement and Concrete Research*, 41(12), pp. 1244–1256.

Modolo, R.C.E., Senff, L., Ferreira, V.M., Tarelho, L.A.C. and Moraes, C.A.M., 2018. Fly ash from biomass combustion as replacement raw material and its influence on the mortars durability. *Journal of Material Cycles and Waste Management*, 20(2), pp. 1006–1015.

Morales, E.V., Villar-Cociña, E., Frías, M., Santos, S.F. and Savastano Jr, H., 2009. Effects of calcining conditions on the microstructure of sugar cane waste ashes (SCWA): Influence in the pozzolanic activation. *Cement and Concrete Composites*, 31(1), pp. 22–28.

Nochaiya, T., Wongkeo, W. and Chaipanich, A., 2010. Utilization of fly ash with silica fume and properties of Portland cement–fly ash–silica fume concrete. *Fuel*, 89(3), pp. 768–774.

Oner, A. and Akyuz, S., 2007. An experimental study on optimum usage of GGBS for the compressive strength of concrete. *Cement and Concrete Composites*, 29(6), pp. 505–514.

Oner, A., Akyuz, S. and Yildiz, R., 2005. An experimental study on strength development of concrete containing fly ash and optimum usage of fly ash in concrete. *Cement and Concrete Research*, 35(6), pp. 1165–1171.

Pandey, V.C., Abhilash, P.C. and Singh, N., 2009. The Indian perspective of utilizing fly ash in phytoremediation, phytomanagement and biomass production. *Journal of Environmental Management*, 90(10), pp. 2943–2958.

Patankar, S.V., Jamkar, S.S. and Ghugal, Y.M., 2013. Effect of water-to-geopolymer binder ratio on the production of fly ash based geopolymer concrete. *International Journal of Advanced Technology in Civil Engineering*, 2(1), pp. 79–83.

Ramaswamy, H.S., Ahuja, B.M. and Krishnamoorthy, S., 1983. Behaviour of concrete reinforced with jute, coir and bamboo fibres. *International Journal of Cement Composites and Lightweight Concrete*, 5(1), pp. 3–13.

Verma, D. and Gope, P.C., 2015. The use of coir/coconut fibers as reinforcements in composites. In: Faruk, O. and Sain, M. (Eds), *Biofiber Reinforcements in Composite Materials* (pp. 285–319). Woodhead Publishing: Sawston.

Verma, D., Gope, P.C., Shandilya, A., Gupta, A. and Maheshwari, M.K., 2013. Coir fibre reinforcement and application in polymer composites. *Journal of Materials and Environmental Science*, 4(2), pp. 263–276.

Ye, Q., Yu, K. and Zhang, Z., 2015. Expansion of ordinary Portland cement paste varied with nano-MgO. *Construction and Building Materials*, 78, pp. 189–193.

20

Challenges to Ensure Healthy Living through Sanitation and Hygiene Coverage: Study on Narail District, Bangladesh

Moshiur Rahman and Parimal Kumar Roy
Bangladesh Public Administration Training Centre

CONTENTS

20.1 Introduction

Water and sanitation are fundamental to human development and well-being. Access to safe water and sanitation is also a human right, as recognized in 2010 by the United Nations General Assembly. Ninety-one percent of the global population now uses an improved drinking water source (WHO, 2015). Safe water sources reached 97% coverage in the early 1990s. But the discovery of arsenic contamination in tube wells reduced this figure to about 74% (UNICEF, 2006).

The report said that Bangladesh has achieved significant coverage in water supply during the past decades primarily due to the availability of suitable groundwater aquifers at shallow depths. Nevertheless, the effective coverage of safe drinking water is 73% (Minnatullah and Vidal, 2010). One of the major causes of it is poor management of drinking water from collection to consumption. Germs may incorporate with drinking water due to the lack of proper maintenance in spite of water options being safe. Due to a lack of knowledge about water safety, community people are found very much unaware regarding applying proper methods while collection, transportation and preservation of drinking water. Thus, a huge gap forms between water coverage and effective water coverage due to lack of domestic water management of people. A similar scenario was found in the case of sanitation coverage in Bangladesh. Bangladesh has made significant progress in reducing open defecation, from 34% in 1990 to just 3% of the national population in 2015. However, the sanitation coverage with improved latrines is only 54% by population (WHO-UNICEF, JMP-2010). However, the current rate of improved sanitation is 61%, growing at only 1.1% annually. Still, the quality of sanitation coverage is an emerging area of concern, with more than 40% of all latrines classified as "unimproved." Drinking water access is widespread, but half of the drinking water consumed fails to meet water safety standards. In urban areas of Bangladesh, piped water supply reaches only about one-third of the population, and there is no systematic sewer disposal and treatment system. Only Dhaka, Bangladesh's capital city, has a sewer system, and it serves just 18% of the city (World Bank, 2016).

The government of Bangladesh has launched a massive campaign and, as of today, achieved over 70% coverage through pit latrines. Ensuring the safety of drinking water in Bangladesh through a conventional monitoring and surveillance system in the context of scarce resources and a predominantly rural population presents an enormous challenge. Despite significant progress in water and sanitation, much remains to be done. Study on ensuring hygiene and sanitation to some extent is related to ensure the healthy living of people in Bangladesh.

20.2 Review of Literature

Most of the South Asian countries are fighting to ensure five basic demands of their people, whereas thinking to have 100% water and sanitation facilities are to some extent a lavish thinking considered two decades before for those countries. Access to adequate water, sanitation and hygiene (WASH) is essential for the health, well-being and dignity of all people. The World Health Organization South-East Asia Region has made considerable progress in WASH provision during the past two decades. However, compared with increases in coverage of improved drinking water, in some parts of the region, access to adequate sanitation remains low, with continued prevalence of open defecation (Chakravarty et al., 2017). In early 2019, the World Health Organization (WHO) and United Nations Children's Fund (UNICEF) reported that globally 26% of health facilities did not have access to an improved water source on the premises, 16% had no hand hygiene services at points of care and 21% did not have sanitation services (WHO, 2018). What is more important is that knowledge about WASH is poor at the community level of these countries. Its ultimate impacts cause fragile health of both mother and children as Benova (2014) cited Insufficient WASH services and compliance to hygiene have important implications for maternal and newborn health. Inadequate sanitation facilities have been significantly associated with higher maternal mortality. An estimated 139,000 newborns (0–27 days) died in East Asia and the Pacific in 2018, representing 49% of deaths in children under the age of 5 years (UNICEF, 2019). Therefore, coverage of WASH services for maternal and newborn care must be improved to reduce risks of maternal and newborn morbidity and mortality (Manava et al., 2019), though Bangladesh made mentionable progress in this regard. However, effective water coverage and knowledge on sanitation is still very poor in the rural areas of Bangladesh.

20.3 Research Question

To what extent and how the challenges of ensuring healthy living through sanitation and hygiene coverage in Bangladesh are manageable?

20.4 Objectives

The main objective of the study is to find out the challenges to help in ensuring healthy living through sanitation and hygiene coverage in Bangladesh.

The specific objectives of this study are as follows:

i. to know the common practices of community people regarding hygiene maintenance;
ii. to find out the challenges of healthy living through providing proper hygiene and sanitation coverage;
iii. to explore a way forward to address the challenges to help in ensuring hygiene and sanitation facilities.

20.5 Study Area

The study area of this research is located in the Southwest region of Bangladesh within the Khulna division covering the districts of Narail (Figure 20.1). It has an area of 990.23 km². The total population of the area is 721,668 with a sex rate of 105 males per 100 female (Census 2011). The population density is 915 per sq. km which is higher than national average (835 per sq.km). The average household size in the sub-project area is 5.6 which is slightly above the national average of 5.5. It is a severely arsenic affected district. Most of the tube well water of Lohagora and Kalia Upazila (sub-district) was found with high arsenic contamination during 2002–2003 blanket arsenic screening under DPHE-UNICEF commissioned community based arsenic mitigation water supply project conducted by Environment and Population Research Centre (EPRC). Five unions of Lohagora sub-district (Itna, Mallikpur, Lahuria and Lohagora) and five unions of Kalia sub-district (Bornal Iliasabad, Pohordanga, Khasial Hamidpur and Joynagor) are selected for data collection. Until now various kinds of alternative and arsenic removal safe drinking water devices have been installed in the area. However, the installation of deep tube wells (DTWs) and pipe water was not affordable for most of the population. Also, a substantial part of the area is not suitable for DTW.

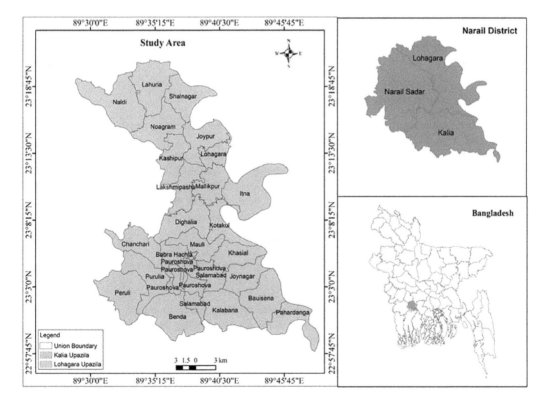

FIGURE 20.1 Study area Lohagora and Kalia sub-district of Narail District.

20.6 Materials and Methods

Both primary and secondary data have been used in conducting the study. Primary data has been collected through an extensive questionnaire survey. Focus Group Discussion (FGD) and Key Informant Interview (KII) also have been arranged in collecting primary data. Secondary data has been also collected through pursuing different reports of government, non-government organizations, web materials, various articles, journals, published and unpublished thesis, reports and books. The collected data with the questionnaire survey has been analysed, mainly with simple descriptive statistics in the graph and tabular form, while the qualitative mode of analysis is mainly in narrative form. In some cases, collected data has been analysed by using statistical techniques like correlation.

20.6.1 Sample Size and Selection Procedure

The total household of 10 Unions is 6457 of which 3476 is at Lohagora sub-district and 2981 is at Kalia sub-district (EPRC-2016). As the population is large in sizes due to short of time and budget, it was not possible to cover all the households of 10 Unions. For that reason, Yamane's mathematical formulae have been followed to determine the sample size.

Taro Yamane's (1970) formula:

$$n = \frac{N}{1 + Ne^2}$$

where N=total households=6,457, n=sample size and e=acceptable sampling error=5%.

$$\text{Therefore, the sample size, } n = \left[\frac{6,457}{1 + 6,457 \times (0.05)^2} \right] = \frac{6457}{1 + 16.14} = 376.7 = 377$$

20.6.2 Tools and Methods of Data Collection

Data have been collected from both primary and secondary sources. For primary data questionnaire surveys, interviews, FGD and KII methods are also used.

Questionnaire surveys were done among 377 households in the selected 10 Unions of two Upazilas. The information was collected on related multi-disciplinary variables, such as social, demographic, water access and use, sanitation and selected disease prevalence. The quality of water was tested on water samples collected from randomly selected about 40 households. Water quality analyses were done to determine thermotolerant coliform bacteria (TTC), manganese, conductivity and arsenic of stored water samples from those options.

20.7 Result and Discussion

It is found that the levels of knowledge related to safe drinking water, sanitation and hygiene maintenance of people in the study area are poor. Most of the respondents had nil knowledge about the carcinogenic impacts of drinking arsenic-contaminated water. Drinking water sources are mainly tube well and deep tube well. Apart from that, people use rainwater, pond water and river water for domestic and irrigation purposes. Tube well water is highly contaminated with arsenic, iron, manganese, lead and cadmium. As the main occupation of 82% of local people is agriculture, they have to depend on groundwater for irrigation during the winter season for harvesting IRRI rice, wheat and vegetable. Almost all the farmers extract groundwater through deep tube well and tube well to support irrigation in the winter period. Thus, a huge amount of water extracts from underground which implies declination of the water

TABLE 20.1

Drinking Water Status of Lohagora and Kalia Sub-District as per the Survey of Environment and Population Research Centre (EPRC) in August 2015

Variable	Kalia (N − 188) %	Lohagara (N − 189) %	Overall (N − 377) %
a) Drinking Water Source			
Deep Tube well	45(23.9)	35(13.9)	80(20.4)
Green Shallow tube well (STW)	36(18.1)	46(18.0)	82(21.7)
Red STW	48(25.5)	51(17.3)	99(26.2)
Not Tested STW	59(31.5)	57(50.8)	116(30.7)
b) Tube well Was Tested after Installed			
Yes	99(52.6)	113(59.8)	212(56.2)
No	89(47.4)	76(40.2)	165(43.8)
c) If Yes Then Is It Coloured			
Green	100(53.2)	90(47.6)	190(50.4)
Red	59(31.3)	51(26.9)	110(29.2)
Not colour	29(15.5)	48(25.5)	20(.4)
d) Is Sufficient Water Available Whole Year			
Yes	78(41.3)	71(37.5)	149(39.5)
No	48(25.5)	58(30.6)	106(28.1)
Sometimes	62(33.2)	60(31.9)	122(32.4)

table. Therefore, a lack of sufficient water was found in the tube wells from November to February. Consequently, the scarcity of drinking water stays behind in the study area. As the main sources of drinking water are contaminated and inadequate supply found in the winter season; thus, people have to depend on other sources like rivers, rainwater and even pond. Therefore, health hazards are often outset in the study area. Arsenic contamination is highly recorded, and tube well water screening has been done in 188 households of Kalia and 189 households of Lohagara Upazila. The status of drinking water found is mentioned in Table 20.1.

Data of Table 20.1 reveals that out of 377 tubewell in two sub-district in Narail, the arsenic contamination is at high risk levels. In the Kalia sub-district, 23.9% of household depends on deep tube well, whereas in Lohagara sub-district, only 13.9% household uses deep tube well water for their drinking purpose. Moreover, 26.2% of tube well in both sub-district are red marked indicating arsenic contamination, while 21.7% of tube well are green colour marked as its water is safe to drink. However, 30.7% of tube well water is not tested.

In the case of sufficient supply of tube well water, 28.1% of respondents replied that they got inadequate water during winter in their tube well, whereas 32.4% replied that they got sufficient water for a period only.

20.7.1 Quality of Water Supply Options

All the installed tube wells provided in the study area with arsenic-safe drinking water table and had higher than 150-m depths (Table 20.2). Concentrations of manganese, TTC, conductivity and arsenic were tested in randomly selected 40 tube wells.

Approximately 100% of tube wells met the manganese and arsenic standards of Bangladesh. Microbiological quality (thermotolerant coliform bacteria concentration; TTC in cfu/100 mL) was determined on water samples from randomly selected 40 DTWs (depth reported by users) during baseline survey. TTC (cfu/100 mL) was not detected in 96% of the samples. The rest showed the presence of <9 TTC cfu/100 mL, except one sample (56 cfu/100 mL). The presence of TTC in DTWs has been rarely reported in earlier studies. Although microbiological quality of the installed DTWs was tested from 40 installed DTW, it is most likely that the water was TTC safe at the tube well site.

TABLE 20.2

Quality of Water Samples from Installed Tube Wells

Variables	Depth (m)	Arsenic (mg/L)	Manganese (mg/L)	TTC (cfu/100 mL)
Sample size	150	40	40	40
Statistics				
Mean	235.8	0.0017	0	3
Median	233.2	0.001	0	0
Minimum	174.7	0.001	0	0
Maximum	293.5	0.028	0	13

TABLE 20.3

Sanitation Status of Lohagora and Kalia Upazila as per the Survey of Environment and Population Research Centre (EPRC) in August 2015

Variable	Kalia (N – 188) %	Lohagara (N – 189) %	Overall (N – 377) %
a) Type of Latrine			
Septic tank	88(46.8)	89(47.2)	177(47.0)
Ring slab	98(52.2)	96(50.3)	194(51.5)
Fixed hole hygienic latrine	2(1.0)	4(2.5)	6(1.5)
b) Ring Slab			
Water seal	68(62.9)	69(36.5)	137(36.4)
Broken slab	38(14.2)	39(20.8)	77(20.5)
Leakage	18(0.9)	17(0.9)	35(0.9)
Septic tank	11(.5)	10(.5)	21(.5)
N/A	53(28.2)	54(28.6)	107(28.3)

20.7.2 Sanitation Status

The sanitation condition of the study area is comparatively better than the water situation. Most households have a latrine. However, there is still a query on the hygienic environment as most of the toilets were found unhygienic. Both the Upazila area was found nearly free from open defecation. Having a toilet in each household is a common scenario except for one or two ultra-poor families. Hygienic use of latrines (ring slab with water seal) was found significantly higher in Lohagara Upazila than in Kaila Upazila. Over 94% of households in the study used sanitary latrines (Table 20.3).

It is found from the above table that only 1.5% of households contain a fixed hole hygienic latrine which is usually cheaper cost and dug on soil, whereas 51.5% of the households have a ring-slab latrine and the rest of them has a septic tank which is costly to make and manage. On the other hand, 36.4% of the family found water seal and 20.5% family found broken slab.

20.7.3 Overall Health Situation of the Study Area

The overall health situation of the study area is not as good as water-borne diseases are widely recorded in whole the year round. Water-related diseases like diarrhoea, dysentery, influenza, cough/cold, typhoid and common fever, pneumonia, pox, gastric, measles and jaundice skin diseases are still common in the study area. Modern medical facilities are not available to a larger percentage of the population. They depend mainly on traditional practitioners. Tube wells are the major source of drinking water supply in the study area. More than 98% of the population is dependent on groundwater for its water supply. The main problem is the presence of arsenic in a large number of the existing tube wells. However, the lowering of the groundwater table during the dry season beyond the withdrawal depth of shallow tube wells is also considered a problem. About 42% of the households have their tube wells, 13% of the households use

TABLE 20.4

Level of Knowledge about Water and Sanitation among Common People of Study Area

Issues ($N=30$)	Responses during Baseline Survey (%)	Responses during Questionnaire (%)
Safe drinking water	5	75
Diseases related to drinking of contaminated water	10	85
Contamination of tube wells	5	80
Health impacts of drinking arsenic-contaminated water	3	75
Sources of arsenic safe water	5	90
Hygienic latrine	2	75
Hand washing	1	75

pucca toilets and 26% semi-pucca, but more than 54% use not hygienic toilets and 6% of the households still use open fields. These poor set up of toilet and insufficient water and sanitation coverage address the second objective of this study as it reveals the challenges of the study area in case of maintaining 100% health hygiene. It can be depicted from Table 20.4 that the knowledge level of community people regarding health and hygiene is not adequate which addressed the first objectives of this study.

20.8 Conclusion

Recognition of the links of socioeconomic and health developments to access safe drinking water and hygienic sanitation is not new. The recent declaration by the UN General Assembly about rights to safe drinking water and sanitation further strengthened and focused its inevitable scopes for improvements in proper terms. Poor water and sanitation conditions in developing and LDCs tarnish the dignity of human being. Lack of knowledge, appropriate technology, institutional and governance, financial, programme and other issues have been hampering the effective improvement in water and sanitation in the least developed countries. This study has opened up opportunities to increase awareness among rural people regarding the most important issues of healthy living water and sanitation. However, the existing drinking water and environmental health conditions inside the study areas are found poor. There has been a strong demand from the local communities for access to safe drinking water and sustainable sanitation. The study reveals many important and glass ceiling factors responsible for inadequate water and sanitation coverage, which addressed the third objective of the study. For obvious reasons, the aims of the study are hiccupped are unveiling all the hidden causes as those are related to the dignity of the community. However, this study tried to sketch out the real scenario of five villages by applying its all-possible tools and technique to ensure the healthy living of thousands of rural people.

20.9 Recommendations

The following recommendations have been proposed based on the experiences obtained through this study:

1. A wider-level awareness programme should be taken regarding health, hygiene, sanitation and water.
2. Safe water devices should be installed in all emergencies in the (>80% contaminated) area.
3. Involvement of the community should be ensured in maintenance and management level works that have been done by the government and NGOs or development partners in providing water and sanitation options to ensure 100% effective coverage.

4. A training programme should be taken for the local people to look after and manage water devices such as rainwater harvesting, pipe water systems, water transportation and arsenic treatment technologies to enhance the quality of life.

5. Further research should be done in the study area with the proper intervention of the government in ensuring the standard livelihood of the local people.

Appendix: Environment and Population Research Centre (EPRC)

Arsenic-Safe Model Villages in Narail District Project

Questionnaire (Baseline survey) on WASH Knowledge assessment among community People

Dear respondent, I am a programme coordinator of EPRC working in a project entitled Creating Arsenic-Safe Model Villages in Narail District. This questionnaire is for social research purposes only. Your information will remain secret. I seek your earnest cooperation.

Thank you-------

Dr. Md. Moshiur Rahman

A. Demographic Profile

1. Name of the respondent:

2. Age---- ---- 4. Religion.....................

3. Gender---------- 5. Occupation...............

4. How many family member of your family?

5. How many members in your family?

6. How many female persons in your family?

7. What is the educational status of other member of your family?

Educational Status	Number of Family Member
Primary	
S.S.C.	
H.S.C	
Graduation	
Post-graduation	
Illiterate	

B. Health Status

Note: Please give tick (√) mark where appropriate.

8. Do you think proper health facilities remain in your locality?

9. If no what are the reasons?..................

10. From where you take treatment usually?
    a) Hospital, b) Community clinic, c) Homeopathy, d) NGO run clinic, e) Other

11. How often you suffer in diseases?
    a) Very often, b) Once a year, c) Once a month, d) Regular, e) Other

12. How often you go for treatment?
    a) Very often, b) Once a year, c) Once a month, d) Regular, e) Other

13. How is the quality of drinking water?
    a) Pure, b) Polluted, c) Arsenic contaminated, d) Not drinkable
14. Do you know when hand washing is necessary?
    a) Yes, b) No
15. Have you any toilet in your home?
    a) Yes, b) No
16. If yes then which type?
    a) Sanitary, b) Katcha, c) Semi-pucca, d) Other
17. If no then which place you use for toilet
    a) Nearby filed, b) Open space, c) Near river, d) No fixed place
18. Have you got any sanitation service from Govt.?
    a) Yes, b) No
19. Have you got any sanitation service from NGO?
    a) Yes, b) No
20. If yes then in what condition?
    a) Free of cost, b) Paying money, c) In lieu of work, d) Another condition

C. Impact of Water and Sanitation on Children Health

21. Are your children habituated to use toilet?
    a) Yes, b) No
22. How much time your family members suffer in disease?
    a) Regularly, b) Sometime, c) Very few time, d) Not at all
23. Which types of disease suffers?: diarrhoea/dysentery/belly pain/fever/other
24. Who suffers in disease mostly?: child/youth/old/women
25. Is there any medical centre located in your locality?
    a) Yes, b) No
26. Where you generally treat the contaminated person? Hospital/clinic/homeopathic/quack
27. Average annual treatment expenditure: <1,000/1,000 – 2,000/2,000 – 5,000/5,000+taka
28. Do you think that inadequate water and sanitation service impose an impact on child health?
    a) Yes, b) No
29. If yes then what should you do to get rid of this problem?
30. Do you think that human should take special care for health and sanitation service?
    a) Yes, b) No
31. Have there any campaign to increase awareness about health and sanitation in your locality?
    a) Yes, b) No
32. If yes then who are the organizer?
    a) Govt, b) NGO, c) Local people, d) Foreign agency, e) Other

D. Citizen Facilities related

33. Do you think citizen facilities are sufficient in your locality?
    If no
34. What are the reasons?
35. What is your demand about water and sanitation facilities to the Govt.?
36. What is your demand about water and sanitation facilities to NGOs?
37. Have any function of INGO in your locality?
    a) Yes, b) No

38. Do you think that the INGO work for a better society?
    a) Yes, b) No, c) No comments

39. What kind of facilities mainly offer INGO
    a) Provide education, b) Medical, c) Citizen amenities, d) Employment

40. What facilities you have achieved from INGO?
    a) Education, b) Treatment, c) Money, d) Employment, e) Others

41. Do you think that government fulfils all your rights?
    a) Yes, b) No, c) No comment, d) Others

42. What is your expectation to government?
    a) To get all facility like a citizen, b) To be freedom as separate nation, c) To access them to be leave country, d) Others

43. What is your expectation to NGOs/INGOs?

44. What we expect from donor agencies?

45. What is your assessment about sanitation service and hygiene for your community?
    a) Satisfactory, b) Strongly satisfactory, c) Dissatisfactory, d) Strongly dissatisfaction

Any suggestion to solve health service inadequate problem
Thanks for your kind cooperation
Researcher Dr. Md. Moshiur Rahman
Programme Coordinator
EPRC, Dhaka

REFERENCES

Benova, L., Cumming, O., Campbell, O.M. (2014). Systematic review and meta-analysis: Association between water and sanitation environment and maternal mortality. *Tropical Medicine & International Health* 19:368–87. doi: 10.1111/tmi.12275.

Chakravarty, I., Bhattacharya, A., Das, S. (2017) Water, sanitation and hygiene: The unfinished agenda in the World Health Organization South-East Asia Region. *South East Asia Journal of Public Health* 6(2):22.

Manava, P., Murry, J., Kim, R., Sobel, H. (2019). Status of water, sanitation and hygiene services for childbirth and newborn care in seven countries in East Asia and the Pacific. *Journal of Global Health* 9(2). doi: 10.7189/jogh.09.020430.

Minnatullah, K.M., Vidal, O. (2010). Achieving the water and sanitation MDGs in Bangladesh. Dhaka.

The World Bank Report, WHO-UNICEF, JMP (2010). Progress on sanitation and drinking water. World Health Organization and UNICEF.

The World Bank Report. (2016). Bangladesh: Improving water supply and sanitation. https://www.worldbank.org/en/results/2016/10/07/bangladesh-improving-water-supply-and-sanitation.

United Nations Children Fund (UNICEF). (2006) Levels and trends in child mortality: Report 2019. Available: https://www.unicef.org/reports/levels-and-trends-child-mortality-report-2019. Accessed: 6 November 2019.

World Health Organization Report. (2015). Progress on sanitation and drinking water: 2015 update and MDG assessment. Geneva 27, Switzerland, ISBN 978 92 4 150914 5 UNICEF-2006. Rural Sanitation, Hygiene and Water Supply.

World Health Organization, United Nations Children's Fund. (2018) WASH in health care facilities: Global baseline report 2019. Available: https://apps.who.int/iris/bitstream/handle/10665/311620/9789241515504-eng.pdf?ua=1. Accessed: 6 November 2018.

21

Circular Economy of Agriculture Wastes in Plastic Alternatives for Food Packaging

**Manal Abdel Rahman Sorour, Marwa Mohamed Helmy,
Amira Sabet El-mahrouky, and Aml Salah Elnawawy**
Food Technology Research Institute, Agriculture Research Center

Salwa Raafat Mostafa
Cairo University

CONTENTS

DOI: 10.1201/9781003231608-25

21.1 Introduction

21.1.1 Plastic Packages

So many plastic items, such as water bottles, milk jags, plastic bags to hold food, forks, knives, coffee cups and shopping carts, are used throughout the world.

Plastic products from the petrochemical industry are not environmentally friendly because of their high carbon footprint (Mohamed et al., 2019). Nowadays, people are more aware of the adverse environmental consequences of plastic materials produced from petrochemical products (Table 21.1).

In addition to these conflicts, plastics are still used indiscriminately in industries for example food industry because they have flexible properties, i.e. strength and rigidity, gas barriers, humidity, clarity and grease, resistance to attacks on food components and flexibility (Avella et al., 2001).

21.1.2 Alternatives of Plastics

Bio-plastic, instead of traditional plastic derived from petroleum, is a type of plastic made of biomass from renewable sources. It can be divided into two categories: biodegradable plastic or bio-based plastic.

There are a couple of terms when talking about alternative plastics that are often misunderstood or used interchangeably when they shouldn't be: bio-plastics and biodegradable plastics. Biodegradable plastic comprises fossil materials, while biomass or renewable resources synthesize bio-based plastics.

The driving parameters for the manufacture of these goods include:
- Petroleum and natural gas are examples of conserving fossil fuel reserves.
- Reduce the emissions of greenhouse gas that cause climate change.
- Helping to decrease the amount of organic waste going to disposal such as food and yard debris (Figure 21.1).

21.1.3 Bio-plastics

- Bio-plastics are plastics that are manufactured from organic materials such as maize, starch as well as polylactic acid (PLA).
- PLA has polypropylene (PP) and polyethylene (PET)-like properties and looks just like ordinary plastic. Theoretically, it degrades in a matter of weeks without leaching toxic chemicals back into the soil. However, this may be a bit misleading, as it is not a great idea to simply let waste of any kind break down in landfills. Like other organic material, its decay produces methane gas, a greenhouse gas many times more powerful than carbon dioxide. Instead, PLA-based bio-plastics may be broken down in a controlled environment by microbes, so they could be taken to a commercial composting facility (Ibrahim et al., 2017).

TABLE 21.1

Share of Most Relevant Types of Plastics in Production and Waste

Polymer	Production (%)	Share in Waste (%)
Polyethylene	30–34	40–47
Polypropylene	12–15	12–19
Polystyrene	8–10	13–17
PET	2–5	5–12
PVC	19–22	10–11
Polyurethane	5–7	4–8
Other thermoplastic	7–12	8–9
Other plastic	7–20	6–11

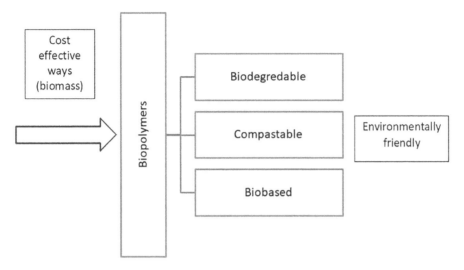

FIGURE 21.1 Basic idea of research.

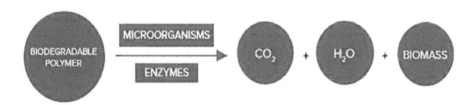

FIGURE 21.2 Biodegradable polymer.

- Bio-plastics are plastics that originate from plant sources such as sweet potatoes, sugarcane, soya bean oil, hemp oil, and maize starch. The activity of microorganisms such as bacteria, fungi and algae naturally degrade these polymers. Bio-plastics may help relieve the energy crisis and also reduce our society's reliance on fossil fuels. They have several remarkable characteristics that are suitable for various applications (Figure 21.2).

21.1.4 Biodegradable Plastics

On the other hand, it refers to plastics based on petroleum blended with an additive that makes them break down quickly. This term is quite vague and used interchangeably with bio-plastics, but the two are quite different chemically (Šprajcar, 2012).

21.1.5 Bio-Based Plastics

The term bio-based consists of both biodegradable and bio-based plastics, meaning those some content has been derived from natural resources or biomass. They may or bio-based plastics: The term bio-based consists of both biodegradable and are bio-based plastics, meaning some content is derived from natural sources or biomass (Figure 21.3). They might or might not be biodegradable but recyclable. The mechanical characteristics are similar to those manufactured from fossils, for example, bio-PVC and bio-PE derived from sugarcane (Braskem) might not be biodegradable but recyclable (Figure 21.4) (Songi et al., 2009).

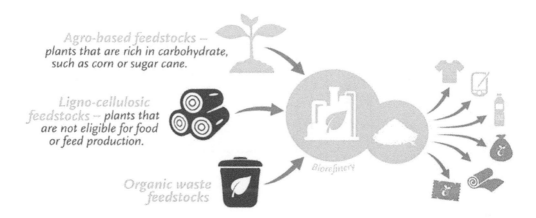

FIGURE 21.3   Bio-based plastic manufactured from a wide range of renewable bio-based feed stocks.

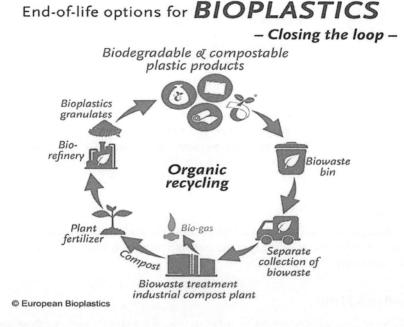

FIGURE 21.4   Bio-plastic.

21.1.6 Why Bio-plastic?

Plastics have become an essential component of our lives. It takes decades for traditional plastic to deteriorate in nature and is produced by non-renewable sources such as petroleum, coal or natural gas. Environmental, economic, and safety challenges have led many scientists to partially replace petrochemical-based polymers with biodegradable one, i.e. bio-plastics. The production of bio-plastics worldwide is <200,000 tons per year, dwarfing over 30 million tons of oil-based plastics. Studies also show that bio-plastics are eco-friendly because of their fewer greenhouse gases emission such as carbon dioxide, which is one of the primary sources of air contaminant sand that contributes to environmental problems such as global warming, climate change, etc.

21.2 Methods of Manufacturing of Bio-plastic

21.2.1 Manufactured from Microorganisms

The presence of polyhydroxy alkanoic acids (PHA) in prokaryotic cells as the storage of polymers cells is currently considered to be highly spread. They are water-insoluble substances. In relatively large amounts, many bacteria generate an intracellular carbon and energy storage compound poly-ß-hydroxy butyric acid (PHB). Development of high-yielding mutant strains resulted in 65% conversion rates for PHB and 71% dry weight eventual PHA yields (Satish & Thakur, 2017).

21.2.2 Manufactured from Plants

The high price is the most important constraint associated with the manufacture of bio-plastic from bacteria compared to petroleum-derived plastic. In exchange, the plant theoretically provides an alternative solution to synthesizing these bulk commodity products at low cost. Although the manufacture of PHA in bacteria and yeast involves an expensive fermentation process with external energy sources such as electricity, it is considerably less expensive in plant systems because it relies on water, soil nutrients and CO_2. In addition, a plant manufacture system is much more environmentally friendly. Although PHB synthesis and its accumulation are limited in the cytolysin bacteria, PHB maybe produced in several subcellular compartments such asystole, plastids, mitochondria and peroxisomes in plants.

21.2.3 Types of Bio-Plastic Are Based On

Cellulose-based plastics are commonly made from wood pulp and can be used to produce wrappers. The most used bio-plastic is thermoplastic starch which accounts for around 50% of bio-plastics drug capsules development by pharmaceutical industry. A plasticizer, such as sorbitol or glycerin, has been added to make it even more stable and produce several features. It is commonly obtained from crops like maize or potato, etc.

- **PLA** is a translucent plastic whose features resemble common plastic based on petrochemicals such as (PE) and (PP). Equipment that is already exists to produce traditional plastics which may be processed by fermenting starch from crops (maize and sugarcane), PLA is formed into lactic acid which will be then polymerized. The blends are used in a broad variety of applications including computer and cell phone cases, foil, biodegradable medical implants, tins cups, bottles, as well as other packaging material.
- **PHB** is somewhat like PP and is used in a wide range of fields including packaging, ropes, bank notes and automotive parts. It is a translucent film, that is biodegradable. The interest within PHB is now extremely high with companies and is biodegradable that have also been used as water resistant worldwide looking to increase their existing production capacity. On an industrial scale, the south American sugar industry has been committed to develop PHB in the cutting edge of poly hydroxy alkanoate (PHA) materials based on bio-plastic technology. These are obtained from the transformation of natural sugars and oils using microbes. They could be processed with in several materials including molded goods, fiber and film.

21.2.3.1 Advantages of Bio-Plastics (Srikanth, 2011; Richard & Kirwan, 2011; Chen, 2014; Ying, 2014)

i. **Reduced CO_2 Emissions**: One ton of bio-plastics produces between 0.8 and 3.2 of CO_2 less the one tons of plastic based on petroleum.

ii. **Cheaper Alternative**: with the fluctuations in oil prices, bio-plastic are becoming more viable.

iii. **Waste**: Bio-plastics decrease the amount of toxic runoff produced by the oil-based alternatives.

iv. **Benefit to Countryside Economy**: Crops' prices (e.g., maize) have risen dramatically in the wake of world interest in the development of biofuel and bio-plastics, and many countries are looking for alternatives to petroleum for the preservation of the environment and energy security.

v. **Reduced Carbon Foot Print**: As main raw material, oil-based plastics require fossil fuel. Moreover, as opposed to bio-plastics, oil-based plastics such as PP and PS need more energy during the plastic production. For standard PP or PS plastics, a life cycle analysis reveals a carbon footprint of approximately 2.0 kg CO_2 equivalents per kg of plastic equivalents (from cradle to factory gate). These emissions of CO_2 are four times greater than PLA resin.

vi. **Several End-of-Life Options**: it is possible to recover and reuse useful raw materials into new products, to minimize the need to have a new virgin material and to substantially reduce the negative environmental effect of 'used' plastic products.

21.2.3.2 Disadvantages of Bio-Plastics

Since nothing is full in this universe, there are a few disadvantages to biodegradable plastics, which are favored over the traditional ones, such as (i) biodegradable plastic is not intended to be reused with other kinds of plastics. (ii) If biodegradable plastic has not been properly disposed of, the results are an inefficient breakdown of the plastic, that can release pollutants into the environment (carbon dioxide, methane, etc.).

21.2.4 Dynamic Growth and Sustainable Development

The bio-plastics industry is a young, creative sector with tremendous economic and ecological potential for a more resource efficient, low-carbon, circular bio-economy. The EU has begun to consider the many benefits of the bio-based economy and is now allocating funds and services to the sector for research and development. Over the next few years, the worldwide demand for bio-plastics is anticipated to rise steadily based on the latest industry data gathered by European bioplastitium. Based on the latest trade information collected by European bio-plastics in partnership with nova-institute, it is expected that global bio-plastic production capacities shall rise from 2.11 million tons in 2018 to 2.62 million by 2023 approximately. Non-biodegradable, bio-based plastics, such as bio-based PE and bio-based PET, are at the forefront, while PLA and PHA will be the key drivers of growth in the bio-based and biodegradable plastic market.

21.2.5 Bio-plastic Processing

Bio-plastics are commercially available and mostly show processing characteristics concerning injection molding that are comparable to their petroleum-based counterparts. It is necessary to be consistent with the petroleum-based predecessors in the cavity design and the processing properties of the new material. Choosing the proper process parameters (particularly temperature) is also crucial for the optimum processing of bio-plastics, as some bio-plastics have a slightly narrow process window, the more sensitive reaction to thermal loading compared with the traditional plastic is the reason for this. From a manufacturing point of view, in injection molding, bio-plastics offer a fascinating alternative to petroleum-based plastics and can replace them. The process and cavity parameters have to be adapted to the bio-plastic as a new material – exactly the same when transitioning from one petrochemical polymer material to another petrochemical polymer material.

21.2.6 The Influences of Biodegradable Bio-Plastic (BDP) (Tobias et al., 2019)

The waste stream is treated by existing available options (recycling, incineration, and landfill) as briefly shown below. Special attention will be given to compost biopolymers, as BDPs allow a potential alternative for waste management by composting as a way of recycling the materials and creating a useful commodity as compost.

21.2.6.1 Recycling

Biodegradable plastics entering the public waste stream may cause some problems for existing recycled plastic systems.

21.2.6.2 Combustion with Energy Recovery

Gross calorific values (GCVs) are equal or greater than coal in commodity plastics. Combustion with energy utilization represents a potentially good choice for petrochemical carbon, which has already a high-value application.

21.2.6.3 Landfill

In the UK waste hierarchy, landfilling of waste plastics is the least favored choice. Historically, it was desirable as it was extremely simple and cheap without the need for separation, cleaning or treatment. In 1999, 65% of the total plastic to be recovered in domestic waste (8.4 million tons per year) has been sent to landfills in Western Europe. However, suitable landfill sites across Europe are running away and public worries about the effect of landfills are growing.

21.2.6.4 Biological Waste Treatments

Anaerobic or composting or digestion like traditional petrochemical-based polymers, biodegradable bio-plastic polymers could be composted. This may be achieved by aerobic waste-management systems such as composting to produce compost rich in carbon and nutrient for addition to soil. There are now more than 300 composting locations in the UK that collectively compost about 2 million tons of waste annually (approximately75% of which is household waste 5% municipal non-household waste and 20% industrial waste).

21.2.7 Agriculture Wastes Used in Production of Bio-Plastic

Agricultural waste, as the primary ingredient for substituting or eliminating the use of plastic, is one of the most effective ways to reduce this possibility. To achieve resource conservation and to maintain the quality of the environment, agricultural waste management is thought to be a critical strategy. Owing to their wide applications within food packages and biomedical sciences, bio-fibers have been attracting increasing interest in recent years. Owing to their safety, low manufacturing costs, and biodegradability, these eco-friendly polymer materials reduce the use of petroleum-based artificial polymers (Mostafa et al., 2018).

Bio-plastics have a lighter environmental load than traditional ones. Today, maize is the most used material for its processing. Although the amounts at stake are pale compared with the world production of this cereal, this still raises a significant question: is it correct to use a food crop for production of bioplastics,? Maize prices are increasing; most of the materials used for food-packaging applications are non-biodegradable such as petroleum-based plastics (Cecilia, 2017). Thus, these products do not satisfy the increasing need for sustainability in society and are not environmentally friendly (Othman, 2014). There is also a need to find alternative food-packaging materials to overcome the problem, and among these, the edible film appears to be the most favorable choice. The reason for this is that edible film is eco-friendly in which the film can be eaten together with the product packaged and consequently no food packaging must be disposed of (Shit & Shah, 2014).

21.2.8 Recommendations Regarding Bio-plastics

- Bio-based statements" should be accompanied by sound measurements based on accepted standards and certification, preferably by third parties. They can be reproduced as a percentage of overall carbon content of the material balance of a material/product showing either the bio-based mass content or the bio-based carbon content (Figure 21.5).

- Claiming that a product is biodegradable without any further specification is classed as vague. Determining the applied testing standards is essential and includes detailed information about the test environment and timeframe.

- Certification in compliance with EN 13432 or requirements standards should be acquired.

- A claim concerning a particular end-of-life alternative may be made only if the equivalent facilities may be accessed by a "fair proportion of users" (European Commission). End-of-life claims should comply with the essential requirements of the relevant European legal systems; the EU packaging and Packaging Waste Directive is especially relevant in the case of packaging. As a supplement to the European legal system, manufacturers should check national waste laws and regulations in place for bio-plastic.

- While a bio-based material's carbon footprint (cradle to gate) can be neutral or even negative (compared to fossil-based materials), this is rarely the case (full life cycle–cradle to grave) for a bio-based consumer product. It is, therefore, better to claim a "reduced carbon footprint".

FIGURE 21.5   Products made of 100% bio-plastics as a replacement for plastics.

21.3 Conclusion

The chapter has reviewed bio-plastic as a plastic alternative for food packages, its types, processing, advantages and disadvantages. Also, there is the possibility of producing bio-plastic to be used for food packages from agriculture wastes.

REFERENCES

Avella, M., Bonadies, E., Martusscelli, E., Rimedio, R., 2001. "European current standardization for plastic packaging recoverable through composting and biodegradation", *Polymer Testing*, 20: 517–521.

Cecilia, C., 2017. "Bioplastic made from upcycled food waste. Prospects for their use in the field of design", *The Design Journal*, 20(1): S1596–S1610.

Chen, Y.J., 2014. "Bioplastics and their role in achieving global sustainability", *Journal of Chemical and Pharmaceutical Reserch*, 6(1): 226–231.

Haider, P., Carolin, V., Johanna, K., Katharina, L., Fredrik, R., Wurm, A., 2019. "Plastics of the future? The impact of biodegradable polymers on the environment and on Tobias", *Angewandte Chemie International Edition*, 58: 50–62.

Ibrahim, M.S., Jafar, A.J., Abubakar, S.A.S., Saleh, Y., Mahmud, L., Ibrahim, A., 2017. "Bioplastics as better alternative to petro plastics and their role in national sustainability: A review", *Advances in Bioscience and Bioengineering*, 5(4): 63–70.

Mohamed, S., Zeinab, K., Taha, A., Waleed, M., Amr, F., Sara, A., Nourhan, S., Alaa, S., Mohamed, G., Menna, M., 2019. "Bioplastics production from agricultural crop residues", *Agricultural Engineering International: CIGR Journal*, 21(3). http://www.cigrjournal.org.

Mostafa, N.A., Awatef, A.F., Hala, M.A., Aghareed, M.T., 2018. "Production of biodegradable plastic from agricultural wastes", *Arabian Journal of Chemistry*, 11: 564–553.

Othman, S.H., 2014. Bio-nanocomposite materials for food packaging applications: Types of biopolymer and nano-sized filler. *Agriculture and Agricultural Science Procedia*, 2: 296–303. doi: 10.1016/j.aaspro.2014.11.042.

Richard, C., Kirwan, M.J,, 2011. *Food and Beverage Packaging Technology*, 2nd edn. John Wiley & Sons: New York.

Satish, K., Thakur, K.S., 2017. "Bioplastics: Classification, production and their potential food applications", *Journal of Hill Agriculture*, 8(2): 118–129.

Shit, S.C., Shah, P.M., 2014. Edible polymers: Challenges and opportunities. *Journal of Polymers*, 2014(427259): 1–13. doi: 10.1155/2014/427259.

Songi, J.H., Murphy, R.J., Narayan, R., Davies, G.B.H., Phil, T.R., 2009, "Biodegradable and compostable alternatives to conventional plastics" *Philosophical Transactions of the Royal Society B*, 364: 2127–2139.

Šprajcar, M., 2012. *Biopolymers and Bioplastics, Ljubljana*. European Regional Development: Slovenia.

Srikanth, P., 2011. *Handbook of Bioplastics and Biocomposites Engineering Applications*. Wiley-Scrivener Publishing LLC: Beverly, MA.

Ying, J.C., 2014. "Bioplastics and their role in achieving global sustainability", *Journal of Chemical and Pharmaceutical Research*, 6(1): 226–231.

22

Meta-Analysis of Studies on Solid Waste Management in the Philippines

Lynlei L. Pintor
ERDB College

Arlen A. Ancheta
University of Santo Tomas

CONTENTS

22.1 Introduction

Since the 2000s, the accelerating volume of waste generated in the Philippines has caught the government's serious attention (Sapuay, 2016, 2019). The country's fast paced industrialization and rapid population growth has made waste management a daunting task (Castillo & Otoma, 2013).

As the country's population rose from 27 million in the 1960s to 97 million in 2012, the amount of solid wastes generated per day also inflates at a constant rate (NSO, 2011; Castillo & Otoma, 2013; Atienza, 2011). This is the consequence of the aggressive industrialization in the country. The manufacture, distribution and use of plastic products add to the generation of wastes (Geyer, Jambeck & Law, 2017). In fact, the National Solid Waste Management Commission (NSWMC) reported a constant increase in the wastes generated by the country from 37,427.46 tons per day in 2012 to 40,087.45 tons in 2016 (SEPO, 2017).

A report from the McKinsey Center for Business and Environment revealed the severity of waste management issues in the country as the Philippines ranks third among the world's highest contributors of plastics leaking into the ocean (Ranada, 2019). This scenario leads to several pressing concerns such as improper wastes disposal, inefficient wastes collection, and lack of disposal facilities (SEPO, 2017). All of these put the country at greater risk of environmental degradation, climate change, and public health emergency.

In response to the alarming issue of waste management in the Philippines, the government designed various laws and policies to reduce waste pollution. These policies include Clean Water Act of 2004, Renewable Energy Act of 2008, Philippine Clean Air Act of 1999, and Ecological Solid Waste Management Act of 2000.

The Republic Act (RA 9275), known as "Clean Water Act of 2004", prevents dumping of solid wastes to rivers and streams and requires filtering of wastewater to preserve the quality of water reservoirs. Meanwhile, the Philippine Clean Air Act of 1999 was created to mitigate greenhouse gas (N20) emissions from wastes. The Renewable Energy Act of 2008, on the other hand, required the utilization of the country's renewable energy such as biogas (Martinez, 2017). Finally, in 2000, the government also crafted the Ecological Solid Waste Management Act of 2000 (RA 9003) which demanded for a "systemic administration of activities which provide for segregation at source, segregated transportation, storage, transfer, processing, treatment, and disposal of solid waste and all other waste management activities which do not harm the environment" (Castillo and Otoma, 2013).

The laws mentioned, especially the RA 9003, provided the legal framework and institutional mechanisms for an organized, comprehensive and ecological solid waste management program in the country. Under RA 9003, the NSWMC was established to execute all solid waste management plans upon accomplishing the objectives of the Act. At present, the country's solid waste management system uses engineered sanitary landfills and waste-to-energy technologies for a more efficient collection, transport, and disposal of solid wastes (Yi et al., 2019).

However, despite these technologies and innovations, solid waste management remains a challenge in the country. For example, studies showed that the volume of solid wastes exceeds the capacity of available sanitary waste disposal facilities and landfills in the Philippines. The scarcity of new landfills paves the way for the use of open dumps which pose threats to public health (Ferronato & Torretta, 2019; Sapuay, 2019). This explains why 790 out of 1,610 municipalities in the country do not comply with the minimum standards of landfills as stipulated by RA 9003 (GIZ, 2012).

Furthermore, experts have also expressed their concern regarding the need for gas-cleaning equipment that will prevent air pollution from the RDF or refuse-derived fuel technology being used in recycling residual wastes (Saupay, 2016, 2019). Ultimately, similar to other developing countries, the Philippines still lacks sustainable waste management models in academic programs, promotional activities on solid waste management (SWM) technologies and services, institutional capacity, and access to public information on solid waste management (ISSOWAMA, 2020).

The aforementioned problems evidently show the existing gaps in the body of knowledge and practice on solid waste management, especially in developing countries like the Philippines. The issues of solid waste management in the Philippines cover diverse and complicated aspects that must be given careful consideration backed with science. This makes research crucial in investigating the institutional, social, legal, and financial aspects of solid waste management (ISSOWAMA, 2020).

There is a call for continuous conduct of research studies that would address scientific and technological barriers that impede better solid waste management in the Philippines. Most importantly, further research on solid waste management would serve as a crucial baseline for policy formulation, innovation, technological breakthroughs, and institutional capacity enhancement. Thus, this study aims to analyze research studies from 2010 to 2020 regarding solid waste management in the Philippines. The objective of the analysis is to bridge the gap in the existing body of knowledge by providing recommendations based on the results.

22.2 State of the Art

As solid waste management issue becomes one of the biggest challenges in the Philippines, ample research studies ventured in analyzing the current system of the country regarding solid waste management. Significantly, most research studies fall under the ten components of the Philippine Solid Waste Management Strategy which include policy gaps, capacity development, social marketing and advocacy, sustainable financing, economic opportunities, knowledge management on technologies and innovation, organizational development and enhancing interagency cooperation, compliance monitoring, enforcement and recognition, good governance, vulnerable groups, and disaster and climate change risks (Acosta et al., 2012).

Several researchers have explored the policy gaps in SWM. These include Premakumara, Gilby and Kataoka (2016), Acostal et al. (2012), and Ragma et al. (2017). Among the popular studies on SWM capacity development were researches conducted by JICA (2005) and Acosta et al. (2012). Various researchers such as Bigornia (2008), USAID (2008), and Atienza (2011) also showed interest in SWM social marketing and advocacy sustainable financing, while Ancheta and Kariuki (2008), Banna et al. (2014), and ADB (2017) ventured in studies on sustainable financing. GIZ (2008), on the other hand, looked into economic opportunities. Knowledge management on technologies and innovation seemed to be the most popular topic among SWM researchers including Antonio (2010), Senoro and Saiyari (2012), Sapuay (2015, 2016, 2019), Martinez (2017), GAIA (2019), Abuag (2019), and ISSOWAMA (2020).

There were also significant studies that addressed issues on organizational development and inter-agency cooperation such as the study of Acosta et al. (2011), compliance monitoring by ADB in 2017, vulnerable groups by Zortea et al. (2019) and Castillo and Otomain (2013), and lastly, enforcement and recognition by Atienza (2011) and Atienza (n.d.). Recently, good governance has been the focus of a number of researchers such as Rola (2019), Castillo and Otoma (2013), and Pagunsan and Shimada (2012). Most importantly, the disaster and climate change risks reduction concerns were also given attention by more than a few researchers including Paul (2009), ADB (2017), Premakumara et al. (2018), NSWMC, DENR and IGES (2019), and Azodo and Ismaila (2016).

The overwhelming amount of new SWM studies published every day highlights the need for a meta-analysis. A review of previously published studies helps provide integrated results and important trends that influence future research and policymakers' decisions (Pedneault, 2018). Clearly, the Philippine SWM research displays the need to integrate findings from the many aforementioned studies. Yet, rarely can one find a meta-analysis of research studies on solid waste management especially in the Philippines. Most of the published meta-analytic studies on solid waste management are foreign studies. Among the few foreign meta-analysis on SWM research conducted includes Morris et al.'s (2012) meta-analysis of 82 studies on End-of-Life (EOL) management methods for source-separated organics and Oke's (2015) meta-analytical review on workplace waste recycling behavior. The rest of the studies conducted were meta-analyses on solid waste management system.

The lack of meta-analytic research contributes to the unabated problem on waste accumulation in the country such as non-compliance to landfill standards, potential pollution from refuse-derived fuel technology, and the of lack sustainable waste management models (ISSOWAMA, 2020).

Hence, a meta-analysis of SWM studies would address the existing gaps in the research body of knowledge and practice on solid waste management. This method would bridge both scientific and technological gaps toward better solid waste management. Most importantly, the integrated results from the meta-analysis would serve as an essential baseline for policy formulation, innovation, and institutional capacity enhancement.

22.3 Results and Discussion

Table 22.1 summarizes the highlights of selected solid waste management studies conducted in the Philippines for the past 10 years, from 2010 to 2020. A significant trend can be observed as far as research focus is concerned. From studying best recycling practices in 2010, researchers are now moving toward exploring zero-waste practices.

In 2010–2013, researchers investigated the country's SWM waste collection and recycling practices and their effects on poverty reduction, good governance, and environmental preservation. SWM problem of the Philippines is attributed to the country's high production volume of scrap steel, plastic waste, waste paper, and glass cutlets. This observation is the same as the observation mentioned in the study of Castillo and Otoma (2013).

Meanwhile, Antonio's (2010) study significantly found that 44% of household and commercial wastes are recyclable or factory-returnable. This data revealed that waste collection and recycling task can be outsourced. In response to this, the NSWMC opened the waste collection and recycling task to volunteers and small-scale waste collectors to complement the LGUs' SWM efforts. Such action was one of the best SWM practices in the country.

Chui (2010) found that the informal waste sector's contribution to managing wastes was indeed significant. The study found that 25% of the waste diversion rate was actually accomplished by the small

TABLE 22.1

Studies on Solid Waste Management (SWM) in the Philippines

Year	Focus of the Study	Methodology	Findings
2010	Recyclables collection trends and best practices	Review of literatures and document review	High export volume on scrap iron/steel and plastic waste, high import on waste paper and glass cullets, and 44% of household and commercial wastes are recyclable or factory-returnable. Best practices include LGU-led waste collection or voluntary collection. Commercial establishments outsource waste haulers or use large trucks/containers. Junkshop cooperatives or eco-aides collect segregated waste and bring it to the MRF. Recycling technologies and facilities transform wastes to useful raw materials or finished products.
2010	3Rs and poverty reduction	Survey and literature review	Informal waste sector contributes in waste diversion and recycling, at no cost to the LGU. LGUs push to implement their own programs and projects of which some have been successful with waste diversion rates of more than the target of 25% attributed to the contribution of the informal sector. Linis Ganda succeeded in organizing cooperatives of junk shops and formalizing the work of eco-aides.
2011	Review of the waste management system	Literature review and policy review from 1938 to 2001	Passed policies from 1938 to 2001. Implemented measures on the role of the LGUs in the collection and disposal of wastes. Factors on the failure of the past government measures include the lack of cooperation of the community, lack of infrastructure, and politics. Implementation of RA 9003 is already behind schedule since there are still open and controlled dumpsites but there are success stories in some municipalities which conducted IEC campaigns and used low-cost and local technologies.
2012	Recycled plastic aluminates relative to environmental physical stresses as barrier material	Procedures of American Society for Testing and Materials and International Standards Organization, Fourier Transformed Infrared (FTIR) approach ANOVA, Dunnett's procedure ($P < 0.05$) and Friedman test	Impregnation of organoclay (OC) improved the water vapor permeability and thermal insulation property. Recycled Plastic Aluminates (RPA) with 10% OC is the best performing RPAs for thermal insulation. All RPAs batches are considered water and leachate vapor barrier, gas vapor retarder, and moderate thermal insulator. Fourier Transformed Infrared (FTIR) showed that all pristine RPAs regardless of thickness and flexible plastic aluminum laminate trimmings (FPALT) particle size, are environmental stress cracking (ESC) resistant.
2014	Characterization of temperature profile in solid waste decomposition for harnessing thermal energy	Experiment on absorber for heat recovery	Temperature profile of heat absorber output was similar to landfill waste temperature profile, a remarkable growth in the initial period (first 100 hours, 37°C), then reaching and maintaining a certain peak condition, and finally goes down gradually after a long period(>40°C, 27 days). No indication of complete organic decomposition after 27 days due to high temperature. The 15.24 m long and 0.0127 m diameter heat absorber design can recover waste energy from the organic waste decomposition.
2016	Resource recovery through refuse-derived fuel	Key informant interview and secondary data gathering	Refuse-derived fuel (RDF) is an option to divert a large portion as an alternative fuel in urban area with high population growth. RDF is used as an alternative fuel raw material for cement coprocessing. It is an environment-friendly facility and complies with 3Rs of waste management

(Continued)

TABLE 22.1 (*Continued*)

Studies on Solid Waste Management (SWM) in the Philippines

Year	Focus of the Study	Methodology	Findings
2019	Synergy in the urban solid waste management system	Case study approach using system dynamics modeling	Constructed system dynamics model demonstrated the synergy in the urban solid waste management system by exhibiting the effect of waste diversion and public participation on the volume of disposed waste. Application of the model quantified the value coupled with public participation. Impact of allocating budget for technical improvements can be reinforced by allocating budget to increase public participation toward solid waste management practices.
2019	Municipal solid waste management through recycling	Literature review and observation	Recycling improves the management of municipal solid wastes and helps reduce the volume of waste disposed of in the landfill and divert it to recycling companies. Recovery can divert a portion of waste away from the municipal solid waste stream that ends up in landfills.
2019	Ocean's plastic waste problem and policy developments	Literature review	"Coasean solution" applicable for small economic activities of small groups. Private institutions moved into promoting "eco-friendly", "recyclable", and "organic" products. No extensive data to measure the production of plastic wastes (GAIA, 2019) but only data on solid waste generation per geo-political region, sources of waste, and broad types of solid waste (SEPO, 2017). There are data showing that solid waste generation can be correlated to increase in economic activity and population density. Some cities banned the use of plastics. Recent discovery on the "plastic-eating enzyme" (Austin et al., 2018 as cited in Dockrill, 2018) as a solution to plastic accumulation. Not easy to clean up the oceans. Need to consider opportunity costs and economic implications.
2019	Waste assessments and brand audits	Waste assessment and brand audit (WABA) tool	Organic wastes comprise more than 50% of generated wastes. Policy actions and implementation of plastic bag regulations lower plastic bag use. Around 59.7 billion pieces of sachets yearly, more than 50% of all unrecyclable residual waste discarded are branded wastes, only 10 companies are responsible for 60% of all the branded wastes.
2020	Streamlining waste-handling approaches to zero waste	Interview, FGD, thematic analysis	Zero-waste practices in selected barangays in San Fernando, Pampanga anchored on RA 9003 with ordinances which strictly implemented segregation at source and segregated collection.
2020	Zero-waste practices and its socio-economic benefits	Structural equation modeling (SEM) and path analysis	Aware citizenry and barangay-centered ESWM implementation in San Fernando, Pampanga. Collaborative approach in the ten-step approach predicted the community perceived socio-economic benefits and zero wastes practices
2020	Promoting environment at grassroots	Interview and case cross case analysis	Model barangays have clear relationship with institutional and noninstitutional support from outside its solid waste management space. Model barangay goes beyond solid waste management but also do network management.
2020	MRF: A living space and community asset	Interview, observation, review of secondary data, and cost-benefit analysis	Material recovery facility (MRF) is beneficial to the community as it provides livelihood, is cheaper in the long run, and results in high diversion of wastes from the landfill.

and medium waste collectors. Having recognized their valuable contribution to the country's SWM endeavor, the NSWMC incorporated the informal sector into the national framework of SWM where they established cooperatives of junk shops and formalized the work of eco-aides.

Commercial establishments also played roles in the SWM project. They outsource waste haulers or use large trucks/containers for their wastes while the junkshop cooperatives or eco-aides collect segregated wastes and deliver them to a material recovery facility (MRF) which transforms wastes to useful raw materials. In this process, SWM has not only increased the waste diversion rate in the country, but it has also opened sources of income; consequently, contributing to poverty reduction.

Good governance, on the other hand, was another important aspect associated with efficient solid waste management. Since 1938, numerous policies have already been enacted to address the country's insistent surge in solid waste production. The most recent law related to SWM is the Solid Waste Management Actor RA 9003 which was enacted in 2001. However, the execution of these regulations remains a challenge. Given the role of the Local Government Units in implementing proper collection and disposal of wastes, Atienza (2011) argued that the current SWM status is a mere reflection of weak political will. Other significant factors affecting the government's progress toward effective SWM, such as poor cooperation from the community and scarcity of available infrastructures, are, in fact, governance issues. This is evident in the delay of the government's full implementation of RA 9003. After almost 20 years of enactment, the use of open and controlled dumpsites remains rampant. Clearly, the government has exhibited weak political will in eliciting the strict compliance of its LGUs. There were, however, few municipalities that displayed strong political will in their SWM practices. These include Los Banos, Laguna; Bagumbuhay, Quezon City; and Caloocan City. This led to Atienza's (2011) conclusion that the application of good governance through conduct of intensive IEC campaigns and the establishment of low-cost and local technologies can lead to sound and effective SWM.

As recycling plastics becomes a useful trend, Senoro and Saiyari (2012) examined the effects of the process to the environment. Results of their study showed that recycled plastic aluminates have the potential to cause environmental stress. Using the American Society for Testing and Materials and International Standards Organization procedures and Fourier Transformed Infrared (FTIR) approach, the study investigated the ESC resistance. The results of their study provided a significant contribution to improving the water vapor permeability and thermal insulation property which could reduce environmental stress caused by recycled plastics.

In 2014–2018, research focus moved toward converting waste to thermal energy. Figueroa (2014) initiated the characterization of the temperature profile in solid waste decomposition to demonstrate the potential for harnessing thermal energy. Upon experimenting with the absorber for heat recovery, the study successfully contributed substantial information on recovering waste energy from the organic waste decomposition. While the study was not able to provide the complete organic decomposition process, it was able to provide the recommended heat absorber design (15.24 m long and 0.0127 m diameter) for an effective waste-to-energy process.

In line with Figueroa's (2014) work, Sapuay (2016) recommended refuse-derived fuel as an alternative fuel raw material for cement coprocessing of cement plants of Lafarge company since RDF facility is environment-friendly and uses the 3Rs of waste management. In addition, a recent discovery on the "plastic-eating enzyme" was published by Austin et al. (2018) which can be a remarkable solution to plastic accumulation. These enzymes are intended to decompose PET. Wax worms are also seen to decompose plastics as found in the study of Bombelli (2017).

Studies conducted in 2019, on the other hand, focused on public participation in the SWM programs, especially on recycling. Using a case study approach, Tinio, Rollon and Moya (2019) constructed a system dynamics model that demonstrated the synergy in the urban solid waste management system which includes public participation in the volume of disposed waste. Their model was able to quantify the value of public participation. The higher the budget allocated for technical improvements, the higher the public participation in the municipality's solid waste management practices.

Sapuay (2019), in her study, added the crucial input of private recycling companies in reducing waste volume in the landfills. The study found out that recycling can serve as an opportunity to help change the mindset of people. In 2019, GAIA published a study about waste assessments and brand audits. The study showed that when brand companies were informed that were part of the only ten brand companies that are accountable for 60% of all unrecyclable residual wastes discarded in the landfills, there was a shift to "eco-friendly", "recyclable", and "organic" products. Such change in practice produced dramatically significant results in lowering plastic bag use.

Finally, current studies conducted in 2020 are now concerned with zero-waste practices at the barangay level. Among the best zero-waste practices in the barangay level found by researchers include the strict compliance to RA 9003 (Ancheta et al., 2020), citizen awareness (Garcia et al., 2020), network management (Castillo, 2020; Acosta et al., 2020), and use of MRF (Garcia et al., 2020).

22.4 Research Gaps

For the past 10 years, numerous researchers have investigated the SWM issues in the country. Topics commonly focused on waste generation, collection cost, and tipping fee based on the linear waste stream model as well as waste characterization and GIS-based mapping. Many researchers have used GIS to track land areas for landfill selection, collection routing, realistic population densities. Spatial researches have been common to generate maps for landfills and MRF. Some also use cost-benefit analysis (CBA) to translate the impact of zero-waste practices into monetary value and enable benefits and costs to be included in the analysis of the decentralized SWM. Most of these reports are classified as government data and are funded by development organizations. Hence, most of the literature reviews conducted in the Philippines are also based on secondary data from government reports.

No studies used A'WOT, a hybrid of SWOT (Strengths, Weaknesses, Opportunities, and Threats) method and only a few utilized qualitative methods, cost-benefit methods, and mixed methods. Quantitative studies are also the most common approach used by researchers to study solid waste management. Survey interviews are commonly used for quantifying SWM practices while using inferential statistics, correlation, and structural equation model are commonly used for analyzing the data. Most researchers in the past have focused on the exploration of the best practices for recycling (Antonio, 2010; Senoro & Saiyari, 2012; Saupay, 2019; Abuag, 2019), clean technology, and zero-waste economy (Sapuay, 2015). Only a significant number of researchers have investigated the implementation of SWM policies (Ragma et al., 2017).

Today, there is an emerging method for studying SWM. The use of mixed methods includes the use of both quantitative and qualitative approaches to come up with a more holistic perspective in SWM. This approach highlights the importance of both the numbers and figures along with the contexts that go with them. More researches that will use mixed methods are expected to provide greater understanding of the gaps present in solid waste management.

Qualitative studies are also not as popular compared to quantitative studies. Only a few studies have used qualitative methods of analysis (Ancheta, 2006) because of its intricacies and complexities in terms of data collection and encoding. However, meta-analysis of studies on solid waste management in the Philippines shows that there is a need for more extensive qualitative studies on this aspect. This is expected to fill the gaps in the social dimension of the issue such as investigating the differing experiences of stakeholders and actors in SWM. Hence, future researchers should focus on the use of qualitative approaches and/or mixed methods to fill in the research gaps and come up with a more holistic view of the issue of solid waste management in the Philippines.

22.5 Conclusion

Solid wastes have evolved through time. As material economy improves, more natural resources are being utilized. As different types of waste emerge through time, solid waste management also becomes a more complex matter. The continuous changes in the composition of solid waste coupled with increased solid waste disposal, overflowing landfills, and shift to non-biodegradable requires a more systematic, comprehensive, and cohesive solid waste management plan. To achieve this, there's a need for researches that will help address the complex and interrelated issues of solid waste management.

Meta-analysis of studies on solid waste management in the Philippines showed that there are several methods that were crafted as well as numerous instruments that were developed to discover and find solutions to our pervasive problem of solid waste. Different theories have already emerged that will help humans further understand the complex issue. There had been shifts and development from a quantitative study of the nature of wastes into qualitative researches that could help further understand the figures and shed light on the issue of solid waste especially those that are non-biodegradable.

One of the new paradigms that have been emerging is the concept of zero-waste economy. Research studies have shown that zero waste is the track that the countries need to take. However, while zero waste has already gained advocates worldwide, there's still a need for large-scale implementation as it is only commonly implemented in the local communities.

The issue of solid waste management still requires more innovative studies and new technologies. Researchers are still confronted with many challenges including new technologies, decentralized practices, professional workers, and right mindsets to be transformed into more coherent solid waste management systems. The country needs more system-based approach models that can be adopted by our communities. Now, more than ever, require for more inquisitive and creative approaches to find more long-lasting solutions to our pervasive problems on solid wastes.

22.6 Scope for Future Research Work

In the near future, the authors plan to conduct qualitative research studies focusing on the narratives of men and women on solid waste management toward the circular economy in urban, peri-urban, and rural areas. Further, the authors will explore research methods that have not yet been explored by other researchers who conducted studies on solid waste management.

Acknowledgements

The authors would like to acknowledge Prof. Belinda de Castro, Ph.D., Associate Professor Moises Norman Z. Garcia, Ph.D., Associate Professor Maria Rosario Virginia Cobar-Garcia, Ph.D., Associate Professor Ronald M. Castillo, MA, Mr. Zosimo Membrebe, MA, Mother Earth Foundation, City of San Fernando, Pampanga, and Global Alliance for Incinerator Alternatives for sharing information on solid waste management.

REFERENCES

Abueg, L. (2019). A survey of the ocean's plastic waste problem, and some policy developments of the Philippines. Munich personal RePEc Archive. MPRA Paper No. 96263. Retrieved from https://mpra.ub.uni-muenchen.de/96263/1/MPRA_paper_96263.pdf.

Acosta, V.L., Paul, J.G., Ricana, M.V.F., Maceda, L.R., Aguinaldo, E.C. (2011). Implementing regional ecology centers in the Visayas region, Philippines: Networking to enhance sector development and solid waste management. In *Proceedings of the International Conference on Solid Waste 2011: Moving Towards Sustainable Resource Management*, Philadelphia, PA, pp. 75–161. Retrieved from https://www.iswa.org/uploads/tx_iswaknowledgebase/05_Waste_Management.pdf.

Acosta, V.L., Paul, J.G., Ricana, M.V.F., Maceda, L.R., and Aguinaldo, E.C. (2011).

Ancheta, A. et al. (2020). Streamlining waste handling approaches to zero waste: Community-based experiences in the city of San Fernando, Pampanga, Philippines. *Journal of Nature Studies*, 19(1), 49–65. Retrieved from http://journalofnaturestudies.org/files/JNS19-1/49-65_Ancheta_Streamlining%20Waste%20Handling.pdf.

Ancheta, C.R. and Kariuki, R.M. (2008). Cost sharing framework for solid waste management. Retrieved from http://nswmc.emb.gov.ph/wp-content/uploads/2017/05/cost-sharing-framework-for-swm.pdf.

Antonio, L.C. (2010). Tudy on recyclables collection trends and best practices in the Philippines. In: Kojima, M. (ed.), *3R Policies for Southeast and East Asia* (pp. 40–70.). ERIA Research Project Report 2009–10, Jakarta: ERIA. Retrieved from eria.org/uploads/media/Research-Project-Report/RPR_FY2009_10_Chapter_3.pdf.

Asian Development Bank (ADB) (2017a). Integrated solid waste management for local governments: A practical guide. Retrieved from https://www.adb.org/sites/default/files/institutional-document/324101/toolkit-solid-waste-management.pdf.

Asian Development Bank (ADB) (2017b). Mainstreaming integrated solid waste management in Asia: Technical assistance completion report. Retrieved from https://www.adb.org/sites/default/files/project-documents/46248/46248-001-tcr-en.pdf.

Asian Development Bank (ADB) (2017c). Report on climate change vulnerabilities and mitigation options in solid waste management. Retrieved from https://events.development.asia/system/files/materials/2017/02/201702-climate-change-vulnerabilities-and-mitigation-options-solid-waste-management-philippines.pdf.

Atienza, V.A. (2011). Review of the waste management system in the Philippines: Initiatives to promote waste segregation and recycling through good governance. Chosakenkyu Hokokusho, Institute of Developing Economies. Retrieved from http://www.ide.go.jp/library/Japanese/Publish/Download/Report/2010/pdf/2010_431_05.pdf.

Atienza, V.A. (n.d.). A breakthrough in solid waste management through participation and community mobilization: The experience of Los Baños, Laguna, Philippines. Retrieved from https://en.apu.ac.jp/rcaps/uploads/fckeditor/publications/journal/RJAPS_V24_Atienza.pdf.

Azodo, A. and Ismaila, O. (2016). Effective solid waste management for environmental quality and sustainability: Knowledge and practices among Nigerian households. In *Proceedings of the 2016 International Conference on SET: A Driving Force for Sustainable Development*, Federal University of Agriculture, Abeokuta. Retrieved from https://www.researchgate.net/publication/303483817.

Banna, F.M., Bhada-Tata, P., Ho, R.Y.Y., Kaza, S., and Lee, M. (2014). Results-based financing for municipal solid waste. Urban development series knowledge papers, 2(20). Washington, DC; World Bank Group. http://documents.worldbank.org/curated/en/237191468330923040/Main-report.

Bigornia, C.G. (2008). The Philippine Environmental Governance Project (EcoGov2): Report on social marketing on SWM-Jagna experience. USAID. Retrieved from http://faspselib.denr.gov.ph/sites/default/files//Publication%20Files/Report%20on%20Social%20Marketing%20on%20SWM-Jagna%20Experience.pdf.

Castillo, A.L. and Otoma, S. (2013). Status of solid waste management in the Philippines. In *Proceedings of the Annual Conference of Japan Society of Material Cycles and Waste Management*. Retrieved from doi: 10.14912/jsmcwm.24.0_677.

Castillo, R.M. (2020). Promoting environment at grassroots: Barangay institutional mapping of solid waste management. *Journal of Community Development Research*, 13(2). Retrieved from http://www.journal.nu.ac.th/JCDR/article/view/Vol-13-No-2-2020-15-30.

Castillo, A.L. and Otoma, S. (2013). Status of solid waste management in the Philippines. In *Proceedings of the 24th Annual Conference of Japan Society of Material Cycles and Waste Management*. Retrieved from doi: 10.14912/jsmcwm.24.0_677.

Chiu, A.S.F. (2010). The 3Rs and poverty reduction in developing countries Lessons from implementation of ecological solid waste management in the Philippines. Asia Resource Circulation Policy Research Working Paper Series. Retrieved from https://www.iges.or.jp/en/publication_documents/pub/discussionpaper/en/1986/swm_philippines.pdf.

Ferronato, N. and Torretta, V. (2019). Waste mismanagement in developing countries: A review of global issues. *International Journal of Environmental Research and Public Health*, 16(6), 1060. doi: 10.3390/ijerph16061060.

Figueroa, A.I. (2014). Characterization of temperature profile in solid waste decomposition to demonstrate potential for harnessing thermal energy, In: *Proceedings of 7th IEEE HNICEM/ISCIII/ERDT*, IEEE Xplore Digital Library, pp. 1–4. Retrieved from https://ieeexplore.ieee.org/document/7016216/authors.

GAIA. (2019). Plastics exposed: How waste assessments and brand audits are helping Philippine cities fight plastic pollution. Retrieved from https://www.no-burn.org/plastics-exposed/.

Geyer, R., Jambeck, J.R., and Law, K.L. (2017). Production, use, and fate of all plastics ever made. *Science Advances*, 3(7), 1–5. doi: 10.1126/sciadv.1700782.

GIZ (2018). Economic instruments for solid waste management: Case study Bayawan, Philippines. Deutsche Gesellschaft fürInternationale Zusammenarbeit (GIZ) GmbH Concepts for Sustainable Waste Management. Bonn and Eschborn, Germany. Retrieved from https://www.giz.de/en/downloads/giz2012-en-economic-instruments-philippines.pdf.

Global Alliance for Incinerator Alternatives (GAIA) (2019). Plastics exposed how waste assessments and brand audits are helping Philippine cities fight plastic pollution. Retrieved from https://www.no-burn.org/plastics-exposed/.

Integrated Solid Waste Management in Asia (ISSOWAMA) (2020). Evaluation of major gaps in knowledge on solid waste in 9 Asian Countries. Retrieved from http://wasteportal.net/en/system-aspects/institutional-aspects/evaluation-major-gaps-knowledge-solid-waste-9-asian-countries.

Japan International Cooperation Agency (JICA) (2005). Supporting capacity development for solid waste management in developing countries. Retrieved from https://www.jica.go.jp/jica-ri/IFIC_and_JBICI-Studies/english/publications/reports/study/topical/waste/pdf/waste_01.pdf.

Madrigal, D.V. and Oracion, E.G. (2018). Solid waste management awareness, attitude, and practices in a Philippine Catholic higher education institution. *Recoletos Multidisciplinary Research Journal*, Negros, Philippines. Retrieved from https://www.researchgate.net/publication/327177428.

Martinez, R.G. (2017). Waste management and biogas technology promotion in the Philippines. Retrieved from http://wepa-db.net/3rd/en/meeting/20170221/PDF/10_S2-1_5_Philippines_0221pdf.

Morris, J., Matthews, H., and Morawski, C. (2012). Review and meta-analysis of 82 studies on end-of-life management methods for source separated organics. *Waste Management (New York, N.Y.)*. 33. doi: 10.1016/j.wasman.2012.08.004.

NSWMC, DENR, and IGES (eds.) (2019). *National Strategy to Reduce Short-Lived Climate Pollutants from Municipal Solid Waste Sector in the Philippines*. Department of Environment and Natural Resources, Quezon City, Philippines. Retrieved from https://www.iges.or.jp/en/publication_documents/pub/policy-submission/en/6987/National+Strategy+to+reduce+short+lived+climate+pollutants.pdf.

Oke, A. (2015). Workplace waste recycling behaviour: A meta-analytical review. *Sustainability*, 7(6), 7175–7194. doi: 10.3390/su7067175.

Pagunsan, J. and Shimada, K. (2012). Efficiency evaluation of Philippines waste management sector: A two stage approach. Retrieved from https://www.researchgate.net/publication/242343111_Efficiency_Evaluation_of_Philippines_Waste_Management_Sector_A_Two_Stage_Approach.

Paul, J. (2009). Rapid assessment of greenhouse gas emissions from municipal solid waste management in the Philippines. In: *Annual Conference of the Society for Environmental Economics and Policy Studies*. Retrieved from https://www.researchgate.net/publication/281116223_Rapid_assessment_of_Greenhouse_Gas_Emissions_from_Municipal_Solid_Waste_Management_in_the_Philippines.

Pedneault, K.S. (2018). The role of meta-analysis in scientific studies. Retrieved from https://www.verywellmind.com/definition-of-meta-analysis-425254.

Premakumara, D., Gilby, S., and Kataoka, Y. (2016). Barriers for implementation of the Philippine national solid waste management framework in cities. Institute for Global Environmental Strategies. Retrieved from www.jstor.org/stable/resrep02912.

Premakumara, D., Menikpura, S., Singh, R.K., Hengesbaugh, M., Magalang, A.A., Ildefonso, E.T., Valdez, M., and Silva, L.C. (2018). Reduction of greenhouse gases (GHGs) and short-lived climate pollutants (SLCPs) from municipal solid waste management (MSWM) in the Philippines: Rapid review and assessment. *Waste Management (New York, N.Y.)*, 80, 397–405. doi: 10.1016/j.wasman.2018.09.036.

Ranada, P. (2019). Why PH is world's 3rd biggest dumper of plastics in the ocean. *Rappler*. Manila, Philippines. Retrieved from https://www.rappler.com/science-nature/environment/108276-philippines-plastic-pollution-ocean-conservancy-study.

Rola, A.C. (2019). Challenges in governance and implementation of solid waste management programs: Focus on plastics. Retrieved from https://nast.ph/index.php/downloads/category/142-visayas-regional-scientific-meeting?download=577:3-dr-rola-presentation-challenges-in-swm-implementation-march-2019.

Sapuay, G.P. (2016). Resource recovery through RDF: Current trends in solid waste management in the Philippines. *Procedia Environmental Sciences*, 35, 464–473. doi: 10.1016/j.proenv.2016.07.030.

Sapuay, G. (2019). Improving MSW management through recycling. *ACTA Scientific Agriculture*, 3(8), 169–176. Retrieved from https://actascientific.com/ASAG/pdf/ASAG-03-0581.pdf.

Senate Economic Planning Office (SEPO) (2017). Philippine solid wastes at a glance. Manila, Philippines. Retrieved from https://www.senate.gov.ph/publications/SEPO/AAG_Philippine%20Solid%20Wastes_Nov2017.pdf.

Tinio, M.R., Rollon, A.P., and Moya, T.B. (2019). Synergy in the urban solid waste management system in Malolos City, Philippines. *Philippine Journal of Science* 148(1), 73–97. Retrieved from http://philjournalsci.dost.gov.ph/images/pdf/pjs_pdf/vol148no1/synergy_in_the_urban_solid_waste_management_with_APPENDIX.pdf.

USAID (2008). A guide to conducting focus group discussions: A community-based social marketing tool to increase participation in solid waste management. Retrieved from http://faspselib.denr.gov.ph/sites/default/files//Publication%20Files/SWM%20FGD%20GUIDE_Final%20Copy.pdf.

Yi, M.K., Yu, K.W., Sheng, L.L., Li, T.Y., and Yen, K.H. (2019). Air pollution characteristics of Reclamation of Refuse Derived Fuel (RDF) recovered from cutting oil waste. *Aerosol and Air Quality Research*, 19(11): 2576–2584. doi: 10.4209/aaqr.2019.09.0481.

Zortea, M., Bonis, M.D., Pupa, F., Ripaldi, G., and Cucculelli, F. (2019). Community-based and integrated solid waste management: Experiences from metro Manila's Tondo district. *Journal of Waste Management & Xenobiotics*, 2(1), 1–5. doi: 10.23880/oajwx-16000117.

23

Management of Urban Flood-Induced Solid Waste in Some Metropolitan Cities: An Overview

Tanmoy Majumder, Paromita Chakraborty, and Bitanjaya Das
KIIT Deemed to be University

CONTENTS

23.1 Introduction

Floods are natural disasters that occur as a result of either intense rainfall or overflow of river banks or due to high tides caused by strong wind forces in the coastal areas (Tingsanchali, 2011). Flood causes massive damage to infrastructures and several life forms, but the aftermath of a flood is worst as it leaves behind a vast amount of solid waste all around the affected area to be disposed of. The risk of the flood being encountered globally is increasing within the context of climate change, specifically in the Asia and Pacific regions. Urban flood is not the same as a river flood; it was learned after the massive flooding in Mumbai (India) in 2005. (National Disaster Management Authority, 2010). The after-effect of solid waste in the city, that too in big metropolitan cities due to flood, is significant. Proper management of flood waste enhances the rapid recovery of the affected area and improves the sanitation and safety of the affected people.

It is necessary to dispose of tons of waste brought along with the flood water or have been dislocated from their dumping site to control the spread of various types of diseases that can result due to the decomposition of waste in the city area. It is a fatal situation as this may lead to contamination of drinking water. The contaminated water serves as a breeding ground for insects like mosquitoes and other pests and can cause the spread of malaria, cholera, typhoid, and other diseases in the area.

Situations in leading metropolitan cities in Asia are continuously reeling under flood, primarily due to very high intensity and short duration rainfall, and not having proper storm drainage facilities. Even some of the cities are located very near to the sea coast or rivers making the situation very complicated. Besides, these urban areas are more affected by floods because of the increase in impervious surfaces (Kundzewicz et al., 2010).

The rapid growth of the population in urban areas is one of the significant factors for the urban flood. Due to urbanization, there is progressive development of waterproofed surfaces (like roads, pavements, roofs); as a result, infiltration of water is reduced during flood events (Texier, 2008). Reduced infiltration, along with weak drainage systems with improper waste management, often results in flooding during the wet period of the year. After the flood, the deposition of solid waste creates a blockage in sewers and becomes a source of toxins and harmful diseases (Illangasekare et al., 2006). Leaching of these toxins may pollute the groundwater. It is often observed that wastes are being disposed of in drains causing blockage of the system, which results in local flooding (Kundzewicz et al., 2010).

Solid waste is generally defined as the waste obtained from local houses, commercial complexes and government enterprises, that includes food waste, clothing, packaging and other waste that have a solid form (Yukalang et al., 2018). During a flood, these wastes scatter around the affected area. Therefore, it is an absolute necessity to have appropriate management of this waste to avoid any detrimental effects on the environment and the people residing in that area. A report from the Central Pollution Control Board of India describes that the generation of solid waste in India was approximately 1, 27,486 TPD (Tons per day) in the year 2010–2011, earlier which was 1, 00,000 TPD in the year 2000. The World Bank has estimated that by the year 2025, the quantity of solid waste will be doubled than what it was in 2012 (Hoornweg and Bhada-Tata, 2012). Also, the approximate growth rate of municipal solid waste in developed and developing countries is about 3.2%–4.5% and 2%–3%, respectively (Suocheng et al., 2001). Management of waste is one of the most expensive public services (Memon, 2010). In India, only the accumulation and transportation of waste cost around 80%–90% of the total budget for waste management (Nandan et al., 2017). Table 23.1 shows GDP along with waste generation rates for a few Asian countries. It is seen that countries with lower GDP have lower rates of waste generation (Shekdar, 2009).

Various methods are incorporated for the effective management of waste, such as the collection of waste and its transportation. Then it is treated and recycled, and at the end, the remaining residue is disposed (Memon, 2010). Figure 23.1 shows the waste management technique followed in South Korea (Oh and Kang, 2013.)

The quantity of waste produced in different areas varies, depending upon the socio-economic status of the people living there. For example, according to Pune Municipal Corporation (India), 1,600 Metric ton (MT)/Day of municipal solid waste is generated and is categorized in Table 23.2.

Various treatment methods are used to disintegrate these wastes such as bio-composting for organic waste, recycling for E-waste, burning in the case of biomedical waste, and landfilling in the case of constructional debris and inorganic waste (Soni et al., 2016). Table 23.3 gives an insight into the percentage of the practice of each waste treatment methods in each region of the world (Memon, 2010).

TABLE 23.1

GDP and Waste Generation of Different Countries

Country	GDP per Capita Estimated for 2007 (USD)	Waste Generation (kg/Capita/day)
Hong Kong	37,385	2.25
Japan	33,010	1.1
Singapore	31,165	1.1
Taiwan	31,040	0.667
South Korea	23,331	1.0
Malaysia	12,702	0.5–0.8
Thailand	9,426	1.1
China	8,854	0.8
Philippines	5,409	0.3–0.7
Indonesia	5,096	0.8–1
Sri Lanka	5,047	0.2–0.9
India	3,794	0.3–0.6
Vietnam	3,502	0.55
Lao PDR	2,260	0.7
Nepal	1,760	0.2–0.5

Source: Shekdar (2009: p. 1440).

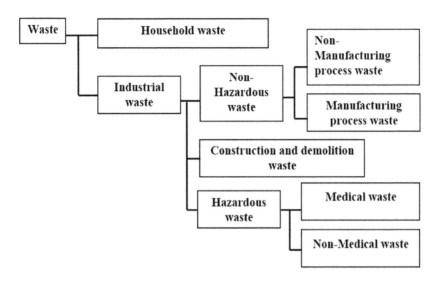

FIGURE 23.1 Waste classification according to South Korea's waste management law (Oh and Kang, 2013: p. 285).

TABLE 23.2

Bio-Degradable and Non-Bio-Degradable Waste Produced in Pune, India

Bio-degradable waste: 720 MT/day	Non-biodegradable: 880 MT/day
Organic waste: 400 MT	Inorganic waste: 560 MT
Households waste: 320 MT	Inert waste: 320 MT

Source: Soni et al. (2016: p. 120).

TABLE 23.3

Waste Treatment Method Followed in Each Region of the World

Region	Sanitary Landfill (%)	Incineration (%)	Open Dumps (%)	Recycling (%)	Open Burning (%)	Others (%)
Africa	29.3	1.4	47.0	3.9	9.2	8.4
Asia	30.9	4.7	50.0	8.5	1.7	4.5
Europe	27.6	13.8	33.0	10.7	11.8	4.4
North America	91.1	0.0	0.0	8.1	0.0	0.0
Latin America	60.5	2.0	34.0	3.2	5.5	2.0

Source: Memon (2010: p. 33).

23.2 Urban Flooding

Urban flooding occurs as a result of locally heavy rainfall and also due to insufficient or poorly developed drainage systems. Floods that occur as a result of overbank river flow are because of the rise in river water levels above the river bank. Land-use changes resulting in large-scale impervious surfaces can considerably affect flooding in urban areas (Suriya and Mudgal, 2012). Few hydrological properties that are affected due to changes in the land-use pattern are alteration in the peak flow properties, total runoff, and quality of water (Leopold, 1968).

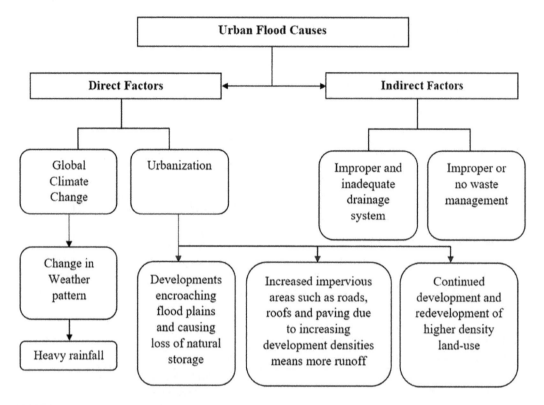

FIGURE 23.2   Factors causing urban flood (Rafiq et al., 2016: p. 725).

23.3 Causes of Urban Flooding

If the rate of absorption of water into the soil, or the runoff of water into the reservoirs, is slower than the rate of rainfall, there may be an occurrence of urban flooding. Other factors such as river flooding, snowmelt or flash flooding also contribute towards urban flooding. Due to the rapid filling of urban lakes, various cities across India are facing frequent urban floods (Figure 23.2). The wetlands that are being encroached upon are considered to be one of the most critical factors in the increase of urban flood as they store flood water and then slowly with time releases it (Melesse et al., 2006). Also, it has been observed that the continuous accumulation of solid wastes and debris on the curb-side of the roads is often washed into the sewers system leading to blockage of sewers and ultimately flooding of that area (Njoku et al., 2015).

23.4 Consequences of Urban Flooding

Urban flooding can cause notable damage to private and public properties, such as residential complexes, railways, roads, ports, and others, besides gathering tons of waste. This damage can cause significant loss to a national economy (Figure 23.3). A study of the previous disasters occurred in the United States found that the quantity of waste produced on regular days was about 5–15 folds less than what occurs as a result of a single disastrous incident (Reinhart and McCreanor, 1999).

FIGURE 23.3   Flood waste being collected (National Institute for Environmental Studies (NIES), 2015: p. 6).

23.5 Classification of Flood Waste

TABLE 23.4

Flood Waste Classification

Types of Waste	Example
Wastes from kitchen	• Left-over food, fruit skins, etc.
Waste generated from paper	• Carton-boxes, paper bags, etc.
Wastes generated from plastic	• Polyethylene bags, food containers, electrical appliances, etc.
Wastes consisting textiles	• Discarded clothes, etc.
Wastes consisting rubber/leather	• Car tires, discarded shoes, etc.
Wastes consisting metal	• Discarded household metal utensils, furniture, etc.
Green waste	• Branches of trees, leaves, etc.
Construction and demolition waste	• Wooden doors and windows, wooden panels, bricks, concrete piles, reinforcement bars, etc.
Mixed waste	• Wastes consisting of various materials.
Hazardous waste	• Television, gas cylinder, battery, medicines, etc.

Source: Agamuthu et al. (2015: p. 133).

23.6 Flood Waste Management

An effective flood waste management system should incorporate appropriate strategies (Table 23.4) such as mitigation, preparedness, after the flood and before flood phases (NIES, 2015). The mitigation phase starts after the complete recovery from the previous flood and ends at the warning of the next flood. The preparedness phase starts after the mitigation phase. During the flood, the land may be covered by water, and collecting waste might be very minimal. After the flood water recedes, flood waste management system may start operating fully to clear out the waste (Figure 23.4).

Table 23.5 shows the problem causing flood in major cities and measures taken to mitigate it.

TABLE 23.5

Problem Causing Flood in Various Cities and the Actions Implemented

City	Problem	Community Awareness	Municipal Action	Community Action
Marikina (Philippines)	Clogging of waste in the river contributes to flooding	Program conducted to make the residents aware	Penalties for dumping waste in the river and working for river dredging	Though the residents comply, waste still arrives from further upstream
Lagos (Nigeria)	Blocked drainage results in flooding	High awareness in the community	New city-wide strategy for the disposal of waste	Less number of actions taken by the community
Managua (Nicaragua)	Disposal of wastes in the river contributes towards flooding	Very poor awareness among the community	Collection of silt and garbage through a network of micro-dams	Flooding continues as a result of low compliance by the community
Mexico City (Mexico)	Flash flooding due to blockage of drains by waste	Fairly low awareness	Programs prioritizing other issues	Very less or no community action
Jakarta (Indonesia)	Flooding due to blockage of channels	Residents are aware	Dredging of channels is on halt due to informal settlement	Though residents comply but waste arrives from further upstream
Mumbai (India)	Major factor for flooding is plastic bags	Community is poorly aware	Plastic bag usage is banned	Low compliance by the community
Bangkok (Thailand)	Poor drainage network causes a slow discharge of storm water	Residents aware of the problem	Post-flood actions such as cleaning of clogged drainage pipelines	Community-based solid waste management (CBM) programs successful

Source: Lamond et al. (2012: p. 200).

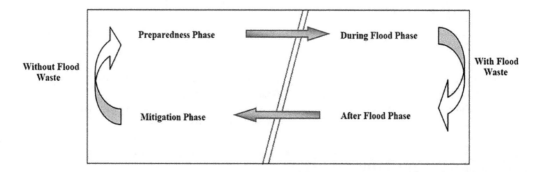

FIGURE 23.4 Flood waste management cycle (National Institute for Environmental Studies (NIES), 2015: p. 3).

Landfilling and open dumping is the most usual procedure of waste management in several developing countries (Karunasena et al., 2012). Of the total disaster management cost, around 27% of the cost accounts for waste disposal alone (FEMA, 2007). But the disposal of wastes can be reduced by recycling; also, it provides a way to acquire some revenue (Agamuthu et al., 2015). It has been observed that construction and demolition (C&D) waste consists of the most amount, as compared to other wastes, in any disastrous event (Figure 23.5). Recycling these wastes such as bricks, metals, concrete, and others for the construction and maintenance of roads, will, in turn, reduce the usage of raw materials and energy consumption (Karunasena et al., 2012). Due to the 2015 flood, an approximate potential amount of USD 1.62 million, which could have been acquired by recycling, was lost due to improper flood waste management in the city of Kelantan (Malaysia) (Agamuthu et al., 2015).

An effective flood waste management system should be established in cities prone to flood. As a response to the flood, the Bangkok Metropolitan Administration (BMA) and the participation of public,

FIGURE 23.5 Collection of waste caused due to urban flooding (National Institute for Environmental Studies (NIES), 2015: p. 9).

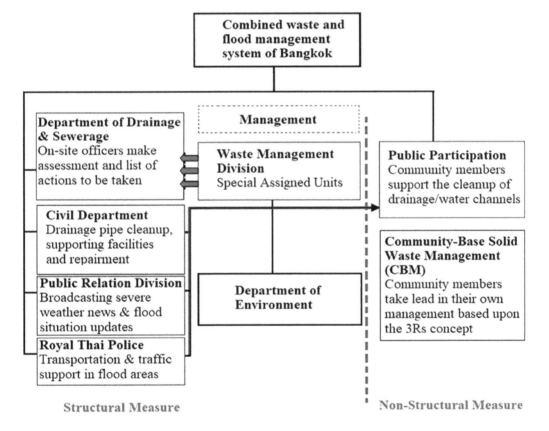

FIGURE 23.6 Combined waste and flood management system (Bangkok) (Pokasoowan and Khamrin, 2015: p. 19).

helped in reducing waste generated during the flood in Bangkok. The city has integrated a combined waste and flood management system, as shown below (Pokasoowan and Khamrin, 2015).

South Korea, after facing severe flood events in recent years, has established an effective flood waste management system (Figure 23.6). The system first estimates the amount of waste generated after the

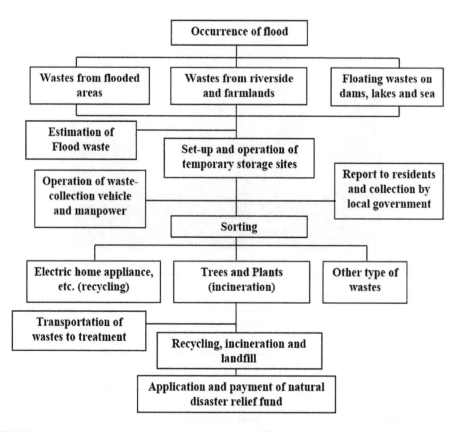

FIGURE 23.7   Collection and treatment of flood waste strategy in South Korea (Oh and Kang, 2013: p. 285).

flood event. It then identifies temporary storage sites in a way such that the location of these sites is not at risk of re-flooding. Also, there should not be any lack of labor forces or vehicles for the collection and transportation of waste by the municipality. Proper disinfection has to be carried out in these sites frequently to avoid the breeding of insects like mosquitoes or flies (Oh and Kang, 2013). Figure 23.7 describes the collection and proper treatment of flood waste that is followed in South Korea.

23.7 Flood Disaster Risk Management

To prevent the spread of several infectious diseases, it is crucial to manage the scattered flood waste. The most important aspect of flood risk management program is to minimize human losses and economic damage. Measures such as risk assessment need to be performed (i.e. identification of any possible hazards). Development of appropriate mitigation strategies should be of high priority, and the following steps can be incorporated (Parkinson, 2003):

  i. Enhancement of local-scale, state-scale and national-scale advocacy partnerships.
 ii. Standardization of hazard risk management tools and practices.
iii. Structural adaptations to construction work to reduce the impact of inundation.
 iv. Initiation of appropriate research and development projects as per the needs.
  v. Appropriate communication protocols should be established.
 vi. Real-time rainfall warning system for taking effective early precautions.

TABLE 23.6

Possible Causes of Flood in Various Indian Cities

City (India)	Possible Causes
Chandigarh	Overloading of the existing drainage system and encroachment of vacant lands by slums.
New Delhi	Storm water drains in poor shape, the sewage flowing into them. Water bodies encroached upon
Jaipur	Amanishah Nullah, the main water channel, now just a dirty drain due to sewage flow. New drainage plan for city pending
Bhopal	Poor drainage system. No trenching ground
Indore	Streams reduced to filthy channels. Poor solid waste management system
Bengaluru	Only 17 out of the 51 lakes, which were there in 1985, remains today. Remaining water bodies pumped with sewage
Puducherry	Illegal construction in open spaces and catchment area of water bodies
Amravati	Large stretches have paleo-channels construction, which may cause floods in future. Absence of an integrated storm water management plan
Bhubaneswar	No sewage treatment plant. Low-lying areas get waterlogged
Kolkata	Threat from rising sea level owing to climate change. East Kolkata wetland being gradually filled
Chennai	Inadequate storm water drainage system and lack of maintenance. Canals degraded and heavily silted. Urban lakes filled and encroached
Varanasi	125 year old sewage system fails to check waterlogging. New storm water drainage system set up 1 year back, but a lot remains to be done
Allahabad	Unplanned development, construction on Ganga and Yamuna floodplains. Poor drainage, encroachment

Source: SANDRP (2015).

  vii. Ban on illegal dumping of construction and excavation debris.

 viii. Pre-monsoon de-silting of major drains.

   ix. Removal of sediments from drains regularly.

   x. Encouraging local authorities' involvement.

India receives a considerable amount of rainfall in the monsoon season, and almost every year, several Indian cities suffer from the flood. Several factors, such as poor drainage network, lack of awareness among people, and illegal dumping, result in a flood when heavy rainfall occurs. Table 23.6 highlights the Indian cities that are vulnerable to flood and also consequential possible solid waste problems.

23.8 Design and Management of Urban Drainage System

Floods occur due to weather phenomena that cause heavy precipitation; as a result, water exceeds that of the storage limit of the basin. Urban floods can have a wide or local origin. Due to the high densities of population and infrastructure, there has been a rise in impermeable surfaces, thereby reducing the infiltration and natural storage. This results in more discharge through the drainage system, thereby putting more stress on them. Therefore, an adequate and well-functioning drainage system needs to be developed. The main objective of the urban drainage system is to provide adequate sizing and design for the conveyance of water, based on peak flow rates (National Disaster Management Authority, 2010).

The drainage system should be well covered to prevent illegal littering, which results in blockage of drains. Plastic bags are not easily decomposed and get clogged easily in the drainage network, limiting the use of plastic may reduce this problem. A well cover drainage system prevents the spread of harmful disease to the surrounding. Establishment of a real-time satellite-radar rainfall-based warning system may be a major help in implementing early precautionary measures during a flood and the data recorded can be used to develop the discharge capacity of new drains and redevelop the old ones (National Disaster Management Authority, 2010). A proper storm drainage system is necessary to mitigate the issue of urban flooding.

23.9 Conclusions

Due to a rise in urbanization, floods that used to be limited in rural areas have now become persuasive in urban areas (Dewan, 2013). Cities, especially those that belong to developing countries, face considerable difficulties in managing flood-induced solid waste management and incurs enormous expenses. Reducing the accumulation of waste, in general, may reduce flooding and its impact. Another primary reason for increasing floods in urban areas is due to an increase in impervious surfaces (like pavement roofs, roads, and parking lots). Creating holes that act as pervious openings by the side of roads and pavements may reduce the effect to some extent, by increasing the seepage of water into the ground. Lack of proper drainage is also a major factor contributing to urban flood. In many Asian counties, due to inferior methods of disaster preparedness system, the capacity to cope with extreme events are reducing. Involvement of both central and local organizations is essential for effective management of flood waste. A proper national guideline is to be prepared specially for the planning and management of flood waste. Recycling is one of the best methods to reduce the impact of the disposal of waste as well as to reduce its cost.

REFERENCES

Agamuthu, P., Milow, P., Nurul, A.M.N., Nurhawa, A.R. and Fauziah, S.H. (2015). Impact of flood on waste generation and composition in Kelantan. *Malaysian Journal of Science*, 34(2): 130–140.

Dewan, A.M. (2013). *Floods in a Megacity: Geospatial Techniques in Assessing Hazards, Risk and Vulnerability.* Springer Geography. ISBN 978-94-007-5875-9.

FEMA (2007). Public assistance debris management guide. https://www.fema.gov/pdf/government/grant/pa/demagde.pdf [Accessed 1st May 2020].

Hoornweg, D. and Bhada-Tata, P. (2012). What a waste: A global review of solid waste management. World Bank. https://openknowledge.worldbank.org/handle/10986/17388 [Accessed 1st May 2020].

Illangasekare, T., Tyler, S.W., Clement, T.P., Villholth, K.G., Perera, A.P.G.R.L., Obeysekera, J., Gunatilaka, A., Panabokke, C.R., Hyndman, D.W., Cunningham K.J., Kaluarachchi, J.J., Yeh, W.W.-G., van Genuchten, M.T., Jensen, K. (2006). Impacts of the 2004 tsunami on groundwater resources in Sri Lanka. *Journal of Water Resources Research*, 1–42: 9.

Karunasena, G., Rameezdeen, R. and Amaratunga, D. (2012). Post-disaster C&D waste management: The case of COWAM project in Sri Lanka. *Australisian Journal of Construction Economics and Building*, 1(2): 60–71.

Kundzewicz, Z.W., Hirabayashi, Y. and Kanae, S. (2010). River Floods in the changing climate: Observations and projections. *Journal of Water Resources Management*, 24(11): 2633–2646.

Lamond, J., Bhattacharya, N. and Bloch, R. (2012). The role of solid waste management as a response to urban flood risk in developing countries, a case study analysis. In: Proverbs, D., Mambretti, S., Brebbia, C.A., de Wrachien, D. (Eds) *Flood Recovery Innovation and Response (III)*, pp. 193–204. WIT Press: Ashurst Lodge.

Leopold, B.L. (1968). *Hydrology for Urban Planning: A Guide Book on the Hydrologic Effects of Urban Land Use.* US Geological Survey: Washington, D.C.

Melesse, A.M., Oberg, J., Beeri, O., Nangia, V. and Baumgartner, D. (2006). Spatiotemporal dynamic of evapotranspiration and vegetation at the glacial ridge prairie restoration. *Hydrological Processes*, 20(7): 1451–1464.

Memon, M.A. (2010). Integrated solid waste management based on the 3R approach. *Journal of Material Cycles Waste Management*, 12: 30–40.

Nandan, A., Yadav, B.P., Baksi, S. and Bose, D. (2017). Recent scenario of solid waste management in India. *World Scientific News*, 66: 56–74.

National Disaster Management Authority (2010). *National Disaster Management Guidelines: Management of Urban Flooding.* Government of India. ISBN: 978-93-80440-09-5.

National Institute for Environmental Studies (NIES) (2015). *Flood waste Management Guidelines for Bangkok.* Center for Material Cycles and Waste Management Research. https://www.apn-gcr.org/resources/files/original/e167b334df13f8c6288f8c009f71369c.pdf [Accessed 1st May 2020].

Njoku, N., Lamond, J., Everett, G. and Manu, P. (2015). An overview of municipal solid waste management in developing and developed economies: Analysis of practices and contributions to urban flooding in Sub-Saharan Africa. *Proceedings of 12th International Post-Graduate Research Conference*, Media City, UK, 10–12 June 2015. University of Salford Manchester, pp. 200–212.

Oh, G.-J. and Kang, Y.-Y. (2013). The status of flood waste treatment and future tasks in South Korea. *Journal of Material Cycles and Waste Management*, 15: 280–289.

Parkinson, J. (2003). Drainage and stormwater management strategies for low-income urban communities. *Environment and Urbanization*, 15(2): 115–126.

Pokasoowan, C. and Khamrin, T. (2015). Solid waste management as a response to urban flood: A case study of Bangkok city. *Naresuan University Engineering Journal*, 11(1): 15–20.

Rafiq, F., Ahmed, S., Ahmad, S. and Khan, A.A. (2016). Urban floods in India. *International Journal of Scientific & Engineering Research*, 7(1): 721–734.

Reinhart, D.R. and McCreanor, P.T. (1999). Disaster debris management-planning tools. US Environmental Protection Agency Region-IV.

Shekdar, A.V. (2009). Sustainable solid waste management: An integrated approach for Asian countries. *Waste Management*, 29: 1438–1448.

Soni, A., Patil, D. and Argade, K. (2016). Municipal solid waste management. *Procedia Environmental Sciences*, 53: 119–126.

South Asia Network on Dams, Rivers and People (SANDRP) (2015). Dams, Rivers and People (DRP), *News Bulletin* (14 December, 2015). https://sandrp.in/2015/12/14/drp-news-bulletin-14-dec-2015-your-city-can-be-next-chennai/ [Accessed 3rd May 2020].

Suocheng, D., Tong, K. and Yuping, Y. (2001). Municipal solid waste management in China: Using commercial management to solve a growing problem. *Utilities Policy*, 10: 7–11.

Suriya, S. and Mudgal, B.V. (2012). Impact of urbanization on flooding: The Thirusoolam Subwatershed: A case study. *Journal of Hydrology*, 412–413: 210–219.

Texier, P. (2008). Floods in Jakarta: When the extreme reveals daily structural constraints and mismanagement. *Disaster Prevention and Management*, 17(3): 358–372.

Tingsanchali, T. (2011). Urban flood disaster management. *Journal of Procedia Engineering*, 32: 25–37.

Yukalang, N., Clarke, B. and Ross, K. (2018). Solid waste management solutions for a rapidly urbanizing are in Thailand: Recommendations based on stakeholder input. *International Journal of Environment Research and Public Health*, 15: 1302.

24

Study of Some Polymeric Coagulants Using Spent Tea Leaves as a Source of Natural Dye Supporting Circular Economy Concepts

Ritwija Bhattacharya, Alokesh Mridha, Richa Sen,
Aniruddha Mukhopadhyay, Debasish Das, and Pritha Bhattacharjee
University of Calcutta

CONTENTS

24.1 Introduction

The synthetic dyes discharged from the effluent to the water bodies are harmful to the aquatic environment. They are not biodegradable; therefore, they affect the aquatic ecosystem by interfering with the photosynthesis of plants and subsequently affecting the food web. Besides, their by-products enter the human system through the consumption of aquatic food, which brings health risks as the derivatives contain carcinogenic aromatic amines (Tahir et al., 2010). Recently, the researchers attempt to re-establish the use of natural dyes to reduce the adverse impact of synthetic dyes on the environment. Less availability, poor reproducibility of shades, inadequate exhaustion capacity into the fibre, and poor fastness property are the demerits of using natural dyes. A huge amount of biomass is exploited to avail the adequate amount of dye required for processing (Křížová, 2015). In our study, STL has been used as an inexhaustible source of natural dye. In the current study, polymeric coagulants have been used to improve dye uptake. Polyferrous sulphate and polyaluminium sulphate are the new generation coagulants that have rapid flocculating ability and the capacity to form flocks at a relatively wider pH range. Polyaluminium sulphate and polyferrous sulphate have a high complex-forming ability (Chairat et al., 2007). These coagulants have been used previously for the removal of dyes from the textile effluent

(Sanghi et al., 2016; Choo et al., 2007). Earlier, a study using the extraction of outer onion skin as the natural dye, reported that different fashion hues could be developed using the same dye with different mordants where ferrous sulphate was reported to be the most efficient mordant (Uddin, 2014). In our study, a comparative assessment was carried out using ferrous sulphate, polyferrous sulphate, aluminium sulphate, and polyaluminium sulphate as mordants to improve the dye uptake capacity.

Literature studies revealed that agro-industrial waste materials processed or unprocessed possess a great potential to be valorised rather than being disposed of in a landfill. These materials are commonly composed of lignocelluloses and can be precursors of valuable products (Acevedo-García et al., 2020). The idea of circular economy has gained importance in recent years defeating the traditional idea of the linear economy of 'take-make-waste' as it had adverse environmental and economic consequences over the years. The circular economy, on the other hand, has elicited a virtuous circle where the products are converted to resources at the end of their service by reusing, repairing, remanufacturing, or recycling the goods (Jawahir and Bradley, 2016; Panwar and Niesten, 2021). Therefore, in a simplified way, it is the ability to recover resources rather than introducing the resources from outside of the system (Barros et al., 2020). According to Chapter 4, i.e. waste hierarchy of 'Directive 2008/98/EC of the European parliament and of the council', a priority of waste hierarchy should be pursued in the following order: "(a) Prevention; (b) preparing for reuse; (c) recycling; (d) other recovery, e.g. energy recovery; and (e) disposal". Although, according to the Directive (EU) 2018/851 of the European Parliament and of the council,the definition of food waste is needed to be included in Directive 2008/98/EC to develop further clarification and a clear concept on its scope (https://eur-lex.europa.eu/homepage.html). The circular economy has restructured the framework of food waste management creating new opportunities to generate high-value products from the waste. Teigiserova et al. (2020) highlighted that developing a database for the hierarchy of food surplus, losses, and waste (FSLW) could augment the efficacy of waste recycling. A matrix was created to categorize FSLW into six distinct types in terms of their edibility and avoidance. For example, despite the quality of the food, some parts of the food are unavoidably transformed into wastes such as bones and pits. Tea leaves were included under category III where the waste generation cannot be eliminated, though it can be minimized by optimizing the processes. It suggests that this category of wastes, when recycled can generate economic benefit and environmental values as they are always available in the system (Teigiserova et al., 2020). Tea factories generate a huge amount of tea wastes. The tea wastes could be a substitute source of income for the workers and the owners of the factory (Chowdhury et al., 2016). According to the "Tea Waste (Control) Order, 1959; in exercise of the powers conferred by Sub-Sections (3) and (5) of Section 30 of the Tea Act, 1953", certain policies for tea waste management are followed. It states that there should be a minimum volume of tea waste production at the ratio of 2:100 kilograms. Tea wastes should be only disposed of by exports or they could be used for caffeine or instant tea production or the production of bio-nutrient or fertilizers. It can be converted to compost or destroyed by burning following the procedure laid down under the law. But in no way, tea wastes can be acquired or sold, or held in stock in any form or whatsoever, without a proper licence procured from the Central Government of India. Concerning the policies and strategies of circular economy concepts in utilizing tea wastes, mention must be made of the initiative by *Nestlé* to launch a zero-waste tea-based bio-fertilizer, which has been declared as a heartening addition to a circular economy (EconPapers. repec.org). It encourages the idea of the establishment of STL as the source of natural dye that could be included in the legislative framework in the future in the light of the circular economy.

24.2 Materials and Methods

24.2.1 Extraction and Preparation of Dye from STL

STL was collected from home, Kolkata (India). STL was soaked in a beaker (500 mL) at a material-to-liquor ratio of 1:25. It was boiled in the presence of 2 g/L sodium carbonate and 2 g/L alum at a temperature of 100°C for 2 hours. Chemicals were of analytical grade, obtained from local distributors, Kolkata. The extraction so obtained was cooled before application on the substrate.

24.2.2 Degumming of Silk Fibre

Silk gum from the loom state of silk was removed. It was degummed in an aqueous solution containing 6 mL/L of non-ionic detergent and 2 g/L of sodium carbonate at fabric-to-liquor ratio of 1:50 (w/v) at 90°C for 1.5 hours. After degumming, the fabric was washed at 70°C for 10 minutes, washed in cold subsequently, and finally dried (Samanta and Konar, 2011).

24.2.3 Mordanting and Dyeing of Silk Fibre

The silk fabrics were mordanted with a 10% solution of ferrous sulphate, aluminium sulphate, polyferrous sulphate, and polyaluminium sulphate. The silk samples were individually soaked in the mordant. The control sample was not soaked in the mordant. The silk samples were dried at room temperature before the application of dye. Dyeing was carried out using a 4% aqueous solution of STL in a water bath. Dye bath temperature was kept at 900°C for 1 hour. Thereafter, the fabrics were rinsed in cold water and subsequently dried in the air (Samanta et al., 2012).

24.2.4 Assessment of Dye Receptivity in Terms of *K/S*

To measure the dye uptake into the substrate, the *K/S* value of the fabric samples dyed with STL was examined in Mcbeth 2020+reflectance spectrophotometer using the Kubelka–Munk equation (24.1), where *R* is the reflectance value of the dyed fabric:

$$K/S = (1-R)^2/(2R) \qquad (24.1)$$

where *K* is the absorption coefficient and *S* is the scattering coefficient. The more the *K/S* value, the better the dye receptivity of the substrate.

24.2.5 Determination of Colourfastness

A standard procedure as described in IS: 3361 (1984, ISO-II) was followed to determine the colourfastness of fabrics to washing. The test samples were washed at 500°C with a 5 gpl of soap solution at a matter liquid ratio of 1:50 for 45 minutes in a Launderometer (MAG Solvics, India, Washfast 1208). The difference between the standard and the sample was measured by the grey scale. Visual assessment was carried out under D 65 illuminant in a standard colour-matching cabinet. To measure the colour fastness to rubbing [IS: 766 (1984)], the test samples were placed at the base of a crock meter, TESTEX, China; Model: TF410 and rubbed with a white piece of silk fabric clamped on the tip of the device. The colour transferred to the cloth was measured by a grey scale. Samples were exposed in an MBTL lightfastness tester and MBTL lamp along with eight blue wool reference standards to measure the colour fastness [IS: 2454 (1984)]. The colourfastness was assessed by matching the change in colour of the specimen with that of the reference (Kumaresan et al., 2011).

24.2.6 Estimation of Exhaustion and Concentration of Dye in the Effluent

The exhaustion of the dye from the aqueous solution to silk substrate was estimated from the difference in the initial concentration of dye in the dye bath at the initiation of the dyeing process and the final concentration of the dye after completion of the process. The concentration of dye in the bath was measured by spectrophotometer (UV-Visible absorbance spectrophotometer, Hitachi, China, U-2000).

24.3 Results and Discussion

24.3.1 Dyeing of Silk with Natural Dyes and Dye Uptake Measurement

The effect of using different salts as mordants on silk after the application of STL on silk was assessed in terms of K/S as shown in Figure 24.1a. The results revealed that the ability of the inorganic salts to add colour to the silk substrate followed the sequential order: polyferrous sulphate, ferrous sulphate, polyaluminium sulphate, aluminium sulphate, and no mordant (Figure 24.1a). Figure 24.1b shows the surface colour perceived by the naked eye after pre-mordanting with different metal salts and dyeing with STL.

The increase in K/S values due to mordanting reveals that the dye molecules form complexes with positively charged metal ions of the mordants. Coordinate bonds are formed between dye-metal chelates and amine ($-NH_2$) groups of silk. One molecule of dye binds with one site of the fibre, while two or more molecules of dye can bind with one molecule of a mordant (Räisänen et al., 2002; Jothi, 2008). As a result, the mordant molecule can hold two molecules of dye with it when it binds to the fibre, therefore enhancing the dye uptake capacity. Ferrous sulphate and polyferrous sulphate are iron salts that are capable of forming a large octahedral complex with the dye molecules. Therefore, when some unoccupied coordination sites of these compounds come in contact with the fibre, they get occupied by functional groups such as amino and carboxylic groups of the silk fibre (Shabbir et al., 2017; Uddin, 2014). Thus, ferrous sulphate salt binds to one site of the fibre, and the other site of the dye molecule forms ternary complex ions (Bhattacharya and Shah, 2000). This resulted in the development of strong bonds between the dye and the fabric, giving a better result for ferrous sulphate and polyferrous sulphate mordants. On the other hand, aluminium salts comparatively form weak coordination complexes with the dye as they strongly bind with the dye molecules but show reduced interaction with the fibre.

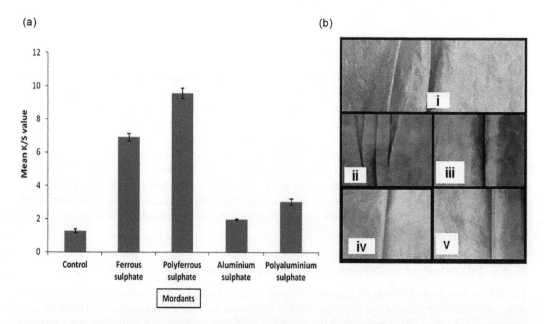

FIGURE 24.1 (a) Comparative assessment of colour uptake of dye assessed in terms of K/S and (b) surface colour perceived by naked eyes (1) control, (2) ferrous sulphate, (3) polyferrous sulphate, (4) aluminium sulphate and (5) polyaluminium sulphate used as mordants.

TABLE 24.1

Colourfastness of Silk Sample to Washing, Rubbing, and Light

Mordants Used	Washing	Rubbing	Light
Control	4	4	4
Ferrous sulphate	5	5	5
Polyferrous sulphate	5	5	6
Aluminium sulphate	5	5	5
Polyaluminium sulphate	5	5	5

24.3.2 Effects of the Mordants on Colourfastness

The results of colourfastness to washing, rubbing, and light are shown in Table 24.1. The rating of colour-fastness to washing appears to be four when dyeing was done in the absence of the inorganic salts. The rating of colourfastness to washing was five when dyeing was achieved in the presence of four inorganic salts. The rating of colourfastness to rubbing and light showed similar trends. In the case of colourfast-ness to light, a maximum rating of six was observed in the silk substrate mordanted with polyferrous sulphate (Kumaresan et al., 2011).

Aluminium and iron have a good complex-forming ability (Mahale and Sunanda, 2003; Bhattacharya and Shah, 2000). They can create insoluble complexes by holding two or more dye molecules together. Such complex formation of coloured compounds within the fibre leads to oligomerization of dye mol-ecules which is responsible for improved fastness to light and rubbing (Samanta and Konar, 2011; Khan et al., 2010). Polyaluminium sulphate and polyferrous sulphate have been identified to be compounds capable of forming large polynucleic complexes with the dye molecules that reduces the surface of expo-sure of the dye-metal complex to light (Verma et al., 2012). This property makes the colourfastness to light better than those of ferrous sulphate and aluminium sulphate. The presence of iron and aluminium helps to hold the dye molecules by forming an insoluble complex that helps to increase the rate of colour-fastness to washing.

24.3.3 Exhaustion and Concentration of Dye in the Effluent

The overall dye uptake of the silk fibre and the residual concentration of dye in the effluent in the absence and presence of different inorganic salts are shown in Figure 24.2. Transfer of dye from the dye bath appears to be very high for the samples where polyaluminium sulphate and polyferrous sulphate were used. Transfer of dye followed the sequential order: polyferrous sulphate, polyaluminium sulphate, ferrous sulphate, aluminium sulphate and no mordant. The residual concentration of dye in the effluent followed the order: no mordant, aluminium sulphate, ferrous sulphate, polyaluminium sulphate, polyferrous sul-phate. The dye transfer achieved in the samples that were mordanted with polyaluminium sulphate and polyferrous sulphate appears to be comparatively high. Polyaluminium sulphate and polyferrous sul-phate caused the reduction of dye concentration in the effluent. Polyaluminium sulphate and polyfer-rous sulphate are capable of forming a large amount of polynucleic complexes such as $(Fe_2(OH)_3)^{3+}$, $(Fe_2(OH)_2)^{2+}$ and $(Fe_8(OH)_{20})^{4+}$, which cause flocculation (Verma et al., 2012). Their fast flocculating ability and broad pH compatibility help in the formation of insoluble complexes within the silk more efficiently as compared to ferrous sulphate and aluminium sulphate. It resulted in the improvement of dye transfer, thereby reducing the concentration of the dye in the effluent.

24.3.4 Policies, Strategies and the Concept of Circular Economy in Utilizing STL in Textile Industries

The legislative and regulatory framework has an essential role in the sustainable growth of the textile industry. Some notable strategies have been adopted by the EU association for the textile sector. The intro-duction of EU Ecolabel 11 is one of the significant policies. Ecolabel is the voluntary label appropriate

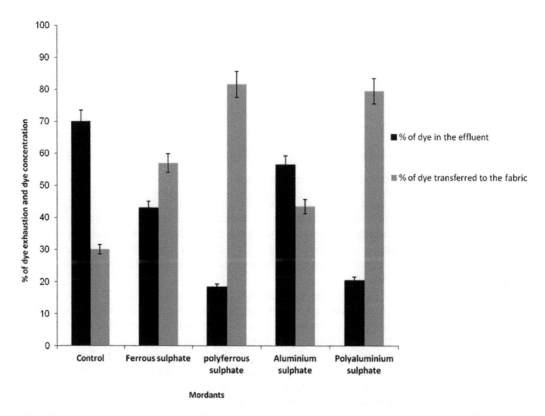

FIGURE 24.2 Exhaustion percentage of dye in the fibre and the total concentration of the residual dye in the effluent.

for the products whose life cycle has a low environmental impact. Another noteworthy regulation for the fashion and textile sector is the REACH Regulation 14 which deals with the manufacturing, market and end-use of the products. The "2010 Directive on industrial emissions (integrated pollution prevention and control, IPPC)" applies to textile industries that combine seven legislative acts on industrial emission and it controls atmosphere, water, and soil pollution. "Waste Framework Directive (Directive 2008/98/EC)" has the most important role in regard to waste. It introduced the principle of the waste hierarchy and also established the idea that waste disposal has to be borne by the owner of the wastes (Jacometti, 2019). This concept clearly connects to the circular economy. Several studies indicated the necessity to introduce circular economy in textile industries to replace the practice of "fabrication, use and dispose of" (Beyer and Arnold, 2021; Brydges, 2021; Huang et al., 2021; Coste-Maniere et al., 2019). Along with the refurbishment of the fabric used in the industry, the substitution of input materials such as the replacement of chemical dyes with natural dyes has been considered a crucial step towards a circular economy (Shirvanimoghaddam et al., 2020). For example, Natsai Audrey Chieza, a designer, has introduced a cost-effective and eco-friendly method by using the bacterial culture of *Streptomyces coelicolor* to produce natural pigmentation on the fabric. The fabric was soaked in the culture, and a unique pattern of gradient colour was developed which also saved a lot of water (Coste-Maniere et al., 2019). Our study highlighted the promising role of STL in the circular economy due to its high availability and various applications. Several experiments revealed the use of STL as an excellent source of biosorbent for the removal of pollutants from effluent water and also its application as fertilizers or bioenergy production (Verrillo et al., 2021; Khayum et al., 2018; Augustine, 2021; Nasar, 2021). Hence, the residues after dye extraction can be potentially utilized in the process of adsorption of pollutants from industrial effluent. Hence, it seeks to close loops in the industrial ecosystem enabling particularly the textile industry to capture value from reuse, recycling and repurposing of the products for multiple generations as shown in Figure 24.3.

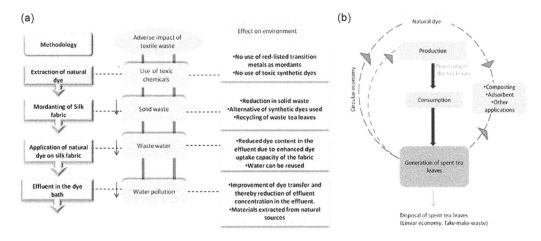

FIGURE 24.3 (a) Effect of using STL as an alternative to synthetic dyes in textile industries. (b) Utilization of STL based on the circular economy.

24.4 Conclusion

The study deals with the potential use of natural dye for a large-scale application in textile industries, making it economically acceptable and environmentally favourable. It focuses on enhancing the dye uptake capacity in silk substrate using STL of *Camellia sinensis*. This was achieved by using ferrous sulphate, polyferrous sulphate, aluminium sulphate, and polyaluminium sulphate as mordants. The maximum colour yield by the silk fibre was acquired when mordanted with polyferrous sulphate. Reduction of the effluent concentration and less depletion of dye could be successfully achieved by the technique without sacrificing the economic demand of the textile industries. Hence, from the perspectives of environmental sustainability and circular economy, STL could be successfully utilized as a source of natural dye for textile colouration.

REFERENCES

Acevedo-García, V., Rosales, E., Puga, A., Pazos, M. and Sanromán, M.A., 2020. Synthesis and use of efficient adsorbents under the principles of circular economy: Waste valorisation and electro advanced oxidation process regeneration. *Separation and Purification Technology*, 242 (1), p. 116796.

Augustine, A., 2021. Spent tea waste as a biomass for co-gasification enhances the performance of semi-industrial gasifier working on groundnut shell. *Biomass and Bioenergy*, 145, pp. 105964.

Barros, M.V., Salvador, R., de Francisco, A.C. and Piekarski, C.M., 2020. Mapping of research lines on circular economy practices in agriculture: From waste to energy. *Renewable and Sustainable Energy Reviews*, 131, p. 109958.

Beyer, K., and Arnold, M. G., 2021. Circular approaches and business model innovations for social sustainability in the textile industry. In: Matthes, A., Beyer, K., Cebulla, H., Arnold, M.G., and Schumann, A. (eds), *Sustainable Textile and Fashion Value Chains* (pp. 341–373). Springer, Cham.

Bhattacharya, S.D. and Shah, A.K., 2000. Metal ion effect on dyeing of wool fabric with catechu. *Coloration Technology*, 116(1), pp. 10–12.

Brydges, T., 2021. Closing the loop on take, make, waste: Investigating circular economy practices in the Swedish fashion industry. *Journal of Cleaner Production*, 293, p. 126245.

Chairat, M., Bremner, J.B. and Chantrapromma, K., 2007. Dyeing of cotton and silk yarn with the extracted dye from the fruit hulls of mangosteen, Garcinia mangostana Linn. *Fibers and Polymers*, 8(6), pp. 613–619.

Choo, K.H., Choi, S.J. and Hwang, E.D., 2007. Effect of coagulant types on textile wastewater reclamation in a combined coagulation/ultrafiltration system. *Desalination*, 202(1–3), pp. 262–270.

Chowdhury, A., Sarkar, S., Chowdhury, A., Bardhan, S., Mandal, P., and Chowdhury, M., 2016. Tea waste management: A case study from West Bengal, India. *Indian Journal of Science and Technology*, 9(42), pp. 1–6.

Coste-Maniere, I., Croizet, K., Sette, E., Fanien, A., Guezguez, H. and Lafforgue, H., 2019. Circular economy: A necessary (r) evolution. In: Muthu, S.S. (ed.) *Circular Economy in Textiles and Apparel*, pp. 123–148. Woodhead Publishing: Sawston.

Huang, Y.F., Azevedo, S.G., Lin, T.J., Cheng, C.S., and Lin, C.T., 2021. Exploring the decisive barriers to achieve circular economy: Strategies for the textile innovation in Taiwan. *Sustainable Production and Consumption*, 27, pp. 1406–1423.

Jacometti, V., 2019. Circular economy and waste in the fashion industry. *Laws*, 8(4), p. 27.

Jawahir, I. S., and Bradley, R., 2016. Technological elements of circular economy and the principles of 6R-based closed-loop material flow in sustainable manufacturing. *Procedia Cirp*, 40, pp. 103–108.

Jothi, D., 2008. Extraction of natural dyes from African marigold flower (*Tagetes erecta* L) for textile coloration. *Autex Research Journal*, 8(2), pp. 49–53.

Khan, M.I., Khan, S.A., Yusuf, M., Shahid, M., Mohammad, F. and Khan, M.A., 2010. Eco-friendly shades on wool using mixed mordants with Acacia catechu (Cutch). *Colorage*, 57(8), pp. 81–88.

Khayum, N., Anbarasu, S. and Murugan, S., 2018. Biogas potential from spent tea waste: A laboratory scale investigation of co-digestion with cow manure. *Energy*, 165, pp. 760–768.

Křížová, H., 2015. Natural dyes: their past, present, future and sustainability. In: Křemenáková, D., Militky, J., and Mishra, R. (eds), *Recent Developments in Fibrous Material Science* (pp. 59–71). Czech Republic, Kanina.

Kumaresan, M., Palanisamy, P.N. and Kumar, P.E., 2011. Application of ecofriendly natural dye on silk using combination of mordants. *International Journal of Chemistry Research*, 2(1), pp. 11–14.

Mahale, G. and Sunanda, R.K., 2003. Silk dyed with Acalypha (Acalypha wilkesiana) and its fastness.

Nasar, A., 2021. Utilization of tea wastes for the removal of toxic dyes from polluted water: A review. *Biomass Conversion and Biorefinery*, pp. 41–43. https://doi.org/10.1007/s13399-020-01205-y.

Panwar, R. and Niesten, E., 2021. Advancing circular economy. *Business Strategy and the Environment*, 29(6), pp. 2890–2892.

Räisänen, R., Nousiainen, P. and Hynninen, P.H., 2002. Dermorubin and 5-chlorodermorubin natural anthraquinone carboxylic acids as dyes for wool. *Textile Research Journal*, 72(11), pp. 973–976.

Samanta, A.K. and Konar, A., 2011. Dyeing of textiles with natural dyes. *Natural Dyes*, 3, pp. 30–56.

Samanta, A.K., Konar, A. and Datta, S., 2012. Dyeing of jute fabric with tesu extract: Part II–Thermodynamic parameters and kinetics of dyeing.

Sanghi, R., Bhattacharya, B., Dixit, A. and Singh, V., 2006. Ipomoea dasysperma seed gum: An effective natural coagulant for the decolorization of textile dye solutions. *Journal of Environmental Management*, 81(1), pp. 36–41.

Shabbir, M., Islam, S.U., Bukhari, M.N., Rather, L.J., Khan, M.A. and Mohammad, F., 2017. Application of Terminalia chebula natural dye on wool fiber: Evaluation of color and fastness properties. *Textiles and Clothing Sustainability*, 2(1), p. 1.

Shirvanimoghaddam, K., Motamed, B., Ramakrishna, S. and Naebe, M., 2020. Death by waste: Fashion and textile circular economy case. *Science of the Total Environment*, 718, p. 137317.

Tahir, H., Hammed, U., Sultan, M. and Jahanzeb, Q., 2010. Batch adsorption technique for the removal of malachite green and fast green dyes by using montmorillonite clay as adsorbent. *African Journal of Biotechnology*, 9(48), pp. 8206–8214.

Teigiserova, D.A., Hamelin, L. and Thomsen, M., 2020. Towards transparent valorization of food surplus, waste and loss: Clarifying definitions, food waste hierarchy, and role in the circular economy. *Science of the Total Environment*, 706, p. 136033.

Uddin, M.G., 2014. Effects of different mordants on silk fabric dyed with onion outer skin extracts. *Journal of Textiles*, 2014, Article ID 405626, 8 pages. https://doi.org/10.1155/2014/405626.

Verma, A.K., Dash, R.R. and Bhunia, P., 2012. A review on chemical coagulation/flocculation technologies for removal of color from textile wastewaters. *Journal of Environmental Management*, 93(1), pp. 154–168.

Verrillo, M., Salzano, M., Cozzolino, V., Spaccini, R. and Piccolo, A., 2021. Bioactivity and antimicrobial properties of chemically characterized compost teas from different green composts. *Waste Management*, 120, pp. 98–107.

Index

For Product Safety Concerns and Information please contact our EU
representative GPSR@taylorandfrancis.com
Taylor & Francis Verlag GmbH, Kaufingerstraße 24, 80331 München, Germany

www.ingramcontent.com/pod-product-compliance
Ingram Content Group UK Ltd.
Pitfield, Milton Keynes, MK11 3LW, UK
UKHW050926180425
457613UK00003B/36

* 9 7 8 1 0 3 2 1 3 7 2 5 4 *